Quantitative Risk Assessment in Fire Safety

Quantitative Risk Assessment in Fire Safety

Ganapathy Ramachandran
and David Charters

Spon Press
an imprint of Taylor & Francis

LONDON AND NEW YORK

First published 2011
by Spon Press
2 Park Square, Milton Park, Abingdon, Oxon OX14 4RN

Simultaneously published in the USA and Canada by Spon Press
711 Third Avenue, New York, NY 10017

Spon Press is an imprint of the Taylor & Francis Group, an informa business

© 2011 Ganapathy Ramachandran and David Charters

The right of Ganapathy Ramachandran and David Charters to be
identified as authors of this work has been asserted by them in accordance
with sections 77 and 78 of the Copyright, Designs and Patents Act 1988.

Typeset in Goudy by
HWA Text and Data Management, London

British Library Cataloguing in Publication Data
A catalogue record for this book is available from the British Library

Library of Congress Cataloging-in-Publication Data
Ramachandran, G. (Ganapathy)
Quantitative risk assessment in fire safety / Ganapathy Ramachandran &
David A. Charters.
        p. cm.
    Includes bibliographical references and index.
    1. Fire risk assessment. 2. Quantitative research. I. Charters,
David A. (David Anderson), 1949– II. Title.
    TH9176.R36 2009
    363.37'7—dc22   2009022649

ISBN13: 978-0-419-20790-0 (hbk)
ISBN13: 978-0-203-93769-3 (ebk)

Contents

Figures

Tables

About the authors

Ganapathy Ramachandran, PhD, DSc, FIFireE, MSFPE, held senior scientific appointments for 23 years (1965–1988) at the Fire Research Station, Borehamwood, Herts, UK. Since retiring in November 1988, he has been practising as a private consultant mainly in research problems in the application of statistical, probabilistic and economic techniques to fire risk evaluation, fire protection engineering and fire insurance. He has published several papers on these topics including chapters in the four editions of the *SFPE Handbook of Fire Protection Engineering*. He is the author of *Economics of Fire Protection* (Spon 1998) and co-author of *Evaluation of Fire Safety* (Wiley 2004).

Ramachandran was a Visiting Professor in Glasgow Caledonian University, University of Hertfordshire and University of Manchester. For the past 12 years he is a Visiting Professor in the University of Leeds. For three and a half years (up until July 2010), he was Technical Director, Fire and Risk Engineering at Faber Maunsell/AECOM.

David Charters, BSc, PhD, CEng, FIFireE, MIMechE, MSFPE is Director of Fire Engineering at BRE Global (incorporating the Fire Research Station, FRS) where he has been responsible for projects such as research into the use of lifts and escalators for emergency evacuation and quantitative fire risk assessments for nuclear and non-nuclear fire engineering projects. He has over 20 years of fire safety experience. Prior to joining BRE Global, he was a Director of Arup Fire responsible for the fire engineering of projects including; Channel Tunnel Rail Link (CTRL) tunnels, City of Manchester Stadium and Attatürk Airport. Previously he was Chief Engineer and Manager; Fire and Safety at NHS Estates, Department of Health where he was responsible for policy advice to ministers and the FIRECODE suite of guidance. Prior to this he was a Senior Fire Risk Assessment Engineer with the UK Atomic Energy Authority.

In 2005, David became International President of the Institution of Fire Engineers. He was appointed Visiting Professor in Fire Risk Analysis at the University of Ulster in 2003. In 1994, David was awarded a PhD in Fire Growth and Smoke Movement in Tunnels from the University of Leeds. He is co-author of the chapter on Building Fire Risk Analysis in the *SFPE Handbook on Fire Protection Engineering*. David was also responsible for quantitative fire risk assessments of London Underground stations following the King's Cross fire and Tartan Alpha following the Piper Alpha disaster.

Preface

As viable alternative for prescriptive rules, particularly for large and complex buildings, performance-based fire safety codes and building design methods are being developed and applied in many countries. These rules, codes and methods are mainly based on a qualitative assessment of fire risk supported by experimental data, case studies, deterministic (scientific) models and professional engineering practice. Statistical data provided by real fires are rarely analysed and the results (evidence) produced by such an analysis are seldom included in the risk assessment. Quantitative methods of risk assessment discussed in this book, on the other hand, explicitly consider statistical data on real fires, in addition to experimental data, and take account of uncertainties governing the occurrence of a fire, spread of fire, damage caused in a fire, reliability of passive and active fire protection systems and evacuation of building occupants.

Fire safety regulations, codes and standards do provide some unquantified levels of safety particularly for the occupants of a building, but these levels may or may not be adequate for some large, tall and complex buildings. Also, these levels may not be acceptable to property owners who have to consider also property damage and consequential losses such as business interruption and loss of profits. Given that there are no quantitative criteria for fire risk in buildings, it is not clear whether the safety levels provided by prescriptive rules are acceptable to the society at large. Criteria for determining acceptable safety levels for property owners and the society are discussed in Chapter 4.

Buildings can be designed according to acceptable levels for life safety and property protection by applying quantitative methods of fire risk assessment. These methods are discussed in detail in Chapter 3. Methods applicable to some particular problems in fire safety engineering are discussed in other chapters. These problems include initiation of fires (Chapter 5), design fire size (Chapter 6), flashover and spread of a fire beyond room of fire origin (Chapter 7) and performance and reliability of detection, alarm and suppression systems (Chapter 8). Building designs should sufficiently consider response and evacuation capability of occupants as discussed in Chapter 9.

Prescriptive rules specified in fire regulations, codes and standards should be further verified, validated and improved, if necessary, in the light of quantitative risk assessment. Otherwise, fire safety engineers will not be able to recommend enhanced fire protection necessary for large tall and complex buildings. This is because owners of such buildings are generally reluctant to spend more money than the cost required for complying with prescriptive rules and performance-based codes. The owners may not appreciate at present that enhanced fire protection may be more cost-effective than protection provided by prescriptive rules and codes.

Enhanced fire protection for large, tall and complex buildings, can be economically attractive if, in addition to cost of fire protection, monetary values of the cost per life saved property/business loss are also considered in a cost–benefit analysis. From among alternative fire protection strategies identified by a performance-based code, a building owner may select a strategy which is the most cost-effective in terms of costs and benefits of fire protection measures considered and interactions and synergies between these measures (Chapter 12). The owner would also consider the costs and benefits due to insurance/self-insurance options. A detailed framework for carrying out a cost-benefit analysis as described above has been discussed in Ramachandran's book *The Economics of Fire Protection* (1998). Quantitative risk assessment is an integral part of this framework.

As discussed above, the four major stakeholders or decision makers in the fire safety field are property owners, architects and designers, consultant firms engaged in fire safety engineering and government departments and organisations involved in the development and enforcement of fire safety regulations, codes and standards. Public fire and rescue services constitute the fifth major stakeholder. These bodies have to provide adequate fire cover to properties in their geographical areas by providing a sufficient number of strategically located fire stations, with enough firefighters and other resources. This problem should necessarily consider the fire risk in the properties in each area and the effectiveness of fire-protection measures and fire brigade performance in reducing the risk – see Chapter 10. Taking into account the interactions and synergies (Chapter 12) with fire protection measures, a fire and rescue service can identify an economically optimum fire cover strategy for any area.

The fire insurance industry is the fifth major stakeholder that has to estimate appropriate premiums that can be charged for different types of properties. If an insurance firm underestimates the premiums to be collected, it might face bankruptcy in a market-driven economic environment involving keen competition with other insurance firms. In the national interest the firm should promote fire safety by offering sufficient rebates in premiums for fire protection systems and self-insurance deductibles. Traditionally, most of the fire insurance underwriters adopt semi-quantitative points schemes discussed in Chapter 2, Section 2.2.2.

The underwriters should validate and improve their premium calculation methods by applying statistical models described in Chapter 3, section 3.3 and in Chapter 11 of Ramachandran's book *The Economics of Fire Protection* (1998). The statistical models would provide more accurate estimates of the 'risk premium' and the 'safety loading' to be added to this premium. An insurer firm can add another loading towards expenses and profits to estimate the total premium to be charged.

The authors hope this book will provide most of the methods and tools for quantitative risk assessment needed by stakeholders in the fire safety field.

Acknowledgements

Ganapathy Ramachandran ('Ram') would like to thank Liz Tattersall for word processing his Chapters 5, 6, 7, 8, 9, 10 and 12, and sections in Chapters 2, 3, 4 and 13. Liz coped admirably well with all the statistical and mathematical functions and formulae in these chapters and sections. Ram's wife, Radha, gave him considerable assistance in checking the typescript of the manuscript and printed page proofs for spotting mistakes to be corrected.

David Charters would like to thank Dr Roth Phylaktou for the permission to incorporate content from the MSc Module on Fire Risk Assessment and Management at the University of Leeds. He would also like to thank Dominic Vallely and James Holland (formerly and currently respectively) of Network Rail, Paul Scott, Fermi Ltd and Matthew Salisbury, MSA for their support in the development of the book. David would like to thank BRE Global and all the other people who gave permission for material in this book. He would also like to thank his sons, Jack and Theo for their encouragement and support.

The authors acknowledge the following permissions to reproduce material in this book.

Figures 1.2 and 11.1 are reproduced from D. A. Charters, 'A review of fire risk assessment methods', *Interflam 04*, reproduced with permission of Interscience Communications.

Figures 1.3, 4.1, 2.5 and Tables 2.1, 2.2, 2.3, 2.4, 2.5, 2.6 and 2.7 are reproduced with permission of Agetro Ltd

Figures 2.1, 2.2, 3.24, 3.25, 3.32 and 3.33 are reproduced with permission of BRE Global Ltd.

Table A.2.1. is reproduced from E. C. Wessels (1980) 'Rating techniques for risk assessment', *Fire International*, 67, 80–89. with permission from Keyways Publishing Ltd.

Tables A.2.2. and A.2.3. are reproduced from P. Stollard (1984), 'The development of a points scheme to assess fire safety in hospitals', *Fire Safety Journal*, 7, 2, 145–153 with permission of Elsevier.

Figures 3.1, 3.2 and 3.3 are reproduced from D. A Charters, Fire Risk Assessment and Management MSc module reproduced with permission of the Energy and Resources Research Institute, University of Leeds.

Figures 3.5, 3.6, 3.8, 3.9, Tables 3.7 and 4.2 are reproduced from D. A Charters and G. Ramachandran, committee draft of BSI PD 7974 'Application of fire safety engineering principles to the design of buildings', Part 7 'Probabilistic risk assessment'. Permission to reproduce extracts from PD 7974 'Application of fire safety engineering principles to the design of buildings' is granted by BSI. British Standards can be obtained in PDF or hard copy formats from the BSI online shop: www.bsigroup.com/Shop or by contacting BSI Customer Services for hardcopies only: Tel: +44 (0)20 8996 9001, Email: cservices@bsigroup.com.

Figure 3.7 and Table 4.4 are reproduced from D. A. Charters 'Fire safety assessment in bus transportation', *Fire Safety in Transport*, reproduced with permission of the Institution of Mechanical Engineers.

Figures 3.10 to 3.19, 5.2, 12.1 and 12.2 are repoduced from D.J .Rasbash, G. Ramachandran, B. Kandola, J. M. Watts and M. Law (2004) *Evaluation of Fire Safety* with permission of John Wiley and Sons Ltd (Figures 3.10, 3.11, 12.1 and 12.2 were previously published in G. Ramachandran (1998) *The Economics of Fire Proctection* published by Spon)

Figures 3.20 and 3.21 and Tables 3.14 and 3.15 are reproduced from W. T. C Ling and R. B. Williamson (1985) 'The modeling of fire spread through probabilistic networks', *Fire Safety Journal*, 9, 287–300 with permission of Elsevier, and from D.J. Rasbash, G. Ramachandran, B. Kandola, J. M. Watts and M. Law (2004) *Evaluation of Fire Safety* with permission of John Wiley and Sons Ltd.

Figure 3.28, 3.29, 3.30 and 3.31 are reproduced from D.A. Charters, 'Control volume modelling of tunnel fires' in A. Beard and R. Carvel (eds) *The Handbook of Tunnel Fire Safety*, reproduced with permission of Thomas Telford Publishing.

Figures 4.2, 4.3, 4.4 and 4.5 are reproduced from D.J. Rasbash (1984), 'Criteria for acceptability for use with quantitative approaches to fire safety', *Fire Safety Journal*, 8, 141–158 with permission of Elsevier.

Figure 7.1 is reproduced from BS5950, Part 8, 1990, Section 4 with permission of British Standards Institution.

Table 11.1 is reproduced from D. A. Charters, M. Salisbury and S. Wu, 'The development of risk-informed performance based codes', 5th International Conference on Performance Based Codes and Design Methods, reproduced with permission of The Society of Fire Protection Engineers.

Figure 11.2 is reproduced from D. A. Charters, 'Quantitative fire risk assessment in the design of major multi-occupancy buildings', *Interflam 01*, reproduced with permission of Interscience Communications.

Every effort has been made to seek permission to reproduce copyright material before the book went to press. If any proper acknowledgement has not been made, we would invite copyright holders to inform us of the oversight.

1 Introduction

We all take risks all the time, whether it is crossing the road, driving to work or watching television. The risk may vary from being knocked down, to being in a car accident or suffering ill health due to lack of exercise. The same can be said of fire safety in buildings. As long as we occupy buildings where there is a chance that ignition sources and combustible materials may be present together, there will be a risk of death and injury due to fire in addition to property damage. We need not be fatalistic, however, this simply identifies the need to manage the risk. It also indicates that, although we should work towards reducing risk, the ultimate goal of zero risk is not currently a realistic expectation.

As Benjamin Franklin once said, 'But in this world nothing is certain, but death and taxes'. It follows that whilst we live there is a risk of death and the only way of not dying is not to live in the first place. This may be self-evident, but it is very important when we start to consider specific risks that we consider them in the context of other risks.

There is a practical benefit to looking at risks in context. Society may decide that it would like to dedicate more resources to addressing one risk than another. For example, for healthcare, it may typically cost about £20,000 to save a life, whereas for fire safety in buildings, it may cost more than say £1million to save a life (Charters 1996). Therefore, society (and/or its representatives) could decide to put more resources into healthcare than into fire safety in buildings. Equally, society may be more concerned about the suffering of people killed and injured by multi-fatality fires than it is about the provision of every possible healthcare intervention to all patients, irrespective of need or prognosis.

For fire safety in buildings, the annual fire statistics indicate that there is a finite level of fire risk in buildings (Office of the Deputy Prime Minister 2005). This may also indicate that if a building complies with the appropriate fire safety standards, then its level of fire risk is broadly tolerable (or possibly acceptable). It could also be said that applying the fire standards to a non-standard building could result in intolerable levels of fire risk. However, no criteria for fire risk in buildings have been set in the UK (British Standards Institute PD 7974 Part 7 2003).

For the fire safety engineering of a non-standard building, this means that the level of risk should be designed to be the same or lower than that for an equivalent standard building (Office of the Deputy Prime Minister 2005; British Standards

Institute BS 7974 Code of Practice 2001). Since there are no quantitative risk criteria, this means comparison of a non-standard building with a compliant building. Similarly, since regulations are not generally framed in terms of risk, this results in an assessment of physical hazards and the balancing of a qualitative arguments about risk. So, in conclusion, we could say that, with respect to fire safety in buildings, we are all taking a risk but rarely, if ever, is it calculated.

The risk of an undesirable event can be defined as the combination of:

- the severity of its outcome (its consequences); and
- how often it might happen (its frequency) (British Standards Institute PD 7974 Part 7 2003).

If we simply consider the potential consequences of fires, then we may not be adequately addressing fires with lower consequences but whose risk is much higher due to their frequency. It is easy to imagine very severe fire events in buildings that have no fire precautions. We can identify ignition sources and combustible material, estimate how quickly a fire would grow and how quickly untenable conditions would develop. We can then consider the potential number of occupants, their likely behaviour and how quickly they might escape from the building. It is less easy to imagine very severe events in buildings with many fire precautions, but they can, and do, occur. Severe fire events like these seem to catch us by surprise and are characterised by a series of failures of the many fire precautions. The lower frequency of these events leads to the sense of surprise. The fact that a series of failures had to occur before the severe event occurred indicates that we had 'defence in depth'. Defence in depth is where many systems are present, but only a few need to work to achieve a safe outcome. Finally, this indicates that all fire precautions have a level of unreliability and occasionally many of the fire precautions may fail at the same time, leading to the severe event. Fortunately, these very severe events are relatively rare, but this rarity means that it is very difficult to assess their frequency directly. For example, if we have a thousand buildings of a certain type and we want to address fires that may occur once every million building years, then on average this fire could be expected only to happen once every thousand years. This is a long period over which to collect and analyse data and make a decision about the adequacy of the fire precautions. What makes it even harder is that this 'one in a million building years' event could happen tomorrow and the day after. This makes the direct estimation of high consequence/low frequency events very uncertain. Therefore, we use events that happen more often to estimate the frequency of rarer events. The frequency of ignition is the usual starting point for fire safety in buildings. To estimate the frequency of the severe fire events we need to understand how ignition may, or may not, lead to them. This involves the study of the reliability of fire safety systems. For example, probabilities can be attributed to the following series of events:

- Does the fire grow?
- If yes, is the fire detected early?

- If yes, is the fire extinguished using first aid fire fighting?
- Is the fire suppressed and/or vented by automatic systems?
- Does the fire spread beyond the compartment of fire origin?

In this way we can assess the impact of fire retardant or non-combustible materials, fire detection, extinguishers, sprinklers, vents, compartmentation etc on reducing the frequency of severe fires if, say, all these fire precautions fail.

We can combine the frequencies and consequences of these severe events to estimate their levels of risk. The levels of risk can be expressed as a frequency distribution (or 'Fn' curve) of risks, with lower risk events near the origin and higher risk events to the upper right-hand side. See Section 1.4 on acceptance criteria for risk and the role of Fn curves. Some events may have trivial or no consequences and so can be discounted from the analysis. Other events are high risk and should be addressed.

Because there are no quantitative criteria for the fire risk in buildings in the UK, it is not possible to say in absolute terms when the risk is low enough. However, if we know where a compliant standard building lies on the distribution, it is possible to do it by comparison. So we can say that we are safe enough, when the risk is lower, than a compliant standard building.

1.1 Fire engineering

Fire engineering can be defined as (BS 7974 2001):

> the application of scientific and engineering principles to the protection of people, property and the environment from fire.

Fire engineering can address one or more objectives. These objectives can include:

- life safety;
- property loss prevention/business continuity; and
- environmental protection.

Fire engineering is generally more demanding, technically and in terms of resources, than the application of simple prescriptive fire safety guidance used to support building regulations (ADB 2007). Therefore, fire engineering is generally used where simple fire safety guidance may not adequately address the fire scenarios or issues of concern. Simple fire safety guidance may not adequately address the fire scenarios or issues of concern when the building is large, complex or unique, or when application of the simple prescription conflicts with the function of the building (usually rendering the fire precaution highly unreliable) or when application of the simple rules is not the most cost-effective approach. Other factors that might indicate where fire engineering is typically used include:

- when there is an atrium;
- when there are multiple purpose groups in one building;
- where a highly innovative design is used to facilitate the function of the building; or
- when there are unique or challenging fire hazards.

Examples where fire engineering is typically used include the larger assembly buildings, hotels, hospitals, industrial and commercial premises, transport interchanges and tunnels, landmark, heritage and headquarters buildings, ships and offshore installations.

1.2 Deterministic approaches

Deterministic approaches use quantitative analysis of physical processes, such as smoke movement and evacuation, to aid decision-making during the design process (BS 7974 2001). To understand the deterministic approaches, a more detailed consideration of the design process may be of benefit.

Design can be characterised as an essentially creative, largely intuitive, often divergent process where many design parameters are manipulated in three and sometimes four dimensions, to meet multiple design objectives. Design objectives may include function, efficiency, cost, buildability, safety, durability, reliability, aesthetics etc. Design parameters may include the location of the site, the size and interconnection of different spaces, means of access and egress, the number of levels, the type of construction etc. For fire safety design, the main design objective is life safety (occupants, fire fighters and others in the vicinity), but other objectives such as property protection/business continuity and the environment may also be included. In many respects, the use of simple prescriptive guidance suppresses the design aspect of fire safety and can encourage the perception that there is only one solution and 'this is it'. However, the nature of many modern buildings means that simple fire safety rules are increasingly difficult and costly to apply. In these circumstances alternative fire safety design solutions are required to ensure that an appropriate level of safety is achieved. This is where analysis is used to assess the level of safety and indicate whether the fire safety design objective(s) have been met. Analysis can be characterised as an essentially logical, structured, rigorous process which takes certain input parameters, undertakes a series of operations/calculations and produces one or more output parameters. For fire safety analysis, the input parameters could include the material contents and their burning characteristics, size of a space, the number, width and distribution of its exits, the number of occupants, a walking speed and a flow rate through the exit. Deterministic analysis can be defined as the use of point values for the above variables and a purely physical model. Through the use of deterministic egress calculation methods a single value of the output parameter of time for occupants to move through an exit can be calculated (see Section 3.6 on consequence analysis). This single value is regarded as an exact value, ignoring the uncertainties governing the input and output parameters.

Decision-making can be characterised as an essentially convergent process of identifying an issue that needs to be addressed, gathering information on it (including that from analysis) and reaching a conclusion based on the evidence. For fire safety, the issue could be the number and width of exits required for egress from a large and complex space. Based on knowledge of the number and width of exits for smaller/simpler spaces, the basis of the input data, analytical model and results of the analysis, a decision on the appropriate number of exits for a large and complex space can be made (Charters 2000).

Quantified fire risk assessment may best be used where deterministic fire safety engineering may not adequately address the fire scenarios of concern. This tends to occur when the (life safety, loss prevention and/or environmental protection) consequences of a fire may be very high/intolerable.

1.3 Probabilistic approaches

Fire risk assessment allows the performance and reliability of fire precautions to be explicitly taken into account in the fire safety engineering of a building (PD 7974 Part 7 2003). It can be used to:

*   select fire scenarios for deterministic analysis; and/or
*   quantify levels of fire risk (to life, property, business and/or the environment).

In the selection of fire scenarios, an event tree for a broad class of occupancies and first order data estimates of frequencies and consequences may be used to identify and define an appropriate fire scenario for deterministic fire engineering analysis. It is important that the scenario selected provides a reasonably severe challenge to the fire safety design, yet is reasonably credible in terms of its frequency. Typically, this may include the failure of fire prevention, reaction to fire of materials, natural fire breaks, first aid fire fighting and non-fire resisting construction. Additional scenarios can be analysed to assess the dependence on a particular fire safety system such as sprinklers, smoke control or compartmentation.

Fire risk assessment can also be useful in quantifying the levels of risk for an individual building design:

*   in demonstrating equivalent life safety to a code-compliant building or satisfaction of fire risk criteria;
*   for cost-benefit analysis of property protection/business continuity; and/or
*   in assessing the environmental impact of large fires.

Therefore, circumstances where fire risk assessment could be useful include where:

1   Life safety is affected by buildings:

    *   containing a very large number of people who are unfamiliar with the building;

- containing a large number of sleeping occupants;
- containing a large number of vulnerable people;
- with very high fire growth rates;
- with restricted means of escape; and/or
- with other unusual features which could be significantly detrimental to fire safety.

Examples where fire risk assessment is useful for life safety in buildings and structures may include larger multi-occupancy buildings, assembly buildings, hotels, hospitals, industrial and commercial premises, tunnels, offshore installations and ships.

2   Property protection/business continuity is affected by buildings:

- with a high value;
- with high value operations; and/or
- with high value public images.

Premises where fire risk assessment is useful for property protection/business continuity may include certain landmark buildings, headquarters, leisure, assembly and heritage buildings as well as some industrial, commercial and transport premises (Charters and McGrail 2002).

3   Environmental protection is affected by buildings with a high potential impact on the environment.

Premises where fire risk assessment is useful in addressing risks to the environment from fire include industrial and transport premises, particularly where hazardous goods may be present in significant quantities (Charters and McGrail 2002). Transportation of dangerous and hazardous goods by road, rail and other means can cause significant environmental damage.

Fire risk assessment is essential where deterministic fire safety engineering cannot adequately address the fire scenarios of concern. This tends to occur when the consequences in terms of life safety, loss prevention and/or environmental protection of a fire may be intolerable.

Factors that might indicate where fire risk assessment would be essential include:

- When the reliability and performance of protection systems is critical. Fire risk assessment should be used to assess the 'defence in depth' of a design that relies heavily on a single or limited number of fire safety systems.
- When the variability of input parameters has a significant impact on the results. Fire risk assessment should be used where there are significant variations in variables like the number of people, their characteristics, fire growth rates etc and deterministic analysis shows that credible combinations of the variables are not acceptably safe (Charters and McGrail 2002).

- When a wide range of fire scenarios is deemed to be necessary. Fire risk assessment should be used in complex, high value, mission critical buildings, where high levels of defence in depth (multiple fire safety systems) are required.

1.4 Background to the development of fire risk assessment

Historically, fire has always been one of the major threats to life. In earliest times, the fires occurred in the natural environment of forests and grasslands. More recently, fires involving combustible materials in the built environment have caused the greatest risk to life.

In early civilisations, the buildings were generally small and single storey and so escape to a place of safety for their occupants was usually quite easy. The property lost could then be replaced relatively quickly due to their simple construction. As civilisations became more organised and sophisticated, settlements became larger. Buildings began to have upper storeys and fires in kitchens and living rooms and later candles and oil lamps added to the risks of fire. Throughout history disasters like the Great Fire of London in 1666 have struck such communities causing massive loss of property and life.

The response to various fire disasters has been to learn specific lessons. Over time this has resulted in the development of building regulations, considerations of compartment size, awareness of the need for a coherent package of fire precautions, the importance of the reaction to fire of building contents and human behaviour in fire. However, the more fire precautions that were placed in buildings did not seem to guarantee safety and fire disasters continued to occur even in apparently safe buildings. Equally, it could be argued that fire risk in buildings could have increased in the absence of fire regulations, codes and standards.

The nuclear industry was the first to learn this lesson in the 1950s with the fire at the Windscale reactor. For other building types in the UK, it was the King's Cross underground fire that raised the question of why an apparently safe building type (there had been few life loss fires on the underground) could suddenly suffer such a severe fire. The answer of course was that some of the fire precautions present work in most fires and so they result in a safe outcome (or a near miss). However, very occasionally, all the measures present will fail in such a way as to lead to a fire disaster. Therefore, the way of dealing with the stochastic (or random) nature of fire is through probabilistic fire risk assessment.

In spite of the many fire precautions in modern buildings, fire statistics still show the extent of the challenge. Fires in a range of countries result in the loss of (World Fire Statistics Centre 2007):

- 0.5 to 2 people per 100,000 of population per year; and
- 0.5 to 1.5 per cent of gross domestic product per year.

1.4.1 Corporate governance

Corporate governance is the way in which companies are directed and controlled for the benefit of internal and external stakeholders in a business. It is, therefore, the mechanism through which management is held accountable for performance and through which stakeholders are able to monitor and intervene in the operations of management.

The most widely accepted principles of corporate governance are provided by the Organisation for Economic Cooperation and Development (OECD) and include:

- protect stakeholders rights;
- ensure equitable treatment of stakeholders;
- recognise the rights of stakeholders in law and encourage cooperation between businesses and stakeholders to create wealth, jobs and sustainable enterprises;
- ensure timely and accurate disclosure of material matters regarding the business;
- ensure strategic guidance of the company through effective monitoring of management for the business and the stakeholders.

To further consider the requirement for fire safety, it is worth identifying the many and varied groups of people who have an expectation of fire safety. These groups of stakeholders can be categorised as:

- staff
- public
- customers
- board of directors
- regulators, and
- other external stakeholders.

Staff expect a safe working environment. This expectation may be balanced by the desire for job security. In other words staff have a high interest in fire safety, but not at a ridiculously high cost. An example of a workforce whose safety related expectations had not been met is that of the offshore oil industry. There were many instances of safety related industrial action following the Piper Alpha disaster in 1988.

The public expect safety for themselves whenever they interact in any way with a business. They also expect others to be safe. The public has a strong 'social' awareness of safety and in particular fires affecting a large number of people (known as 'consequence aversion'). An example of the failure to ensure public safety is the King's Cross underground station fire in 1987.

Customers expect to be safe whilst on the business's premises. Here it is not the actual level of safety that is important; it is the perception of safety that matters.

Hence, road tunnels may be statistically safer than the open road, but public knowledge of this fact does not affect public fear of tunnel fires following the multiple fatality fires at Mont Blanc, Tauern and Kaprun in the late 1990s.

Boards have expectation of safety, both as stakeholders and as individuals who are ultimately responsible and liable if things do go wrong. The Health and Safety at Work Act 1974 effectively establishes board members as negligent if risks are not as low as reasonably practicable (ALARP). Criminal law is increasingly being directed at board members to establish corporate liability. Chief Executives, such as that at London Underground Ltd at the time of the King's Cross fire, are being forced to resign.

Regulators try to balance all the different requirements of the various groups of society. The regulators recognise the national interest in a strong economy. The balance between ever-improving safety and reasonable social/economic cost is enshrined in the ALARP principle/legal precedent where risks can be said to be ALARP when the cost of further risk reduction far outweighs the reduction in the risk.

1.4.2 Risk management

Early interpretation of these principles concentrated on financial aspects, however, the Combined Code (1998) and the Turnbull report (1999) widened the emphasis on internal controls to business-wide risk management activity. Not surprisingly, risk management incorporates all the usual features of a management system:

- establish a strategic policy (individual policies such as fire safety may be required to support this);
- put in place an organisational structure, resources, competencies and responsibilities to deliver the policy;
- plan and implement risk assessments leading to the implementation of control measures to reduce risk;
- monitor the use and effectiveness of the control measures and investigate incidents;
- audit and review the system to confirm that it continues to address risks and identify opportunities for improvement.

These risks may cover a range of areas including health and safety, finance, quality, environment, security, IT, resources, reputation, regulations and business continuity. Fire may impact on all of these areas and the key is to identify, analyse and control the risks from fire.

A hazard can be defined as something with the potential to cause harm, for example a pile of combustible waste is a fire hazard. Risk can be defined as the combination of the frequency of an unwanted event, such as a fire and its consequences, i.e. how often it might occur and how bad the outcome might be.

Once the risks have been assessed, a hierarchy of risk reduction measures can be applied:

1   Elimination: remove the hazard;
2   Prevention: reduce the likelihood of the event occurring;
3   Mitigation: reduce the severity of the event;
4   Control of consequences: emergency and contingency planning and crisis management.

The remaining business risk can then either be:

1   Transferred: by means such as insurance, or
2   Absorbed: with a budget allocation to cover the expected level of loss.

1.4.3 Legislation

One of the main drivers for fire risk assessment is legislation specifically relating to fire safety. This varies from country to country but in many countries it typically includes:

* building regulations;
* fire precautions in existing buildings (normally workplaces); and
* other safety legislation.

1.4.3.1 Building regulations

The objectives of building regulations vary tremendously from one country to the next. Typical fire safety objectives could include:

1   to ensure satisfactory provision of means of giving an alarm of fire and a satisfactory standard of means of escape for persons in the event of a fire in a building;
2   that fire spread of the internal linings of buildings is inhibited;
3   to ensure the stability of buildings in the event of fire; to ensure that there is a sufficient degree of fire separation within buildings and between adjoining buildings; and to inhibit the unseen spread of fire and smoke in concealed spaces in buildings;
4   that external walls and roofs have adequate resistance to the spread of fire over the external envelope, and that spread of fire from one building to another is restricted;
5   to ensure satisfactory access for fire appliances to buildings and the provision of facilities in buildings to assist fire fighters in the saving of life of people in and around buildings.

Guidance on how to comply with building regulations can be found in a variety of guides. These guides are intended for the some of the more common building situations. However, many recognise that there are alternative ways of complying with the requirements of building regulations. The means for developing alternative design solutions is commonly known as fire safety engineering. Some guides indicate that a fire engineering approach may be the only practical way to achieving a satisfactory standard of fire safety in some large and complex buildings. Further they may indicate that the factors that should be taken into account include:

- the anticipated probability of a fire occurring; and
- the anticipated fire severity.

These factors clearly imply the use of fire risk assessment.

Furthermore, alternative solutions are accepted on the basis of equivalency, within whose definition is often an explicit reference to the level of fire risk. For equivalency, there is a need to:

> ...demonstrate that a building, as designed, presents no greater risk to occupants than a similar type of building designed in accordance with well-established codes.

Therefore, fire risk and fire risk assessment can play a major role in the design of new buildings. Further information on fire engineering can be found in many guides, for example PD7974 (2001) contains guidance on the application of probabilistic fire risk assessment.

1.4.3.2 *Fire regulations for existing buildings*

The fire regulations for existing buildings are many and varied. Some of the more recent examples include requirements for:

- a fire risk assessment to be undertaken;
- identification and recording of the significant details of those at particular risk;
- provision and maintenance of fire precautions to safeguard those in the workplace;
- provision of information, instruction and training to employees about fire precautions in the workplace.

With many relatively low risk buildings to assess, the fire risk assessment process used to support this kind of legislation relies.

1.4.3.3 *Other safety legislation*

Other safety legislation is based on the development and approval of safety cases. One of the main pieces of evidence in support of a safety case is risk assessment. Some of the legislation that might include a fire risk assessment or a risk assessment that includes fire hazards concerns:

- rail transport;
- major chemical sites;
- offshore oil and gas installations; and
- nuclear power installations.

1.4.4 Fire safety management

1.4.4.1 Fire safety policy

The fire safety policy for an organisation is a document that lays out the commitment of the organisation to fire safety on behalf of its stakeholders. Fire safety policies may contain references to supporting documentation such as legislation, procedures and risk reduction plans. Policy statements are usually endorsed by the Chief Executive of an organisation, to demonstrate commitment from the highest level.

1.4.4.2 Management structure

To support the delivery of the policy, there is need for a fire safety management structure. This structure will range from the Chief Executive and an Executive Director responsible for fire safety to all members of the organisation.

Other roles may include a senior manager who may be responsible for:

- supervising the day-to-day upkeep of the policy;
- ensuring staff participate in fire safety training;
- carrying out fire drills;
- coordinating and directing staff action in accordance with an emergency plan;
- receiving reports of fires;
- preparing reports to the board of directors on fire safety.

Members of organisations also have general duties and responsibilities and these include:

- understanding the nature of fire hazards;
- awareness of fire hazards in the building;
- practising and promoting fire prevention;
- knowing and following procedures for when a fire breaks out.

There may also be a role for a specialist fire safety advisor. This role may include:

- assisting management in the interpretation and application of legislation and guidance;
- advising management of their responsibilities;
- undertaking fire safety audits;
- undertaking fire risk assessments;

- liaising with staff, designers and approval authorities when specifying fire precautions;
- training staff in fire safety.

Specialist fire safety advisors may be internal or external to an organisation. If however, they are internal, their position in the management structure should be adjacent to the normal management lines, so that they are independent of operations managers who are responsible for fire safety in their area.

1.4.4.3 *Audit and risk assessment*

Lord Kelvin said 'you cannot predict what you cannot measure'. Equally, what cannot be measured or predicted cannot be managed. Therefore, the two crucial measurement processes in the feedback loop of the risk management process are audit and risk assessment.

Audit is a structured interview process which includes a range of questions and seeks evidence of compliance. It was first developed in the financial industry for verifying accounts, but is equally applicable to the assessment of any management process such as risk and quality assurance.

Risk assessment on the other hand tends to address the physical situation and seek to measure/predict and assess the acceptability of risk in a particular place. Over time a range of fire risk assessment methods have been developed to address risk in a range of applications. These methods range from simple qualitative fire risk assessment methods for the Fire Safety Order to the complex quantitative and probabilistic methods applied in the nuclear industry. Figure 1.1 shows a breakdown of fire risk assessment methods and terms.

The following subsections outline each of these types of fire risk analysis.

1.4.5 *Qualitative fire risk assessment methods*

Qualitative methods rely on identification of factors that affect risk. These factors may be those that affect the level of hazard, such as combustibles or the level of fire precautions intended to mitigate those hazards such as fire detection. An assessment of all these factors is undertaken against a benchmark set of values

Figure 1.1 Breakdown of different fire risk assessment methods

and a judgement made as to whether the factor is higher, lower or about the same as the benchmark. A review is then undertaken of all the factors to judge how overall the area being assessed compares with the benchmarks. If additional mitigation measures are judged to be required, this is a user input or a choice from a prescribed list. Assessments under the fire regulations for existing buildings usually fall into this category.

1.4.5.1 Unstructured methods

An unstructured method of qualitative fire risk assessment might follow the five-step risk assessment process outlined below:

1   identify potential fire hazards;
2   identify those in danger from fire;
3   evaluate the risks and whether the current fire precautions are adequate;
4   record findings and actions and communicate to employees;
5   keep the assessment under review and revise it when necessary.

The method is unstructured as it has no prescribed standards for what types and levels of precautions are appropriate for various levels of hazard. An example of an unstructured fire risk assessment can be seen in Section 2.1.1.

1.4.5.2 Structured methods

Structured methods of qualitative fire risk assessment use the same process as unstructured methods except that in addition they tend to prescribe:

*   different levels of hazard;
*   different levels of precaution;
*   the acceptability of different combinations of hazard and precaution;
*   appropriate risk reduction alternatives.

A good example of this approach is that of FIRECODE HTM 86 (1994) 'Fire risk assessment in hospitals'. This method identifies different types of fire hazard such as 'smoking' and 'fire hazard rooms' and whether the level is 'acceptable', 'high' or 'very high'. It then goes on to identify different types of fire precaution including 'alarm and detection' and 'compartmentation' and whether the level of provision is 'unacceptable', 'inadequate', 'acceptable' or 'high'. Certain combinations of hazard ratings and precautions ratings are deemed to be acceptable. Other combinations lead to the need to reduce risk by a range of prescribed alternatives.

This kind of qualitative flexible prescription identifies areas of high hazard and/ or inadequate precautions and prescribes one or more prescriptive solutions. This is relatively easy and consistent to apply, taking only a matter of hours to survey and days to report. This method does not suffer from the apparent acceptability of inappropriate fire strategies. However, the method cannot easily address non-standard situations, limits options for upgrade and may not be applicable to new designs.

1.4.6 *Semi-quantitative fire risk assessment methods*

Semi-quantitative fire risk assessment methods place some numerical value on the level of risk, without that numerical value representing a precise value of risk.

1.4.6.1 *Points schemes*

Points schemes use similar factors to other qualitative methods, but inject some numeracy to result in a score. Typically each aspect of fire safety (or factor) is scored from a range and then multiplied by a weighting to provide an overall score for that factor. The overall score for each factor is then added together to provide an overall score for the area being assessed. This overall score can then be compared with a benchmark score to assess the level of risk. Risk reduction can be effected by improving any number of factors and recalculating the overall score for the area.

The 1980s version of FIRECODE HTM 86 (1987) for fire risk assessment in UK hospitals and the ADAC (Allgemeiner Deutscher Automobil-Club) European fire risk assessment of tunnels fall into this category.

Points schemes have the advantages that they are relatively easy and quick to use and provide a degree of resolution of the level of risk by producing a number. Their disadvantages include the relatively arbitrary values given to the weightings, the implicit fire safety strategy that they represent and the freedom to develop high scoring fire safety solutions that do not work in practice.

1.4.6.2 *Matrix methods*

Matrix methods are used where the subjective nature and lack of quantitative information from qualitative methods are likely to lead to poor decision making with significant consequences. Matrix methods usually take the form of a structured brainstorm in a workshop. The Fire HAZard ANalysis (HAZAN) is often the first task in as asset's fire risk assessment. The objectives of the HAZAN workshop are to:

- identify fire hazards and accidents that could occur in the asset;
- estimate the consequences according to an accident severity matrix;
- rate the hazards by using a risk classification matrix;
- screen out minor fire hazards; and
- provide a list of identified major fire hazards.

The major fire hazards will then be subject to a Quantitative Risk Assessment (QRA). An example of a matrix and the output tables can be seen in Section 2.2.1.

The advantages of matrix methods are that they place risks in broad categories by breaking down difficult questions into smaller parts, the judgements can largely be evidence based, they are auditable and mathematically robust and based on consensus. The disadvantages are that the risk categories are broad, judgements are often subjective and dependent on who actually attends the workshop and the dynamic of the brainstorm process. This is why they are often used to prioritise hazards and events as a precursor for full quantitative fire risk assessment.

1.4.7 Quantitative fire risk assessment

Quantitative risk assessment methods predict a discrete level of risk. They tend to use a combination of probabilistic and/or physical modelling to predict the frequency of certain events and their likely consequences. Most methods can be characterised as probabilistic or full quantitative. This approach assumes that event consequences from the asset being risk assessed will be like the others from which the statistics are derived.

1.4.7.1 Probabilistic risk assessment

Probabilistic methods simply use statistical analysis to predict the frequency or probability of an unwanted event. The consequences of the unwanted event are taken directly from the statistical information, e.g. probability of fire spread beyond the item first ignited or probability of area damaged being greater than x.

Probabilistic methods have the advantage that they have a sound theoretical basis and provide a numerical prediction of risk. The disadvantages are that they require a high degree of technical input, are reliant of fire statistics being available and assume that the asset being assessed is the same as a homogeneous set from which the fire statistics were taken.

1.4.7.2 Full quantitative risk assessment

Full quantitative fire risk assessment predicts the level of consequences from unwanted events as well as their frequency. Figure 1.2 illustrates the process.

HOW OFTEN IS A SEVERE FIRE LIKELY TO OCCUR?

One of the main challenges in assessing fire risk is the fact that the kinds of very severe events that we are interested in are relatively infrequent. For example, the King's Cross fire happened after approximately 10,000 station-years of experience.

If severe fire events occurred all the time, their frequency would be easy to predict. Thankfully, they are relatively rare. This rarity means that we are unable to predict these extreme events explicitly from historical data. So to predict the frequency of fire events that have not happened yet, we have to break the event down from ignition to outcome into sub-events which occur frequently enough for there to be meaningful data available. The means for constructing event and/

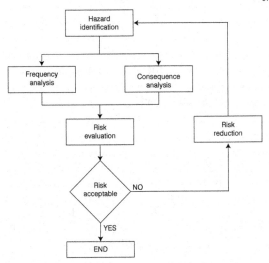

Figure 1.2 Schematic of the full quantitative fire risk assessment process

or fault trees and these were developed for probabilistic safety analysis in the nuclear industry.

HOW SEVERE IS A FIRE LIKELY TO BE?

How severe a fire event is likely to be can be evaluated in a number of ways. The prediction of event consequences tends to rely on modelling which in many respects is more mature and more widely used in processes like fire engineering. For fire risk assessment, this modelling tends to involve predicting:

* how quickly a fire will grow;
* how the heat and smoke will move about;
* how quickly the occupants will be aware of, react to and evacuate;
* how tenable the environment is for occupants before evacuation is complete.

Full quantitative fire risk assessment methods have the advantage that they have a sound theoretical basis and provide a numerical prediction of risk. The disadvantages are that they require a high degree of technical input and are reliant on fire statistics and appropriate models being available.

1.4.8 *Other methods*

A wide range of fire risk assessment methods are available including:

* checklists
* GOFA
* Delphi

- HAZID
- HAZOP.

These methods are described in the technical literature and share many of the same characteristics as the methods described above, with the possible exception of the Delphi method, described below.

DELPHI

The Delphi method is a way of gaining a group opinion without the difficulties inherent in meetings (Linstone and Turoff 1975). The group never meet and all communication is via a group controller. The group controller selects the members of the group, presents the basic problem and informs the individuals in the group of progress. When the main elements of a problem have been described, a consensus is sought on the value to be ascribed to each issue.

The consensus process is via a series of rounds of voting. After each round the controller returns each panel member with their score and a measure of the group score. The members are then asked if they wish to revise their score until the controller judges that consensus has been reached or that consensus is unattainable.

The advantages of the method are the lack of group pressure present in a meeting when operating in an area with no data or a very difficult decision. The disadvantages include extended time implications, greater dependency on the group controller to avoid unconscious bias, empirical and subjective assessment of risk and a range of other doubts about the quality of the outcome. For example, recent work by Carvel et al. (2001) found that there were significant variations between the estimates of a panel and actual experimental values.

1.5 Examples of the adoption of fire risk assessment

1.5.1 Introduction

This section describes several disasters that have shaped the provision of safety from fire in different industries and situations. The nuclear and chemical industries in particular, for example, consider their physical neighbours whilst dealing with complexities which are beyond the understanding of the vast majority of the public and even many of those industries' employees. Prevalent in all the 'disaster led' revisions of safety provision below is the term 'safety case', a term whose semantics express the move away from explaining 'why something is not dangerous' to a documented appraisal of 'why something is safe'.

These industries are not only dealing with the risk from fire, of course, as the 'release' catastrophes of Bhopal (1984) and Chernobyl (1986) illustrate only too well. Indeed, the first recognisable quantitative risk assessment was undertaken over 200 years ago. The concepts of insurance and domestic risk management have a history almost as long as the concept of risk itself, but the real milestone came with the development of probability theory by Pascal in 1657. This seems to have initiated a flurry of activity culminating in the first quantitative assessment of risks by Laplace in 1792.

Laplace also helpfully described the application of probability theory as (1814):

> …common sense with a little mathematics.

1.5.2 Nuclear installations

1.5.2.1 Windscale fire 1957

On 10 October 1957, a fire occurred at the Windscale nuclear power plant. The fire spread to the plant's air filters, there was a loss of containment and a release of radioactive material into the local area which led to a ban on milk produced locally.

The safety philosophy at this time was one of reaction and prevention. For every incident, changes were made to the design and operation of plant so that the recent event could not happen again. This resulted in more and more safety systems being incorporated but still incidents (usually different in nature from the previous one) would occur. To address this problem a different safety philosophy was required – one that recognised that all systems can fail and that combinations of initiating events and system performance on demand are important and can lead to very different outcomes for events. This led to the use of methods for predicting the probability of unwanted events that had been developed in the communications, aviation and electronics industries after World War Two (Green and Bourne 1972).

One of the key findings from the inquiry into the incident was that there was need for 'better communication between the plant's management and local interests'. This, in turn, ultimately led to a way of expressing society's risk expectations in the Farmer curve (later known as Fn curves, see Farmer 1967). Implicitly, the best way of using the Farmer curve was to analyse the level of risk from a plant or activity and compare it with the levels on the curve.

1.5.2.2 Farmer curve

The Farmer curve was the first quantitative expression of risks to society. The graph plots the number of reactor years against the size of release of iodine 131 in curies. The curve indicates that relatively small releases can be tolerated every 100 to 1,000 reactor years but that larger releases cannot be tolerated unless they are every million to 10 million reactor years. This type of curve later became known as Fn or frequency–consequence curves (Allen *et al.* 1992), see Figure 1.3.

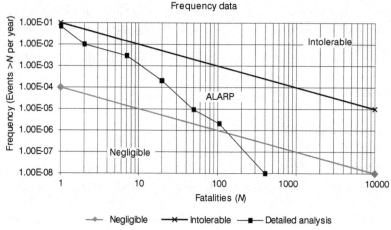

Figure 1.3 An F–n Curve

1.5.2.3 Safety cases and probabilistic risk assessments (PRAs)

The Nuclear Installations Act, 1965 (as amended) (HSE 2002) requires any operator of a defined nuclear installation to be licensed and gives the HSE the powers to 'attach to the licence such conditions as may appear … to be necessary or desirable in the interests of safety'. There are 36 standard Licence Conditions attached to all Nuclear Site Licences of which Licence Condition 14 requires the licensee to:

> make and implement adequate arrangements for the production and assessment of safety cases consisting of documentation to justify safety during design, construction, manufacture, commissioning, operation and decommissioning phases of the installation.

A safety case can be defined as the totality of documented information and arguments which substantiate the safety of the plant, activity, operation or modification in question. It provides a written demonstration that relevant standards have been met and that risks have been reduced as low as reasonably practicable (ALARP).

The safety case for a plant as a whole should be a living document which is subject to review, change and amendment as time elapses. For example, the safety cases may change due to important changes to the plant, its mode of operation, or the understanding of safety-related issues. It may also change in the light of operating experience.

The safety case for a nuclear plant (or modification etc) should be based upon robust design, defence-in-depth and deterministic analysis of normal operation and fault behaviour. The latter should consider faults that are reasonably foreseeable during the lifetime of the plant and for which provisions have been designed into the plant to prevent or mitigate them. This is known as design base accident

(DBA) analysis. In addition, a deterministic analysis should be undertaken of more severe faults and failures which have not been specifically protected against in the design and which, in the extreme, could lead to large releases of radioactivity. This constitutes severe accident analysis.

To supplement and support the deterministic analysis, a probabilistic safety analysis (PSA) may be required. The requirement, depth and level of the PSA should be commensurate with the significance of the hazard. The PSA will provide the means to identify failure scenarios, confirm the effectiveness of defence-in-depth provisions, search for weaknesses in the design, show there is reasonable balance of risk for all hazards and operations, and derive numerical estimates of risk. The PSA provides quantitative input to the ALARP case and it can provide estimates of relative benefits (in terms of risk reduction) of improvements.

This shows how quantitative risk assessment (of all hazards including fire) is intrinsic to, and underpins, much of the safety case regime for the licensing of nuclear installations. The most common methods used are HAZOP (similar to matrix methods), deterministic analysis (similar to consequence analysis) and PSA (similar to full quantitative risk assessment).

The nuclear industry uses safety assessment principles (SAPs) to help judge the acceptability of risk assessments. The SAPs contain some 330 principles, the vast majority of which relate to engineering or operational good practices as well as dose and risk criteria. The overall requirement is that risks should be reduced to ALARP: the criteria are for the guidance of inspectors and are not design targets for licensees. To support SAPs, more detailed internal guidance in the form of Technical Assessment Guides has been produced, some of which are available on the HSE website (HSE 2003).

1.5.3 Chemical plant and offshore installations

1.5.3.1 Flixborough, 1974

On Saturday 1 June 1974, the Flixborough works of Nypro (UK) Ltd were practically demolished by a massive explosion, killing 28 and injuring 36 people. The plant manufactured caprolactam, which is the basic ingredient of Nylon 6. Originally, this was achieved by the hydrogenation of phenol but the process was changed in 1972 and cyclohexane was oxidised to produce cyclohexanone, bringing with it various new problems regarding safety. Any escape of cyclohexane, which was circulated through reactors under pressure at a temperature of 155°C could prove potentially dangerous.

Following the detection of a cyclohexane leakage in Reactor 5, a large crack was discovered in the vessel. It was decided at a management meeting that the reactor should be removed and inspected, and that a by-pass pipe should be installed between Reactors 4 and 6 in order to allow the plant to continue to function. It was inadequacies in the design and installation of this pipe that led to it rupturing and releasing large quantities of cyclohexane vapour which subsequently exploded.

1.5.3.2 COMAH regulations

This section describes the Control of Major Accident Hazards Regulations 1999, which came into force on 1 April 1999 (HSE 1999). They implemented the SEVESO II Directive and replaced the Control of Industrial Major Accident Hazards (CIMAH) Regulations 1984.

COMAH applies mainly to the chemical industry, but also to some storage activities, explosives and nuclear sites, and other industries where threshold quantities of dangerous substances identified in the Regulations are kept or used.

A major accident is defined as an occurrence (including in particular, a major emission, fire or explosion) resulting from uncontrolled developments in the course of the operation of any establishment and leading to serious danger to human health or the environment, immediate or delayed, inside or outside an establishment, and involving at least one dangerous substance.

The COMAH regulations have a regime that is similar in many ways to that of the nuclear industry in that the Competent Authority (CA) sets down its policies, procedures and guidance for the handling and assessment of safety reports. The assessment involves the exercise of professional judgement by inspectors and a framework within which these judgements are made. This is intended as a practical tool for inspectors, to help achieve consistency in the approach to safety report assessment.

A safety report has to contain certain information, which relates to the major accident hazards and how major accidents are prevented or how the consequences of such an accident are limited. The information provided has to be sufficient to meet the purposes of the report and to help assessors gather this information the CA has drawn up assessment criteria. There are about 130 criteria set out in six groups. The predictive criteria deal with the identification of major accident hazards and risk analysis. They cover:

- principles of risk assessment and the use of appropriate data;
- identification of major hazards and accident scenarios;
- likelihood of a particular major accident scenario or the conditions under which they occur including initiating and event sequences; and
- consequence assessment.

The safety report as a whole should enable a view to be taken on the suitability and sufficiency of the risk assessment for drawing soundly based conclusions. It should be clear that the operator's approach to demonstrating compliance with the 'all necessary measures' requirement, is fit for purpose (HSE 2003).

1.5.3.3 Piper Alpha, 1988

At 21:45 on 6 July 1988, one of two condensate injection pumps tripped on the Piper Alpha platform in the North Sea (Cullen 1990). Due to a failure in the

Permit to Work system, the night shift attempted to restart the other pump which had been shut down for maintenance. Unknown to them, a pressure safety valve had been removed from the line and a blank flange assembly had been fitted that was not leak-tight.

The initial explosion caused extensive damage. It led immediately to a large crude oil fire in B Module, the oil separation module, which engulfed the north end of the platform in dense black smoke. This fire which extended in C Module and down to the 68ft level was fed by oil from the platform and by a leak in the main line to the shore, to which pipelines from Claymore and Tartan platforms were connected. The fire spread rapidly, unchecked by the firewater system which had failed to operate partly because the fire pumps were on manual due to diving operations.

At about 22:20 there was a second major explosion which caused a massive intensification of the fire. This was due to the rupture of the riser on the gas pipeline from Tartan as a result of the concentration and high temperature of the crude oil fire. In total 226 persons were aboard the rig, most of these in the accommodation part of the platform. Only 61 people survived the disaster.

1.5.3.4 The Offshore Installations (Safety Case) Regulations, 1992

Prior to 1988, the Department of Energy was responsible for both production of oil and gas and for safety in the North Sea. This potential conflict of interest was resolved by enforcing the above regulations through the Health and Safety Executive (HSE).

The Offshore Installations (Safety Case) Regulations 1992 (HMSO 1992) were in addition to and amended the Offshore Installations (Safety Representatives and Safety Committee) Regulations 1989. The Regulations are applicable to fixed and mobile installations, combined operations and abandonment of fixed installations. Similar to the nuclear and COMAH regimes they require a safety case that is reviewed, in this case by the Offshore Safety Division of the HSE. The risk assessment methods are also similar to the nuclear installations, but modelling pool and jet fires as hazards in their own right rather than simply as mechanisms for the release of radioactivity or as a failure mode of safety critical systems.

A key part of an offshore safety case is demonstrating that all hazards with the potential to cause a major accident have been identified, their risk evaluated, and that measures have been or will be taken to reduce the risks to people affected by those hazards to the lowest level that is reasonably practicable (ALARP) (HSE 2003). Acceptable safety cases will demonstrate that a structured approach has been taken which:

a. identifies all major accident hazards. The identification methods should be appropriate to the magnitude of the hazards involved and a systematic process should be used to identify where a combination or sequence of events could lead to a major accident;

b. evaluates the risks from the identified major accident hazards. Any criteria for eliminating less significant risks should be explained and in deciding what is reasonably practicable, relevant good practice and sound engineering principles should be taken into account. In addition, human factors need to be accounted for and safety critical tasks should be analysed to determine the demands on personnel;

c. describes how any quantified risk assessment has been used, and how uncertainties have been taken into account;

d. identifies and describes the implementation of the risk reduction measures. The reasoning for or against the choice of risk reduction measures to be implemented should be clear.

1.5.4 Transport

1.5.4.1 King's Cross fire, 1987

On Wednesday 18 November 1987, 31 people lost their lives when a fire occurred in escalator 4 at the King's Cross underground station (Fennell 1988). Although there had been a smoking ban in place since 1985, ignition, probably due to smokers' materials, occurred at about 19:25. The fire probably grew from the ignition of detritus before involving grease and the plywood skirting board. Between 19:43 and 19:45, a modest escalator fire was transformed into a 'flashover' that erupted in the ticket halls. A new mechanism, the 'trench effect' (where the flames laid flat against the wooden escalator steps) led to the flaming and a large amount of dense black smoke in the ticket hall that caused horrendous injuries and killed 31 people.

A combination of factors, including poor communications between members of staff and London Fire Brigade, uncertainty about evacuating passengers, confusion about who was in charge, and a failure to operate the water-fog equipment, meant that the fire was able to take hold, eventually propagating violently up the escalator through the 'trench effect'.

The safety ethos of the time, as summed up by Lord Fennell (1988), was that:

* 'there have been smoulderings ... year in year out';
* 'there had been some escalator fires, ... but no passenger had ever been burned ...'; and
* 'a safe environment is not one in which there is an absence or low number of serious injury incidents ...'.

With the benefit of hindsight these statements clearly did not adequately describe the nature of fire safety at that time. The dilemma for the operator was that a significant body of historical experience indicated that such an event was highly unlikely, but on the other hand the potential ignition of combustible materials mean that it is difficult to call this kind of event incredible.

1.5.4.2 *The Railways (Safety Case) Regulations 2000*

Prior to the King's Cross fire, the underground system in London was regulated by the Railway Inspectorate. Company standards were in place and the Inspectorate would inspect and audit these standards and investigate any significant incidents, e.g. derailment.

From a fire perspective, London Underground Ltd had a very good record with relatively few large fires, injuries or fatalities over its 100-year or so history. There were, however, some occasional but large escalator fires and a large number of small fires, usually ignition of rubbish by the third rail arcing.

One of the initial regulatory responses to the King's Cross fire was to extend certification under the Fire Precautions Act 1971 to cover underground stations (the Sub-surface Railway Station Regulations, also known as 'Section 12').

Subsequently, the Railways (Safety Case) Regulations 2000 (HMSO 2003) replaced the 1994 Regulations and made them the sole responsibility of the HSE. Here, a safety case is described as a document produced by a railway operator which describes their operations, analyses the hazards and risks from those operations and explains the control measures, such as procedures and managerial systems that the operator has put in place to manage those risks.

Railways operators are required to prepare and submit safety cases to HSE (HSE 2003). The safety case needs to provide sufficient specific information to describe the nature and extent of the operation and must demonstrate that the operator has undertaken adequate risk assessment for all operations, identified risk control measures, and has systems in place to ensure the measures are implemented and maintained. HSE inspectors form judgements about the completeness of a safety case and the adequacy of the arguments presented to show that risks have been properly controlled. The criteria represent what is currently accepted as good practice. The criteria have been published to make them widely known throughout the railway industry and help develop a common understanding of the requirements for producing safety cases and to make the process by which the HSE assess them transparent. The criteria used to assess risk assessment aspects of Railway Safety Cases are:

- The Safety Case should give details of the duty holder's organisation and arrangements for identification of hazards and assessment of risk.
- The Safety Case should justify the methodologies used for the identification of hazards and assessment of risk with particular reference to any assumptions and data used, together with the methods of calculation.
- The Safety Case should describe the significant findings of the risk assessments and demonstrate that the control measures are adequate to control the risk to a level as low as reasonably practicable.
- The Safety Case should describe the duty holder's arrangements to review risk assessments in the light of new information, new technology, incidents, or other changes that may affect risks, and to ensure that the risk assessments remain valid.

1.5.5 Tunnels

1.5.5.1 Mont Blanc fire

On 24 March 1999 a fire occurred in the Mont Blanc tunnel between France and Italy, killing 38 people (Lacroix 2001). The tunnel is 11,600m long with bi-directional single lane traffic and had a transverse ventilation system (air inlet and exhaust) along the length of the tunnel.

At 10:53 an HGV stopped halfway through the tunnel and was seen to be on fire. The HGV pulled a refrigerated trailer containing 9 tons of margarine and 11 tons of flour and was insulated by polyurethane foam. The driver was unable to reach his cab extinguisher and did not attempt first aid fire fighting. At 10:57 the alarm was raised by tunnel emergency telephone. The fire brigades entered the tunnel at 10:57 and 11:11 from France and Italy (respectively), but were unable to effect extinguishing operations due to thick black smoke.

There were delays in mobilising emergency response on the Italian side of the tunnel and in configuring the ventilation system to manage the smoke. The fire was between 75 and 190MW at its peak and burnt for 53 hours. Many of the dead were found in their vehicles and four cars had attempted U-turns.

1.5.5.2 Tauern fire

On 29 May 1999, in the Tauern Road Tunnel in Salzberg, Austria, a rear-end collision of a lorry caused a large fire that killed eight people (Eberl 2001). The tunnel is 6,400m long with bi-directional single lane traffic and a fully transverse ventilation system.

At 04:48 ignition occurred after a road traffic accident. The lorry contained spray cans and paints and there may have been a fuel spill from one of the cars. The fire grew steadily, the ventilation system was configured in emergency mode with a clear layer through which many people escaped to the adjacent safety tunnel. Ten to 15 minutes after ignition, a series of explosions were seen to lead to a rapid increase in heat and smoke and smoke stratification was lost.

The fire was extinguished by 11:00 the following day and this revealed that four people had died in the original road traffic accident and four people were found to have died from the effects of the fire in a vehicle. They had been seen to evacuate the tunnel and then return to pick up papers from vehicles.

1.5.5.3 The European Tunnel Directive

Standards for road tunnels in the UK are set in a guide BD78/99 published by the Highways Agency. Most new road tunnel projects are now subject to some form of fire risk assessment using either matrix methods, probabilistic analysis and/or full quantification.

In 2004, the EU published a Directive for road tunnels on the Trans-European road network. It remains to be seen how this will be enacted by Member States, what implications it will have for other road tunnels (via the ALARP principle) and how, and to what extent, fire risk assessment will form part of the safety regime in future.

1.5.6 Summary

This section shows how quantitative fire risk assessment was first developed several centuries ago. It again found favour in addressing risk issues in the nuclear industry where many of the technical foundations were established.

As fire disasters have affected other activities, these approaches and their regulatory regime have been implemented in different ways to suit different needs. In the nuclear industry great effort is placed on a robust quantification of fire hazards and risks against a background of approved codes of practice; whereas in the railway industry, fire safety is normally addressed by accepted company standards with qualitative risk assessment to satisfy the fire regulations for large numbers of simple existing buildings and quantitative risk assessment methods used by exception to address particular significant issues.

Buildings generally use qualitative techniques and semi-quantitative techniques, but quantitative approaches are being increasingly used to better inform fire safety decisions in a wide range of buildings such as utility buildings, hospitals, airports and railway stations.

1.6 General principles of fire risk assessment

1.6.1 Introduction

There are a wide range of fire risk assessment models and approaches available (Ramachandran 1979/80, 1988, 2002, Beck and Yung 1994, Fraser-Mitchell 1997, Frantzich *et al.* 1997, Watts 1996, Fitzgerald 1985, Charters and Marrion 1999). However, until recently fire risk assessment was rarely used in the design of buildings and so this section proposes that, for fire risk assessment to become an integral part of fire engineering design, it needs to satisfy at least three requirements. It needs to:

*   be consistent with the regulatory paradigm;
*   be based on engineering and scientific principles; and
*   add value to the design process.

This section also discusses the proposition that there are two things that quantified fire risk assessment does not need to do to become an integral part of the design process. It does not need to:

*   address all aspects of building fire safety; or
*   recreate exact fire events.

These five issues are discussed in more detail below.

1.6.2 Consistency with the regulatory paradigm

To be consistent with the regulatory paradigm means that quantified fire risk assessment needs to address the concerns and objectives of fire regulations. In England and Wales, the legislation that governs the design of new buildings is the Building Regulations 1991. The objectives for fire safety are contained in Part B of these regulations (ADB 2007) and are divided into five areas:

- B1   Means of warning and escape
- B2   Internal fire spread (linings)
- B3   Internal fire spread (structure)
- B4   External fire spread
- B5   Access and facilities for the fire service.

The objectives are stated as functional 'Requirements'. For example, B1 requires that:

> The building shall be designed and constructed so that there are appropriate provisions for the early warning of fire, and appropriate means of escape in case of fire from the building to a place of safety outside the building capable of being safely and effectively used at all material times.

The prescriptive guidance on how to meet this requirement is contained in the ADB (2007). For a specific type of building this guidance provides information on the:

- number of occupants
- number of exits
- overall exit width
- travel distance to an exit
- location of exits.

For fire risk assessment to be accepted as 'evidence tending to show' that a design fulfils the functional requirement of the legislation, it is essential that it addresses the issue or issues of concern (the functional requirement) and uses the same parameters as the prescriptive guidance.

1.6.3 Based on engineering and scientific principles

To satisfy engineering and scientific principles, models for quantified fire risk assessment need to have a good theoretical and evidential basis. General principles such as:

- the conservation of mass and energy;
- that risk is the combination of the frequency and severity of an unwanted event; and

- fire safety is concerned with the time and location of certain events such as evacuation, untenability and/or structural collapse, are essential in judging the theoretical robustness of a model.

The evidence and data on which the model is based is also important. For probabilistic data such as the frequency of fires, reliability of systems and any data on the frequency of the unwanted outcome(s) of concern, we need to assess:

- What is the set of cases that the data is drawn from?
- What case is the data measuring?
- How similar is the building being designed to the cases considered?
- Will variations in statutory controls and design practice skew the data?

Equally, for physical data, we need to ask: is it based on standard tests or ad-hoc tests, large or small scale, new or as used, single or multiple tests, in a similar enclosure etc? The answers to these questions and the nature of the data will affect the way in which the data is used, and more significantly, the way in which the answer is used.

An example of the kind of difficulty that can arise occurred with the use of a well-known points scheme for fire risk assessment. The scheme covered a building occupancy where evacuation of occupants is difficult and often hazardous, so the fire strategy was usually based on a significant degree of compartmentation. However, in attempting to include all fire safety systems in the assessment, the scheme allowed a very low level of compartmentation to be compensated for by a very high degree of fire signs, notices and emergency lighting. Although this 'alternative strategy' may work in some occupancies, it was clearly inappropriate here and this occurred because the theoretical model did not adequately represent the system it was modelling.

This example also implies that there may be fire systems whose need is self-evident, where the impact of their level of performance on the level of risk cannot be predicted using current knowledge. Fire signs, notices and emergency lighting all fall into this category and are, therefore, normally best served by prescriptive system standards.

However, no matter how hard we try to model all aspects, to make the fire risk assessment feasible, simplifications and assumptions are necessary. So, as with all engineering analysis, simplifications or assumptions must be clear and supported by evidence. When undertaking quantified fire risk assessment against absolute criteria and comparative analysis, where the ranking of solution is important, it is possible to make simplifications and assumptions err on the side of safety. However, in comparative analysis, where the quantitative difference between cases is important, care should be taken that the assumptions do not bias the comparison in one direction or another.

There also needs to be a balance in approach to the method of quantified risk assessment. For example, some models incorporate very sophisticated physical sub-models for fire growth, smoke movement, detection, egress, structural response etc and use single point data for highly variable variables. They often also link the physical sub-models together with simple 'yes/no' Boolean logic. This may be an attempt to use the best available knowledge, but there may quickly come a point where the 'span of sophistication' is so great that the higher level of sophistication becomes unhelpful. Equally, other models use very sophisticated statistical models to quantify the risk from fire. However, these methods often make highly simplified assumptions about the physical environment, which mean that they cannot take the particular nature of a specific building into account when assessing the risk.

Both of these types of model have been used in support of national fire safety policy decisions but have not been used extensively on specific projects. Therefore, it is recommended that quantitative fire risk models should treat the frequency and the consequence sides of the fire risks predicted with an appropriate and balanced emphasis, depending on the aspect of fire safety being analysed.

1.6.4 Adding value to the design process

To add value to the design process, the analysis needs to be feasible, usually on several different design options, during the 'design window'. Design is often an iterative process and fire safety analysis needs to be undertaken in a finite period of time for its results to be of use. The design iteration time step can be anything from a few days to a few months depending on the building, but the trend is generally downward and so detailed quantified fire risk assessment that takes weeks to perform is going to be of limited use in the design of most buildings.

Just as crucially, the results of the analysis also need to improve the design decision-making process in terms of the safety, aesthetics, function and/or cost of the finished building. Therefore, as with the regulatory paradigm, the quantified risk analysis needs to have the same currency as the architect. Fortunately, the regulatory paradigm and the design parameters of interest to the architect have a high degree of overlap and so it should be relatively easy to use the same design parameters to address issues of concern to the architect.

The above two requirements combined demand that quantified fire risk assessment be quick enough to use, but sufficiently sophisticated that it can provide the necessary detail and level of accuracy for design decision making.

1.6.5 Not addressing all aspects of building fire safety

A common expectation of quantified fire risk assessment methods is that they should address all aspects of building fire safety. It is certainly true that many fire safety systems impact on the level of fire safety in more than one way. For example, a sprinkler system will suppress a fire, reducing property damage and reducing the risk to occupants. However, independent aspects of fire safety can be addressed using different techniques, e.g. radiation modelling of external fire spread and prescriptive standards for egress, such as travel distances, as long as the resultant level of those particular fire precautions does not impact adversely on a separate fire safety objective.

The same is also true of quantified risk assessment. This does, however, imply that those aspects not addressed by quantified fire risk assessment need to be addressed by prescriptive standards or by deterministic performance-based fire design. It also implies that the resulting package of fire precautions needs to form a coherent whole.

1.6.6 *Not recreating exact fire events*

Another common expectation of quantified fire risk assessment is that it should recreate, exactly, past fire events. Examination of structural analysis shows that it does not exactly model the forces in, or the forms, of a structural member in, say, a bridge. So fire risk analysis, like any other safety critical engineering analysis, only needs to represent actual events to the extent necessary to provide a sound basis for design decision-making.

References

ADB (2007), The Approved Document to Part B of the Building Regulations 2000, As amended 2002, (ADB), Office of the Deputy Prime Minister, The Stationery Office, London.

Allen F R, Garlick A R, Hayns M R and Taig A R (1992), *The Management of Risk to Society from Potential Accidents*, Elsevier, London.

Beck V B and Yung D (1994), The development of a risk-cost assessment model for the evaluation of fire safety in buildings, *Proceedings of 4th International Symposium of Fire Safety Science*, International Association of Fire Safety Science, Ottawa.

BS 7974 (2001), Code of Practice on the Application of fire safety engineering principles to the design of buildings, British Standards Institute, London.

Carvel R O, Beard A N and Jowitt P W (2001), The influence of longitudinal ventilation systems on fires in tunnels, *Tunnelling and Underground Space Technology*, 16,(1) 3–21.

Charters D (2000), What does quantified fire risk assessment need to do to become an integral part of design decision making?, *Proceedings of Conference on Engineered Fire Protection Design*, Society of Fire Protection Engineers, San Francisco.

Charters D and McGrail D (2002), Assessment of the environmental sustainability of different performance based fire safety designs, *Proceedings of the 4th International Conference on Performance-Based Codes and Fire Safety Design Methods*, Society of Fire Protection Engineers, Melbourne.

Charters D and Wu S (2002), The application of 'simplified' quantitative fire risk assessment to major transport infrastructure, Society of Fire Protection Engineers Symposium on Risk, New Orleans.

Charters D, Paveley J and Steffensen F-B (2001), Quantitative fire risk assessment in the design of a major multi-occupancy building, *Proceedings of Interflam 2001*, Interscience Communications, Cambridge.

Charters D A (1992), Fire risk assessment in rail tunnels, *Proceedings of the International Conference on Safety in Road and Rail Tunnels*, Independent Technical Conference, Basle.

Charters, D A (1996), Quantified assessment of hospital fire risks, *Proceedings of Interflam 96*, Interscience Communications, Cambridge.

Charters D A and Marrion C (1999), A simple method for quantifying fire risk in the design of buildings, *Proceedings of the Society of Fire Protection Engineers Symposium on Risk, Uncertainty, and Reliability in Fire Protection Engineering*, Society of Fire Protection Engineers, Baltimore, MD.

Combined Code (1998), *Corporate Governance*, Organisation for Economic Cooperation and Development (OECD), Paris.

Cullen W (1990), *The Public Inquiry into the Piper Alpha Disaster*, Department of Energy, HMSO, London.

Eberl G (2001), The Tauern Tunnel incident: what happened and what has been learned, *Proceedings of the Fourth International Conference on Safety in Road and Rail Tunnels*, Independent Technical Conferences, Tenbury Wells.

Farmer F R (1967), Siting criteria – a new approach, International Atomic Energy Agency (IAEA) Symposium, April.

Fennell D (1988), *Investigation into the King's Cross Underground Fire*, Department of Transport, HMSO, London.

FIRECODE HTM 86 (1987) *Fire Risk Assessment in Hospitals*, Department of Health, HMSO, London.

FIRECODE HTM 86 (1994) *Fire Risk Assessment in Hospitals*, Department of Health, HMSO, London.

Fitzgerald, R W (1985), An engineering method for building fire safety analysis, *Fire Safety Journal*, 9, 233.

Frantzich H, Magnusson S-E, Holmquist B and Ryden J (1997), Derivation of partial safety factors for fire safety evaluation using reliability index β method, *Proceedings of 5th International Symposium of Fire Safety Science*, International Association of Fire Safety Science, Melbourne.

Fraser-Mitchell, J (1997), Risk assessment of factors relating to fire protection in dwellings, *Proceedings of 5th International Symposium of Fire Safety Science*, International Association of Fire Safety Science, Melbourne.

Green A E and A J Bourne (1972), *Reliability Technology*, Wiley-Interscience, New York.

HMSO (1965), Nuclear Installations Act, 1965, HMSO, London.

HMSO (1988), *The Tolerability of Risk from Nuclear Power Stations*, Health and Safety Executive, HMSO, London.

HMSO (1992), The Offshore (Safety Case) Regulations 1992, HMSO, London.

HMSO (2003), The Railways (Safety Case) Regulations 2000 (as amended 2003), HMSO, London.

HSE (1999), The Control of Major Accident Hazards (COMAH) Regulations 1999, Health and Safety Executive, London.

HSE (2003), *Good Practice and Pitfalls in Risk Assessment*, Research Report 151, Health and Safety Executive, London.

Lacroix D (2001), The Mont Blanc Tunnel fire: what happened and what has been learned, *Proceedings of the Fourth International Conference on Safety in Road and Rail Tunnels*, Independent Technical Conferences, Tenbury Wells.

Laplace P-S (1814), *Essai philosophique sur les probabilités*.

Linstone H A and Turoff M (1975), *The Delphi Method: Techniques and Applications*, Addison-Wesley, Glenview, IL.

Office of the Deputy Prime Minister (2005), *UK Fire Statistics*, Office of the Deputy Prime Minister, London.

PD 7974 Part 7 (2003), Application of fire safety engineering principles to the design of buildings, Part 7 Probabilistic Risk Assessment, British Standards Institute, London.

Ramachandran G (1979/80), Statistical methods in fire risk evaluation, *Fire Safety Journal*, 2, 125–154.

Ramachandran G (1988), Probabilistic approach to fire risk evaluation, *Fire Technology*, 24, 3, 204–226.

Ramachandran G (2002), Stochastic models of fire growth, *SFPE Handbook of Fire Protection Engineering*, 3rd Edition, Section 3, Chapter 15, Society of Fire Protection Engineers, Baltimore, MD.

Turnbull N (1999), *Internal Control: Guidance for Directors on the Combined Code*, (The Turnbull Report), The Stationery Office, London.

Watts, J M (1996), Fire risk assessment using multiattribute evaluation, *Proceedings of 5th International Symposium of Fire Safety Science*, International Association of Fire Safety Science, Melbourne.

World Fire Statistics Centre (2007), *Information Bulletin No 23*, TheGeneva Association, Geneva.

2 Qualitative and semi-quantitative risk assessment techniques

This chapter describes some of the qualitative and semi-quantitative fire risk assessment techniques used in fire safety. The qualitative techniques are categorised as unstructured and structured. The semi-quantitative methods described include matrix methods and points schemes.

2.1 Qualitative techniques

Most fire risk assessments of buildings are undertaken to satisfy legislation and in the vast majority of cases that means the fire regulations in existing buildings which addresses life safety in occupied buildings. There can also be many other reasons for undertaking a fire risk assessment of a building. These include:

- life safety
- asset protection
- mission/business continuity
- environmental protection
- heritage
- public image
- post-fire analysis.

The initial qualitative assessment process in fire safety engineering (also known as the Qualitative Design Review (QDR) or Fire Engineering Design Brief (FEDB)) can be likened to a qualitative risk assessment process based on a table-top exercise.

Qualitative methods rely on identification of factors that affect risk. These factors may be those that affect the level of hazard, such as combustibles or the level of fire precautions intended to mitigate those hazards, such as fire detection. An assessment of all these factors is undertaken against a benchmark set of values and a judgement made as to whether the factor is higher, lower or about the same as the benchmark. A review is then undertaken of all the factors to judge how, overall, the area being assessed compares to the benchmarks. If additional mitigation measures are judged to be required, this is a user input or a choice from a prescribed list. Assessments under the Fire Safety Order in the UK fall into this category.

Qualitative fire risk assessment methods can be placed into two categories: unstructured and structured.

2.1.1 Unstructured methods

An unstructured method of qualitative fire risk assessment would follow the five-step risk assessment process outlined in the guidance to the Fire Safety Order 2005 (TSO 2005). These steps are:

1   Identify potential fire hazards.
2   Identify those in danger from fire.
3   Evaluate the risks and whether the current fire precautions are adequate.
4   Record findings and actions and communicate to employees.
5   Keep the assessment under review and revise it when necessary.

The method is unstructured as it has no prescribed standards for what types and levels of precautions are appropriate for various levels of hazard. This can make judging adequate compliance rather subjective.

2.1.2 Checklists

Quite often checklists are used to help identify fire hazards in unstructured qualitative risk assessments and in fully quantitative fire risk assessments. The advantage of a checklist is that it indicates to the non-professional what may constitute a fire hazard and it reminds the professional risk assessor of the range of fire hazards. The disadvantages can be that they are used too literally to record every piece of combustible material and they can lead to unusual fire hazards being missed, because any checklist can never be totally comprehensive. The fire hazard lists below are examples of checklists. Fire hazards may include combustible materials and ignition sources.

Combustible material (Figure 2.1) can include:

*   products based on flammable liquids, e.g. paints, varnish, thinners;
*   flammable liquids and solvents such as petrol, white spirit and paraffin;
*   flammable chemicals;
*   wood;
*   paper and card;
*   plastics, rubber, foam such as polystyrene and polyurethane used in packaging and furniture respectively;
*   flammable gases, such as LPG;
*   textiles;
*   packaging materials;
*   waste materials such as wood shavings, offcuts, dust, paper and textiles.

Figure 2.1 An example of combustible materials

Ignition sources (Figure 2.2) can include:

- matches and smokers' materials;
- naked flames;
- electrical, gas or oil fired heaters;
- hot processes such as welding and grinding;
- cooking;
- engines or boilers;
- machinery;
- faulty or misused electrical equipment;
- lighting equipment such as halogen lamps;
- hot surfaces and the obstruction of equipment ventilation;
- friction from loose bearings or drive belts;
- static electricity;
- metal impacts such as metal tools striking each other;
- arson.

Lists of hazards such as these can form the basis of checklists and be used as part of a hazard identification process.

Figure 2.2 An example of a car fire, most car fires are ignited by electrical faults or arson

A number of factors are important in determining the level of hazard from a combustible material. For combustible materials, the extent of the hazard depends on their:

- ignitability – e.g. certain flammable liquids are easier to ignite than others; thin items tend to be easier to ignite than thick items (see Drysdale 2002 p208);
- reaction to fire – e.g. rate of flame spread, heat release, smoke production, toxicity etc (BS EN 13501-1 2000);
- amount – the higher the fire load the greater the potential rate of heat release and fire severity;
- orientation – vertical, corner and ceiling surfaces and high-racked storage tend to increase the rate of flame spread (see Drysdale 2002 p232);
- location – materials that are grouped together or near exits or large numbers of people may present an increased risk.

A number of factors are important in determining the level of risk represented by an ignition source. For ignition sources the extent of the hazard that they present may depend on their:

- frequency of occurrence – more frequent ignition sources such as cooking and hot works tend to increase the level of risk;

- ignition energy – higher ignition energy may mean increased initial fire growth rate;
- temperature – higher temperate ignition sources may result in increased initial fire growth rate (see Drysdale 2002 pp191 and 212);
- number – multiple ignition sources from, say, arson may be more dangerous than a single source;
- timing – ignition when many people are present or no-one is present may result in increased life and property fire risks (respectively);
- location – ignition at the bottom of an item or near an exit may increase the risk.

Unstructured approaches are simple to apply and may be very general in nature. This may mean that they can be highly subjective and may rely heavily on the expertise and judgement of the risk assessor.

Some guidance for fire risk assessment in existing buildings requires that fire risk assessments are 'suitable and sufficient' without prescribing an approach. This guidance describes a unstructured qualitative form of fire risk assessment as the regulations apply to a wide range of workplaces, many of which are small and simple in nature.

A summary report for an unstructured fire risk assessment can be seen in Figure 2.3. The risk assessment consisted of a survey of the building, locating all hazards and precautions, designed or installed, with relation to fire safety. In addition to this, staff members in the different occupancies were asked some questions in relation to safety procedures and checks were made concerning the maintenance of hand-held fire extinguishers. All these factors were then evaluated to assess whether the risks from fire were adequately addressed.

Where available, the maintenance records for fire safety related equipment were inspected. However, during the survey of the building, no equipment was tested.

There are a number of alternative frameworks for carrying out fire safety risk assessments. The assessment in this building was carried out using the five-step process above. The checklist, which defines the manner in which the assessment was carried out is outlined below:

1   Assess the fire risks in the workplace.
2   Check that a fire can be detected in a reasonable time and that people can be warned.
3   Check that people who may be in the building can get out safely.
4   Provide reasonable fire-fighting equipment.
5   Check that those in the building know what to do if there is a fire.
6   Check and maintain your fire safety equipment.

The building was surveyed on a floor-by-floor and occupancy-by-occupancy basis. Within each area, the hazard to persons within the occupancy and persons in other occupancies were considered.

CFire Solutions		*Fire Safety Risk Assessment Schedule*	
Client:	Bristol Holdings	**Date of assessment:**	21 July 2002 / 13 October 2002
Site:	Febrile House, Umbridge	**Assessment officer:**	Cherry Stone
Location of occupancy:	Ground Floor	**Title off occupancy:**	Fenda
Description of occupancy:	Open plan office		
Fire risks identified	Both escape doors from this space are not remote from each other (i.e. they are in the same space).		
Fire detection and alarm provisions	Manual call points provided.		
Means of escape in the event of fire	Travel distance to alternative escape doors is short (< 18m). No emergency lighting in tenant's area. Fire doors (main escape doors) appear to be FD30 with self closers but door fit was inadequate with a 10mm gap between doors.		
Fire suppression equipment	Manual fire extinguishers (water and CO_2).		
Fire safety procedures and management	No written procedures noted.		
Maintenance of fire safety provisions	Water extinguishers checked March 2002 and CO_2 extinguishers checked Jan 2001. Fire alarm tested weekly.		
Additional notes	In general, due to the area being small, means of escape is considered to be adequate if emergency lighting were installed. The concern is that a fire could grow undetected to a point where manual fire fighting is ineffective and smoke could spread into both escape stairways.		

Figure 2.3 Summary report for an unstructured fire risk assessment

During the survey of the premises, representatives from all available tenants were asked questions, including:

- Do you have an evacuation procedure?
- What do you do when the alarm sounds?
- Have you had fire drills?
- When was the last drill conducted?

2.1.3 Structured methods

Structured methods of qualitative fire risk assessment use the same process as unstructured methods except that, in addition, they tend to prescribe:

- different levels of hazard;
- different levels of precaution;
- the acceptability of different combinations of hazard and precaution; and,
- appropriate risk reduction alternatives.

A good example of this approach is that of FIRECODE HTM 86 (1994) 'Fire risk assessment in hospitals'. This method identifies different types of fire hazard such as 'smoking' and 'fire hazard rooms' and whether the level is 'acceptable', 'high' or 'very high'. It then goes on to identify different types of fire precaution including 'alarm and detection' and 'compartmentation' and whether the level of provision is 'unacceptable', 'inadequate', 'acceptable' or 'high'. Certain combinations of hazard ratings and precautions ratings are deemed to be acceptable. Other combinations lead to the need to reduce risk by a range of prescribed alternatives.

Figure 2.4 shows a summary assessment sheet for a fire risk assessment to HTM 86 (1994). Like the unstructured fire risk assessment shown in Figure 2.3, this risk assessment consisted of a survey of the building, locating all hazards and precautions, with relation to fire safety. Similarly, staff members in the different occupancies were asked some questions in relation to safety procedures and checks were made concerning the maintenance of hand-held fire extinguishers. All these factors were then evaluated to assess whether the risks from fire were adequately addressed.

Again, where available, the maintenance records for fire safety related equipment were inspected. However, during the survey of the building, no equipment was tested.

As before, the building was surveyed on a floor-by-floor and occupancy-by-occupancy basis. Within each area, the hazard to persons within the occupancy and persons in other occupancies were considered.

During the survey of the premises, representatives from all available tenants were asked the following questions:

- Do you have an evacuation procedure?
- What do you do when the alarm sounds?

Assessment record	Meddleston Primary Health Care NHS Trust					
	10/04/1999					
Victoria Ground Floor Pickersgill A Block 10	High Standard	Acceptable Risk/hazard or HTM 85 Standard	High risk/ Hazard	Very high risk/hazard	Inadequate	Unacceptable
1 Patients			X			
HAZARDS						
Ignition sources						
2 Smoking		X				
3 Fire started by patients			X			
4 Arson		X				
5 Work processes		X				
6 Fire hazard rooms				X		
7 Equipment		X				
8 Non patient access areas		X				
9 Lightning		X				
Combustible materials						
10 Surface finishes		X				
11 Textile and furniture		X				
12 Other materials		X				
PRECAUTIONS						
Prevention						
13 Management		X				
14 Training		X				
15 Fire notice and signs		X				
Communications						
16 Observation					X	
17 Alarm and detection	X					
Means of escape						
18 Single direction of escape		X				
19 Travel distance		X				
20 Refuge		X				
21 Stairways						
22 Height above ground level	X					
23 Escape lighting		X				
24 Staff		X				
25 Escape bed lifts						
Containment						
26 Elements of structure		X				
27 Compartmentation		X				
28 Sub-division of roof and ceiling voids						
29 External envelope protection		X				
30 Smoke control						

Figure 2.4 Record sheet from a structured qualitative fire risk assessment method (HTM 86 1994)

- Have you had fire drills?
- When was the last drill conducted?

The main difference of the structured compared with the unstructured fire risk assessment is the evaluation and presentation of the findings. In this case, each assessment sheet was completed from information gained both during the survey and through desktop analysis of drawings and policy statements. This section of the report has further broken down the assessments into three distinct areas:

- adequate assessment areas;
- compensated assessment areas; and
- inadequate assessment areas.

The results of the building survey, review of documentation and records and of ad hoc audit questions are then expressed against the prescriptive standards on the summary assessment sheet in Figure 2.4.

This kind of qualitative flexible prescription identifies areas of high hazard and/ or inadequate precautions and prescribes one or more prescriptive solutions. This is relatively easy and consistent to apply, taking only a matter of hours to survey and days to report. This method does not suffer from the apparent acceptability of inappropriate fire strategies. However, the method cannot easily address non-standard situations, limits options for upgrade and is not applicable to new designs.

The structured fire risk assessment approach was adopted for hospitals because they comprise a more homogeneous estate and in this case, structured methods may be more efficient and provide greater consistency of assessment.

2.1.4 Fire hazard definitions

Another example of qualitative prescription in fire risk assessment can be found in this extract from BS 5588 : Part 11 (1997) 'Code of practice for shops, offices, industrial, storage and other similar buildings', Section 8.1.1.

Examples of different types of hazard are as follows:

- **Normal hazard.** Where any outbreak of fire is likely to remain localised or is likely to spread only slowly, and where there is little hazard of any part of the building structure igniting readily.
- **Low hazard.** Where there are very few flammable and no explosive materials present and where the hazard of a fire breaking out and smoke and fumes spreading rapidly is minimal.
- **High hazard.** Where there are:
    - Materials stored or handled in such quantities or dispositions that they would be likely, if ignited, to cause a rapid spread of fire, smoke or fumes. For example, processes handling large quantities of highly flammable liquids, gases or solids, such as polyurethane foam;

- Unusual circumstances relating to the occupants;
- Certain areas which, due to their function, may present a greater risk of fires occurring and developing than elsewhere.

A similar type of approach is adopted in BS9999: 2008

2.2 Semi-quantitative techniques

Semi-quantitative fire risk assessment methods place some numerical value on the level of risk, without that numerical value representing a precise value of risk.

2.2.1 Matrix methods

The matrix method, so called for its use of matrix-like tables, was developed from the HAZard and OPerablity (HAZOP) method used in the chemical industry. HAZOP is a technique for systematically considering deviations from the design intent of a system and was developed in the chemical industry to identify failure modes of complex process plant (Mannan 1980, Chemical Industry Safety and Health Council 1977). HAZOP studies identify deviations by the application of a series of guide words such as 'too much', 'too little' to the process parameters, such as flow and pressure. The process is best suited to group sessions with multi-disciplinary participants. The technique is completely general and can be applied to processes or procedures of any type and complexity.

The matrix method below incorporates many of the features of HAZOP such as systematic hazard identification with a multi-disciplinary group, but in a much more general way. It then takes the process further by classifying the frequency and consequences of identified events in categories representing an order of magnitude, e.g., one per year to once every ten years etc.

Matrix methods are usually based on a 'structured brainstorm' in a workshop environment (Mannan1980). Rather than a formal 'walking of the halls' with clipboard in hand, they make use of the collective experience and knowledge of people involved in an asset (i.e., the building or other construction under consideration) or people who have related generic knowledge e.g. the organisation's safety officer. Often this includes representatives of the design team and the management team.

It is a method which tries to identify high risk areas (i.e., specific locations or specific activities) so that more precise techniques, which are too costly to apply to the whole facility, can be efficiently targeted. Often these high risk areas are then subject to a quantitative risk assessment.

2.2.1.1 Matrix methodology

The matrix method of fire risk analysis is best illustrated by a flow chart which is specifically designed for a rail facility (see Figure 2.5).

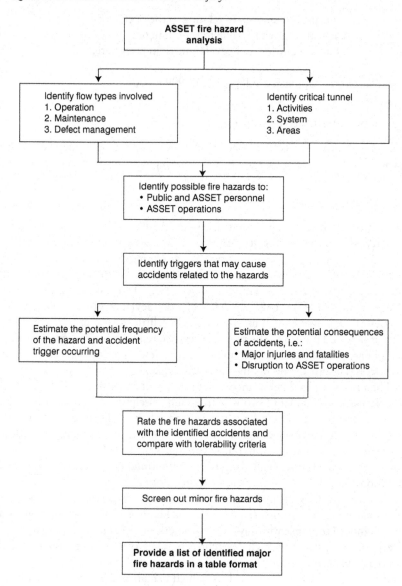

Figure 2.5 An example of a matrix method flow chart

Each of the stages are done in turn. It is a 'one off' and not an iterative approach, i.e., this is very much a snapshot risk assessment method rather than a method of ongoing management. A table is used to record findings as shown in Table 2.1. This table is completed by the workshop attendees.

Tables 2.2–2.7 present typical 'look-up' matrices, which are used by the workshop teams to objectively assess fire hazards. These tables will be created before the matrix workshop. Note that Tables 2.2, 2.3 and 2.5 are industry-specific to different degrees.

Hazards considered can be defined to include a narrow or wide range, such as:

- flammable liquids;
- flammable gases;
- recognised hazardous loads;
- HGV vehicles;
- multi-passenger vehicles;
- cars and light vans.

Other hazards include:

- trespass and arson;
- asset equipment.

Further hazards identified by the workshop team are also included in the analysis. The workshop will normally be limited to fire hazard analysis, however, asset operations and effectiveness of equipment are often the subject of a further HAZOP workshop and evaluation.

Table 2.1 Typical matrix method record sheet

Hazard no	Fire hazard		Accident		Accident severity	Hazard frequency	Trigger probablity	Accident frequency	Risk classification	Tolerability/ remarks
	Potential	Causes	Potential	Trigger						

Table 2.2 Hazard /accident frequency

Category	Description	Definition	Guide frequency	Example
A	Frequent	Likely to occur repeatedly (more than once a week)	> 1 per 100 hrs	Stoppage due to flow congestion Normal operations
B	Occasional	Likely to occur from time to time (once a year or more)	1 per 100 to 1 per 10,000 hours	Temporary loss of control function
C	Probable	Likely to occur once on the system during its operational life (e.g. once in 30 years)	1 per 10,000 to 1 per million hours	Failure of active system, e.g. ventilation fan
D	Improbable	Unlikely to occur during the system design life (i.e., less than one in 100 years)	1 per million to 1 per 10^8 hours	Major chemical spillage in asset
E	Remote	Very unlikely to occur during system design life (i.e., less than 1% likelihood of happening throughout the life of the system)	1 per 108 hours to 1 per 10^{10} hours	Major earthquake
F	Incredible	Extremely unlikely to occur during the life of the system (i.e. once in a million years).	< 1 per 10^{10} hours	Meteor impact

Table 2.3 Accident trigger probability

Category	Description	Definition	Guide frequency	Examples
A	Frequent	Likely to exist repeatedly on the system	> 10%	Traffic driving too fast
B	Occasional	Often exist somewhere on the system	1% – 10%	Maintainer working in the plant room
C	Probable	Likely to exist from time to time on the system	0.01% – 1%	Staff working in asset during running hours
D	Improbable	Likely to exist only on rare occasion on the system	0.001% – 0.01%	Trespasser in asset (possible arsonist)
E	Remote	Unlikely to exist even on rare occasion on the system	0.0001% – 0.001%	Control room unavailable, e.g. due to fire
F	Incredible	Extremely unlikely ever to exist on system	< 0.0001%	Two separate serious fires in asset

Table 2.4 Accident frequency categories

Accident frequency		Hazard cause frequency					
		A	B	C	D	E	F
Accident	A	A	B	C	D	E	F
trigger	B	B	B	C	D	E	F
	C	C	C	D	E	F	F
probability	D	D	D	E	F	F	F
	E	D	D	F	F	F	F
	F	E	E	F	F	F	F

Table 2.5 Accident severity

Category	Description	General description	Examples
I	Catastrophic	Possible multiple deaths, or probable death of passenger or member of the public, or total system loss Uncontrolled incident leading to: Asset unavailable (asset cannot operate)	Vehicle strikes maintenance crew Passenger(s) fall from moving vehicle Collapse or loss of asset Assets needing repair or recommissioning before flow can resume Incidents resulting in asset unavailability for > 24 hours
II	Critical	Probable major injury to one or more passengers or members of the public, or probable major injury and possible death of staff, contractor or trespasser or major system damage Loss of control function Asset cannot operate	Trespasser struck by vehicle Collision of vehicles in asset Assets needing inspection before flow can resume Controlled evacuation from vehicles into asset Incidents resulting in asset unavailability for < 24 hours
III	Marginal	Probable minor injuries to one or more passengers or public or staff, contractors or trespassers, or possibility of major injury to a single individual Major service disruption Third party action requiring halt to operations whilst remedial action (rescue, repair, inspection) takes place	Maintainer strikes head on sharp object. Strain injury to asset staff Controlled evacuation of vehicles from asset Signalling fault reported by TO Fire alarm and FP system operation in Control Centre Control Centre fault requiring investigation
IV	Negligible	Trivial injuries to passengers or staff, with no more than possibility of minor injury Unscheduled maintenance required without interruption to service.	Unavailability of FP system due to maintenance / defect management

Table 2.6 Risk classification

Risk classification		Accident severity category			
		I	II	III	IV
Accident	A	A	A	A	B
frequency	B	A	A	B	C
category	C	A	B	C	C
	D	B	C	C	D
	E	C	C	D	D
	F	C	D	D	D

Table 2.7 Risk tolerability

Category	Description	Definition
A	Intolerable	Risk shall be reduced by whatever means possible.
B	Undesirable	Risk shall only be accepted if risk reduction is not practicable.
C	Tolerable	Risk shall be accepted subject to confirmation that risk is as low as reasonably practicable.
D	Acceptable	Risk shall be accepted subject to endorsement of the supporting hazard analyses.

2.2.1.2 Quantification for matrix methods

Matrix methods rely on each category of frequency or consequence representing an order of magnitude. The workshop attendees, assisted by historical information and their own experience, can then identify a range of fire hazards using a structured brainstorming technique. Having identified a reasonably comprehensive a list of fire hazards, these are rationalised to eliminate duplicates and trivial events. Each fire hazard is then evaluated in turn by the workshop attendees against each of the parameters e.g. frequency, consequence etc.

In the tables, ratings for each parameter are represented by letters and these are then combined using a risk matrix to identify the level of risk using capital roman numerals. These associations are not arbitrary and rely on the basic principles of risk assessment. To illustrate this, we will consider the numerical rating system used in the second example in section 2.2.1.3. Here each category of each parameter represents an order of magnitude in the scale of interest and is represented by an integer number rising as the scale rises.

The total risk rating for life safety is given by:

$$\text{Total life risk} = F + S + P \tag{2.1}$$

where: F is the rating for the frequency of initiating event (usually ignition)
S is the likely severity of the outcome, and
P is the probability that that severity will result from that initiating event.

Clearly there is a range of events that could occur from any particular initiating event. The workshop may feel that a particularly severe outcome is likely from an event, but this may be offset, to a degree, by a relatively low probability and vice versa. This helps guard against unintentional bias in the process. For example, small fires with large consequences may happen frequently and be categorised as 1 and 3 respectively. Equally, large fires may happen infrequently and be categorised as 3 and 1 respectively. In this case both would have a risk category of 4 and so it is important that the frequency and consequence ratings are consistent with each other.

It often helps to define the nature of the events being quantified as most likely or reasonable worst case or both (depending on the objective of the workshop) to help participants converge on a shared view of the ratings.

The process is addition rather than multiplication, because each number represents an order of magnitude:

$$\text{Risk} = 10^F \times 10^S \times 10^P = 10^{(F + S + P)} \tag{2.2}$$

For the purposes of prioritisation we can simply use Equation (2.1).

For the estimation of total business risk it is slightly more complicated because business risk is a combination of asset damage and performance loss 'damage' that may have different values for the same event.

Total business risk is given by:

$$\text{Total business risk} = F + P + (\log_{10}[10^{Cd} \times 10^{Cp}]) \tag{2.3}$$

where Cd and Cp are the asset damage and performance loss ratings respectively. Thus for example:

$$3 + 2 + \log_{10}(10^3 \times 10^3) = 8.3$$

In the example spreadsheet discussed at the end of this section the results can appear to be counter-intuitive at first sight.

2.2.1.3 Matrix workshops

The staged objectives of such workshops are, in sequence, to:

- identify fire hazards and accidents that could occur in an asset;
- estimate the consequences (by making use of 'accident severity matrices');
- rate hazards (by using a risk classification matrix);
- screen out minor fire hazards; and
- provide a list of identified major fire hazards.

Once all the participants have been briefed on the asset under assessment and the process of assessment, a structured brainstorming process is used to fill in the first five columns of the record sheet (Table 2.1). The keys to good brainstorming are recognised as:

- say the first thing that enters your head;
- no discussion; and
- no critical comment.

Once the fire hazards and accidents have been identified, they can be sorted and duplicates amalgamated and irrelevances discarded, i.e. a list of hazards is created. These would be entered into the first column of Table 2.1.

The next stage is to estimate which of the matrix categories each fire hazard best fits. This can best be done with a combination of historical data and experience from the key stakeholders around the table. One of the major beneficial side effects of the matrix workshop is the final column of tolerability where comments on the acceptability of hazards or their mitigation can be recorded for future reference. This often leads to greater mutual understanding of a project and its hazards amongst design and management teams alike.

Once all the fire hazards have been rated, they can be sorted in terms of priority and the minor fire hazards noted and the major hazards addressed further by quantitative fire risk assessment.

Often matrix methods are used to collect information on hazards and key event progressions that are then used in a full quantitative fire risk assessment. In a quantitative fire risk assessment following a matrix method risk assessment:

- Matrix methods are used where the subjective and lack of quantitative information from qualitative methods may lead to poor decision making with significant consequences.
- Matrix methods can be applied to any building or situation where there are hazards with high potential consequences and/or where the risks of an activity may be high. That is why matrix methods are usually a precursor to full quantitative fire risk assessment.

Fires are important in terms of safety risk and commercial cost, both of which need managing effectively. The importance of managing fire risk for both safety and commercial reasons is emphasised by current fire losses and trends:

- There are hundreds of thousands fire events per year in most countries.
- Significant fires have progressively increased over recent years with commercially significant fires accounting for most of the increase.
- The annual cost to national economies is measured in billions of pounds per year.

CASE STUDY

This study summarises the proceedings of a fire safety risk assessment workshop in support of a rail Safety Risk Framework. The workshop was held at a railway station.

The declared objectives of the workshop were to:

1   Provide a company-wide preliminary indication of the nature, scale and source of fire risk, i.e. what proportion of the assets had high, medium and low levels of fire risk.
2   Identify any high consequence and/or high likelihood events requiring immediate attention.
3   Produce a risk ranked list of assets to enable detailed risk assessments to be undertaken.

The workshop used a structured, formal assessment process similar to that of HAZID (HAZard IDentification). It was structured in that the brainstorming and risk rating of hazards followed a process defined by the headings of the matrix table. It was formal in that the matrix table, including comments, is a written record that forms part of the risk register and should be a living document for the system, i.e. it is updated regularly and when aspects of the asset or its operation are changed.

After confirming the assets under consideration, the workshop systematically assessed the assets and the results were recorded on a spreadsheet.

The main parameters that the workshop addressed are described below:

*   asset type;
*   location;
*   failure mode/hazard;
*   cause;
*   consequence;
*   for three scenarios:
    *   unwanted activation of fire alarm;
    *   most likely fire;
    *   worst case fire;
*   to assess for each scenario their:
    *   frequency;
    *   severity;
    *   probability;
    *   asset loss;
    *   performance loss;
    *   current controls;
    *   comments/high priorities;
    *   actions;
    *   asset type.

The workshop focused on the particular railway station. The fire risks contained within the station are complex and varied and are categorised by distinct locations which in general, may include:

- platforms;
- concourses;
- retail;
- non-public areas;
- neighbouring buildings, as appropriate, which may include:
  - offices;
  - hotels;
  - hospitals;
  - construction sites;
  - car parks;
  - connecting underground stations.

The failure modes/hazards under consideration were 'fire and/or unwanted fire alarm activation'.

The workshop identified the main potential causes of the failure mode/hazard for each location. This comprised typical combustibles/ignition sources, e.g. 'overhead line equipment, rubbish, electrical faults, vehicles, arson, smoking'.

The workshop identified the general scale of event in terms of life safety, asset loss and performance loss, e.g. 'death/injury, loss of building, closure of line'.

The workshop then assessed the risks under three main categories:

- unwanted activation of fire alarm;
- most likely fire; and
- worst case fire.

For unwanted activation of fire alarm, the workshop assessed the frequency and consequences of unwanted activation of fire alarm, e.g. an unwanted alarm within a retail area leading to evacuation.

For most likely fire, the workshop considered the frequency and consequences of fires that tend to be more common but not necessarily very severe, e.g. a small fire in a kitchen or office that is confined to the room or item of origin.

For worst case fire, the workshop considered the frequency and consequences of fires that are likely to be much less frequent and probably much more severe, e.g. a fire that involves the whole of a carriage on a platform.

For each of the three event categories, five parameters were assessed:

- F hazard frequency
- P incident probability
- S accident severity
- Cd cost of asset damaged or replacement cost.
- Cp cost of performance penalty or operational loss.

Each parameter was given a rating. Since precise values for these parameters are not available, each rating represents a range of values. For unwanted activation of fire alarm, life safety and business (asset and performance), the total fire risk rating was calculated.

The current controls were identified by the workshop in terms of compliance with main fire legislation and standards, e.g. the fire safety order and/or company standards.

The workshop also identified any high priorities, significant fire scenarios, queries or further comments for the record.

The assessments of the assets under consideration were recorded on a spreadsheet. The basis of the assessment was determined using Tables 2.8 – 2.12.

The following fictional characters may have participated in the workshop:

- Station Manager
- Duty Station Manager
- Fire Systems Manager
- Safety Risk Manager
- Project Sponsor / Company Fire Adviser
- Workshop Facilitator / Fire Risk Engineer
- Project Manager
- Workshop Secretary

The fire hazards for the railway station have been assessed and the results are summarised in Tables 2.8 and 2.9, listing the locations in descending order of risk rating scores in terms of life safety and possible business cost. These overall scores have been determined from the results obtained for the three hazard scenarios considered.

A risk rating of 10 represents a potential maximum level of risk occurrence of 1×10^{-1} fatality equivalents per year which equates to one fatality in every 10 years. Further examples are risk ratings of 9 and 4 which represent potential levels of risk of 1×10^{-2} and 1×10^{-7} fatality equivalents per year respectively. These equate to one fatality in every 100 years for a risk rating of 9, and one fatality in every 10,000,000 years for a risk rating of 4. A 'fatality equivalent' equals one fatality or 10 major injuries. Table 2.13 highlights that the link works, with a risk rating of 10, represents the highest safety risk for this station.

A risk rating of 13 represents a potential maximum loss of £1,000,000 per year (or £10,000,000 once in ten years), 12 represents £100,000 per year and 11 represents £10,000 per year etc.

The figures in Table 2.14 are asset damage and performance penalty costs combined. The performance loss system is a financial penalty system for late trains and so may represent some of these issues. Not explicitly, these may be in the minds of the workshop members when they rate the risks. Table 2.14 highlights that link works, with a risk rating of 13.3, represents the highest business risk. The retail outlets, with a risk rating of 13, are indicated as representing the second highest risk.

Table 2.8 Hazard frequency, F

Occurrence frequency, F	Range	Rating
Never	< 1 in 10,000 years	0
Remote	1 in 1,000 to 1 in 9,999 years	1
Rare	1 in 100 to 1 in 999 years	2
Infrequent	1 in 10 to 1 in 99 years	3
Occasional	1 in 1 to 1 in 9 years	4
Frequent	1 to 10 times per year	5
Common	> 10 times per year	6

Table 2.9 Accident severity (life safety), S

	Rating
None	0
Minor injuries	1
Major injuries	2
One fatality	3
Multiple fatalities	4

Table 2.10 Accident probability, P

Likelihood	No. of fires before accident	Rating
Never	> 10,000	0
Very unlikely	1,000 to 9,999	1
Improbable	100 to 999	2
Occasional	10 to 99	3
Near certain	1 to 9	4

Table 2.11 Cost of asset damage or destroyed, C_d (£)

	Rating
< 100	1
100 to 999	2
1,000 to 9,999	3
10,000 to 99,999	4
100,000 to 999,999	5
1,000,000 to 9,999,999	6
> 10,000,000	7

Table 2.12 Cost of performance penalties or operational loss, C_p (£)

	Rating
< 100	1
100 to 999	2
1,000 to 9,999	3
10,000 to 99,999	4
100,000 to 999,999	5
1,000,000 to 9,999,999	6
> 10,000,000	7

Table 2.13 Life safety risk rating

Location	Risk rating
Link works	10.0
Food preparation basement	9.1
Retail outlets	9.0
East side offices (including station control room)	8.3
Non–public areas – west side offices and south west offices	8.3
Platforms 9–11	8.3
Platform areas 2–8	8.3
Concourse and forecourt	8.0
Platform 1 and access road	8.0
Clothes store (above platforms 9–11)	7.0
Car parks	6.3
Parcel post	6.3
Underground station	5.0
Public highway	5.0
Hotel Way	4.0

Overall risk ratings for the entire station for both business and life safety risk are presented in Table 2.15. The values indicate that the overall score is largely dependent on the greatest single risk due to the logarithmic nature of the process (see above).

As could be expected, the life safety rating is dominated by the risk presented by fires and not unwanted alarm activation. Fires dominate the business risk rating whilst the risk to asset and performance are similar. There may be more scope to reduce risks, cost effectively, for some aspects more than others.

Table 2.14 Business risk rating

Location	Risk rating
Link works	13.3
Retail outlets	13.0
Platforms 9–11	12.5
Platform areas 2 to 8	12.5
Food preparation basement	12.3
Platform 1 and access road	12.1
Concourse and forecourt	11.5
East side offices (including station control room)	11.5
Non–public areas – west side offices and south west offices	11.5
Parcel post	11.0
Underground station	11.0
Car parks	9.3
Clothes store (above platforms 9–11)	9.3
Public highway	9.3
Hotel Way	8.0

Table 2.15 Overall station risk rating

	Risk rating
Overall life safety risk rating	10.1
a. Unwanted activation	8.1
b. Fires	10.1
Overall business risk rating	13.6
2.1 a. Asset	7.9
b. Performance	8.0
2.2 a. Unwanted activation	12.6
b. Fires	13.6

An example of the full results of the workshop can be found in the spreadsheet presented in Table 2.16 (courtesy of Mr J. Holland, Network Rail).

2.2.1.4 Summary

The matrix method can be applied to many assets and the results presented graphically. For example, a histogram of number of areas against ratings can be generated to assess the distribution of risks facing an organisation or project and which risks should take priority for further assessment and/or mitigation.

Table 2.16 Example of part of a risk ranking spreadsheet (most likely fire event)

Ref	Location	Causes	Consequences	F	S	P	Tot	C_d	C_p	Tot	Current controls	Comments/ priority
1	Platforms/ concourse	Rubbish, smoking, trains, electrical vehicles and contents, retail, arson	Serious injury or death	3	2	2	7	3	3	8.3	Company standards, procedures, staff, PA system, means of escape	
2	Storage	Rubbish, smoking, arson, food, packaging, electrical, rats	Serious injury or death	3	2	2	7	3	3	8.3	Controlled access, staff, automatic detection and warning system, compartmentation, extinguishers	Time limit to current occupancy
3	Parking	Vehicles, rubbish, arson	Serious injury	2	1	3	6	1	1	6.3	Highway code, CCTV, staff	Check fuel spill facilities
4	Passages	Rubbish, smoking, arson, gas main, contractors	Serious injury	1	1	3	6	1	1	5.3	Company standards, CCTV, means of escape, staff, manual call points	
5	etc	etc	etc									

Legend: F = Frequency
S = Severity (life safety)
P = Probability
C_d = Asset damage or destroyed
C_p = Performance penalties or loss

Distributions of life safety and business fire risk for a major business can also be generated.

It is important to note that the range of each category is significant and that the risk categories calculated by combinations of several categories represent even wider ranges. These are intended for prioritisation purposes only, based on evidence and views of stakeholders and other experts. Subsequent quantified fire risk assessment will probably predict a level of risk consistent with the matrix method, but any backward comparison is probably only of academic interest.

Matrix methods tend to be used when the risks and/or consequences involved with an activity are potentially more critical. They are often a precursor to full quantitative risk analysis of a subset of the hazards identified. Matrix methods have a more robust technical basis but their results are based on stakeholder judgements and so should be viewed in a comparative rather than absolute way.

2.2.2 Points schemes

Points schemes use similar factors to other qualitative methods, but inject some numeracy to result in a score. Typically, each aspect of fire safety (or factor) is scored from a range and then multiplied by a weighting to provide an overall score for that factor. The overall scores for each factor are then added together to provide an overall score for the area being assessed. This overall score can then be compared with a benchmark score, provided by the regulator and/or their stakeholder group, to assess the level of risk.

2.2.2.1 Gretener method

During the 1960s, the Association of Cantonal Fire Insurance Houses, the umbrella organisation of public legal insurance in Switzerland, felt the need for assessing objectively, on a standard basis, the fire risk and necessary protection measures particularly for industrial and other special types of buildings whose construction and purpose necessitate more than the normal protective measures. The initiative for developing a standard method for this purpose was taken up by M. Gretener, Head of the Fire Prevention Service for Industry and Commerce in Zurich. He presented the first comprehensive paper dealing with this problem at the third International Fire Protection Seminar of the Association for the Promotion of German Fire Protection held in Eindhoven – see Gretener (1968). The basic features of this method of fire risk evaluation are as follows.

Gretener proposed the formula

$$R = \frac{P \cdot A}{N \cdot S \cdot F} \tag{2.4}$$

where R is a numerical measure of fire risk. The parameter P quantifies the inherent or potential hazard or risk of the building and includes factors such as fire load, combustibility, size of fire area with or without fire venting system, building height and tendency to produce smoke, toxic or corrosive gases. The parameter

A is the activation factor representing the tendency for a fire to start and may be quantified by the probability of a fire starting.

The parameter N refers to 'normal' fire precautions such as fire brigade, fire extinguishers, trained personnel, hydrants and water supplies. The parameter S is concerned with the existence of 'special' protective measures such as automatic fire detection systems and sprinklers and F with the fire resistance of the building. The parameters in the numerator of Equation (2.4) are factors enhancing fire risk and those in the denominator are factors reducing the risk.

In the estimation of R, the component excluding A denoted by

$$B = \frac{P}{N \cdot S \cdot F} \tag{2.5}$$

is the 'effective' or 'endangering' fire risk, representing the extent of spread or damage if a fire occurs. For each type of building, a certain 'normal' or 'maximum' value, Bmax, is specified for the risk factor B, taking into account insurance requirements and factors such as the distance of the nearest fire station and the average times taken by the fire brigade to attend and extinguish fires. Depending on the potential fire risk represented by the risk component P under given conditions, fire protection requirements according to the components N, S and F are evaluated such that the value of B does not exceed Bmax. An advantage of this method is that the planner of a building project is not strictly tied to specifically prescribed protective measures but a range of possible measures which can ensure that the risk does not exceed Bmax A disadvantage of this method is that the basic equations are dimensionally meaningless.

Risk to life depends primarily on the number of people in a building. It also depends on the physical and psychological conditions of the occupants of the building. For example, life risk is greater in an old people's home than in a residential building. Hence, a slightly smaller value is specified for Bmax where there is a greater risk to human life. The value of Bmax is also adjusted depending on the presence or absence of factors affecting the parameter A quantifying the probability of fire starting.

Each of the risk parameters P, N, S and F are evaluated by enumerating the factors affecting the parameter and assigning points to these factors. These points are then multiplied to provide the overall risk value for the parameter. For example, as mentioned earlier, the potential fire risk quantified by P depends on factors such as fire load, combustibility, smoke factor, corrosion factor, shape, size and height of a fire compartment, presence or absence of heat and smoke ventilation systems and height of the building or number of storeys. The parameter S depends on the existence of protection measures such as guard service, automatic fire detection systems and sprinklers.

The points to be assigned to the factors are specified in tables developed for the parameters P, N, S and F. Appendix A2.1 is an example showing the values of S suggested by the European Insurance Committee, CEA. The value of N is set equal to unity for a building with 'normal' or 'standard' fire precautions without any 'extra' or 'special' protection measures. Tables and methods for calculating P, N, S and F are discussed in details by Bürgi (1973), Kaiser (1980) and Wessels (1980). Modified points schemes have been suggested by Purt (1971) and Cluzel and Sarrat (1979); in the latter method, different values, one for people safety and another for property protection, are allocated to evaluate the performances of protection measures. As mentioned above, with some modification to suit local conditions, the Gretener method has been applied to fire protection and insurance problems in Switzerland, Austria, Italy, France, Belgium and the Netherlands. A number of commercially available computer models of the Gretener method have been developed in the last decade and these include FREM (Watts 1995a), RiskPro (2000), FRAME (2000) and Risk Design (FSD 2000).

2.2.2.2 *Fire safety in hospitals – United Kingdom*

A more complex points scheme is the fire risk assessment method for hospitals developed by the University of Edinburgh during the 1980s under the sponsorship of the Department of Health and Social Services – see Stollard (1984). The Fire Precautions Act 1971 – A Guide to Fire Precautions in Hospitals (1976/79) provided the basis for identifying 17 factors contributing to the 'norm' against which assessment can be made. Based on a 'Delphi' exercise involving fire safety experts, three more factors were added – staff, patients and visitors, fire brigade – 20 factors in all.

The Delphi group considered a hierarchy of fire safety in order to determine the relative importance of the 20 components of fire safety identified. This exercise estimated contributory values of objectives to policy, of tactics to objectives and of the 20 components to tactics. By the multiplication of these matrices, a vector for the contribution of components to overall policy was produced. Each number in this vector was then expressed as a decimal fraction of the whole to give a set of relative values for the contribution of components. This vector was then modified to give a new set of values of component contribution which took into account the interactions between components evaluated in a matrix form. The 20 components and their relative values are given in Column 1 of the table in Appendix A2.2. Revised values of the components after considering the interactions are given in Column 2 of this table.

The patient areas of a hospital were divided into fire zones or survey volumes. Each zone was then assessed by a surveyor to assess the deficiency in each of the 20 components, in comparison with the 'ideal' or 'norm', on a six-point scale 0–5 with 0 representing 100 per cent deficiency and 5 indicating no deficiency. Work sheets were used for this assessment. The grade for each component was then multiplied by the relative (decimal fraction) contribution of that component to the overall fire safety policy to calculate a single score between 0 and 500. The scores for the components were then added. A total score between 450 to 500 was considered as good, 350 to 450 as acceptable, 280 to 350 as unacceptable and less than 280 as definitely unacceptable. An example of the summary sheet is given in Appendix A2.3. The method was superceded in 1994, because the arbitrary interaction between factors could lead to an acceptable risk score and an inadequate fire strategy (see section 1.4.6.1).

The hierarchical approach described above was further developed at the University of Ulster for application to dwellings – see Shields and Silcock (1986) and Donegan, Shields and Silcock (1989). This approach has been refined and implemented for the assessment of fire risk in telecommunication facilities – see Budnick, Kushler and Watts (1993).

2.2.2.3 Risk assessment schedules

Schedule rating is a risk assessment method similar to a points scheme developed in the USA by the insurance industry for calculating fire insurance premium rates, particularly for industrial and commercial properties. For these properties, unlike occupancies such as dwellings for which class rates are determined, a premium rate is specifically calculated for each property by the application of a schedule or formula. The specific rate is designed to measure the relative quantity of fire hazard present in the property. In the past, many insurance companies in the United Kingdom adopted a tariff system developed by the Fire Offices' Committee which specified premium rates for different classes within an occupancy type, based on fire resistance and fire protection measures such as sprinklers.

Based on various physical hazards, the schedules establish an arbitrary point from which insurance premium rates are built up. A schedule of additions and reductions is computed for a particular property and the difference applied to the arbitrary point. Schedule rating provides a plan for quantifying the fire hazards in any particular property; it is an empirical standard for the measurement of relative quantity of fire risk. Schedule rating takes into consideration various factors such as construction and occupancy with a view to determining which features either enhance or reduce the probability of loss.

Examples of the schedules developed in the USA are the Mercantile Schedule, the Dean Schedule and the Specific Commercial Property Evaluation Schedule (SCOPES) of the Insurance Services Office (ISO). SCOPES is the most commonly used insurance rating schedule according to which a percentage occupancy charge is determined for a building from tabulated charges for classes of occupancy, modified by factors reflecting the specific fire risk in the building. The basic building grade is a function of the resistance to fire of structural walls and floor and roof assemblies. The building fire insurance rate is the product of occupancy charges and building grade modified by factors such as exposure to fire in nearby buildings and protection provided by portable fire extinguishers, fire alarm systems, sprinklers etc. Tabulated values and conversion factors are based on actuarial analyses of fire losses (claims) paid by insurers and reported to the insurance industry.

The Fire Safety Evaluation System (FSES) is a schedule approach, developed in the USA in the late 1970s to determine equivalencies to the National Fire Protection Association (NFPA) 101 Life Safety Code for certain institutional occupancies. The technique was developed at the Centre for Fire Research at the National Bureau of Standards (now National Institute of Standards and Technology) in co-operation with the US Department of Health and Human Services. It was adapted to new editions of the Life Safety Code and published in NFPA 101A: Alternative Approaches to Life Safety (1995). The FSES was developed to provide a uniform method of evaluating health care facilities to determine what fire safety measures would provide a level of fire safety equivalent to that provided by the NFPA Life Safety Code.

The FSES subdivides a building into 'fire zones' for evaluation. A fire zone is defined as a space separated from other parts of the building by floors, fire barriers or smoke barriers. When a floor is not partitioned by fire or smoke barriers, the entire floor is the fire zone. In application, every zone in the facility should be evaluated.

The FSES begins with a determination of occupancy risk factor, calculated as the product of the assigned values for the risk parameters of the following five characteristics: patient mobility, patient density, fire zone location, ratio of patients to attendants and average patient age. Variations of these parameters have been assigned relative weights, as indicated in Table 3–1 of the worksheets for health care occupancies, in Chapter 3 of NFPA 101A. A hardship adjustment for existing buildings is applied to the occupancy risk factor; this modifies the risk in existing buildings to 60 per cent of that for an equivalent new building.

Thirteen fire safety parameters were selected for offsetting calculated occupancy risk by safety features. These parameters and their respective ranges of values are reproduced from Table 3–4 in Chapter 3 of NFPA 101A. This table is designed to be used for selecting appropriate values for each safety parameter by carrying out a survey and inspecting the fire zone.

An important concept of the FSES is redundancy through simultaneous use of alternative safety strategies. The purpose is to ensure that failure of a single protection device or system will not result in a major fire loss. Three fire safety strategies are identified: containment, extinguishment and people movement. Table 3–5 in Chapter 3 of NFPA 101A indicates which fire safety parameters apply to each fire safety strategy. The limited value of automatic sprinklers for people movement safety is adjusted for by using one-half of the parameter value in this column.

Following the procedure described above, the FSES determines if the fire zone considered possesses a level of fire safety equivalent to that of the NFPA Life Safety Code. This is done by comparing the calculated level for each fire safety strategy with stated mandatory minimum requirements. The FSES and other fire risk ranking methods have been discussed in detail by Watts (1995b).

2.2.2.4 Points schemes for road tunnels

An example of this is the application of the ADAC (2002) tunnel fire points scheme to a Victorian road tunnel in the UK. The ADAC points scheme is based on a particular national standard for new road tunnels and awards the majority of its points for ventilation systems and means of escape.

Physical constraints meant that the Victorian road tunnel could not score highly in these categories and so received a low score (i.e. implying a high, if not unacceptable, risk). However, the tunnel operators had implemented a range of alternative arrangements which meant that the tunnel was much safer than implied by the points scheme. These measures were not recognised in the points scheme and so made no difference to the score achieved by the tunnel even though it could be seen from historical data that they would make a significant contribution to the reduction of risk.

In some situations, such as existing hospitals, it has been found in hindsight that the application of the simple rules is not the most cost-effective approach. One notable example of this was a small urban Victorian hospital where the ward corridor was turned into a type of hospital street (with significant implications in terms of compartmentation, fire dampers and stopping) which required a massive investment with (given other fire precautions in the hospital) relatively little improvement in safety.

Points schemes remain popular with those looking for a quick method to apply to a large number of buildings. This is particularly true when their originators are not aware of points schemes' limitations or the availability of equally easy to use alternative methods that are more technically sound.

2.2.2.5 Merits and demerits of points schemes

As discussed by Ramachandran (1982), points schemes follow simple numerical approaches and are found easy to apply in practical problems concerned with the evaluation of fire risk and effectiveness of fire safety measures. They are useful for comparing fire risk in a building against a 'norm' or prescribed 'standard' and identifying components of fire safety which are deficient in the building. Points schemes allow a certain amount of flexibility in building design and in providing comparisons between premises which are not basically dissimilar in general occupancy and use. The latter is of assistance in making decisions on the allocation of scarce resources for fire safety improvement.

Though based on practice, experience and expert opinion, points schemes are empirical methods with somewhat arbitrary or subjective allocation of points, scores or ranks. It might be argued that statistical information provided by real fires and case studies is consciously present in the minds of those who contribute to the allocation. But there is room for argument and serious disagreement between people in the determination of points or ratings for different factors enhancing or reducing fire risk. In the points schemes, equal weights are generally attached to all the factors when adding or multiplying the points allocated to the factors for evaluating the overall numerical measure for fire risk in the whole or part of a building. This is not a satisfactory procedure. Points schemes do not consider sufficiently the interactions between fire safety measures.

The numbers generated by a points scheme are subjective qualitative indicators of relative fire risk. They are not objective quantitative estimates of absolute fire risk and effectiveness of fire protection measures. Hence, points schemes cannot provide quantitative inputs required for any cost benefit analysis of fire protection measures. With a statistical/probabilistic analysis based on factual data provided by real fires, case studies and experiments, it is possible to trace back from the final risk figure the independent quantitative contributions from different factors and their interactions. Scores or ranks can be easily derived from such an analysis in order to facilitate the presentation and application of results.

Points schemes are, therefore, best suited to specific building types because they can be judged on the same criteria and then compared with each other (rather like testing in schools of pupils at specific ages; a benchmarking exercise).

Four limitations of points schemes should be borne in mind. They:

- are based on scoring and weighting systems that may be largely arbitrary;
- contain an implicit fire strategy, so valid alternative fire strategies may not achieve the appropriate score;
- may be applied subjectively which may have an impact if comparison is to be made over several regions with different assessors, i.e. one assessor may generally be more generous or strict in assessing certain components;
- have no structure between factors which means that inappropriate fire strategies may achieve a good score whilst the actual level of risk is not acceptable. The healthcare example of this was that a ward, with no compartmentation and excellent emergency lighting, exit signs and fire notices, would achieve an acceptable score. When the fire safety of patients depends on protection by compartmentation before progressive horizontal evacuation, an acceptable score is dangerously inappropriate.

For fire safety, the main objective is usually life safety, but other objectives such as property protection/business continuity, heritage, image/reputation and the environment may also be important to organisations. In many respects the use of points scheme risk assessment methods suppresses the decision-making aspect of fire safety and can encourage the perception that there is only one solution and 'this is it'. However, the nature of many modern buildings means that points scheme fire risk assessment methods are increasingly difficult to apply. The literal application of the implicit solution in specific buildings may not be the most cost/risk effective and so can have significant cost implications.

Due to the limitations of points schemes, the original version of HTM 86 (1987) for fire risk assessment of hospitals was replaced by a structured qualitative approach (see Section 2.1.2).

For fire safety engineering, conditions where points scheme fire risk assessment methods may not adequately address the issues of concern tend to occur when the building is large, complex or unique, or when application of the simple risk assessment solutions conflict with the function of the building (usually rendering the fire precaution highly unreliable).

Appendix A2.1 Relative S values for active systems

Active System		S
FIRE DETECTION		
Guard service		
Control rounds minimum twice per night		1.03
Control rounds every two hours		1.05
Alarm boxes (manual)		1.04
Automatic fire detection		
Detectors in relation with public fire service	E1	1.20
	E2	1.15
	E3	1.11
	E4	1.07
	E5	1.04
	E6	1.02
Automatic extinction installation		
Detection value		1.10
ALARM TRANSMISSION		
Central station in building		1.03
Automatic transmission		1.04
Simultaneous alarm		1.05
Security line		1.03
INHOUSE FIRE FIGHTING		
Fire-fighting teams		1.05
Voluntary fire service		1.17
Professional fire service		1.28
AUTOMATIC EXTINCTION		
Sprinkler installation, class I, in relation with public fire service	E1	2.40
	E2	2.40
	E3	2.39
	E4	2.38
	E5	2.36
	E6	2.33
Sprinkler installation, class II, in relation with public fire service	E1	1.62
	E2	1.61
	E3	1.60
	E4	1.59
	E5	1.57
	E6	1.55
CO_2 installation		not yet
Halon Installation		determined
FIRE VENTILATION		
PUBLIC FIRE SERVICE		

The efficiency of automatic detection and extinction is considered to be a function of the operation of the public fire service. S-values have therefore been determined in relation to the "E-factor" for the public fire service	E	Time (min)	Distance (km)
	1	≤10	≤1
	2	>10–15	>1–3
	3	>15–20	>3–6
	4	>20–30	>6–10
	5	>30–40	>10–15
	6	>40	>15

Appendix A2.2 Values of components

		1	2
01	Staff	0.0866	0.0889
02	Patients and Visitors	0.0646	0.0643
03	Factors affecting smoke movement	0.0586	0.0656
04	Protected areas	0.0565	0.0555
05	Ducts, shafts and cavities	0.0443	0.0400
06	Hazard protection	0.0676	0.0649
07	Interior finish	0.0500	0.0497
08	Furnishings	0.0592	0.0625
09	Access to protected areas	0.0448	0.0407
10	Direct external egress	0.0436	0.0412
11	Travel distance	0.0478	0.0488
12	Staircases	0.0509	0.0488
13	Corridors	0.0511	0.0509
14	Lifts	0.0356	0.0342
15	Communications systems	0.0487	0.0506
16	Signs and Fire Notices	0.0401	0.0406
17	Manual firefighting equipment	0.0328	0.0302
18	Escape lighting	0.0411	0.0462
19	Automatic suppression	0.0316	0.0329
20	Fire Brigade	0.0445	0.0435
		1.0000	1.0000

Appendix A2.3 Sample of a summary sheet

Health authority:
Building: Victoria Hospital
Survey volume: Ward 1
Date of survey:
Surveyor:
Number of bedspaces: 30

Component		Grade	Percentage Contribution	Score
01	Staff	0 1 2③4 5 X	9	27
02	Patients and Visitors	0 1②3 4 5 X	6	12
03	Factors affecting smoke movement	0 1 2③4 5 X	7	21
04	Protected areas	0 1 2 3 4⑤ X	6	30
05	Ducts, shafts and cavities	0 1 2 3④5 X	4	16
06	Hazard protection	0 1 2 3 4⑤ X	7	35
07	Interior finish	0 1 2③4 5 X	5	15
08	Furnishings	0 1 2③4 5 X	6	18
09	Access to protected areas	0 1 2 3④5 X	4	16
10	Direct external egress	0 1 2 3④5 X	4	16
11	Travel distance	0 1 2③4 5 X	5	15
12	Staircases	0 1 2 3 4⑤ X	5	25
13	Corridors	0 1 2 3 4⑤ X	5	25
14	Lifts	0 1 2 3 4⑤ X	3	15
15	Communications systems	0 1②3 4 5 X	5	10
16	Signs and Fire Notices	0 1 2③4 5 X	4	12
17	Manual firefighting equipment	0 1 2 3④5 X	3	12
18	Escape lighting	0 1②3 4 5 X	5	10
19	Automatic suppression	⓪1 2 3 4 5 X	3	0
20	Fire Brigade	0 1 2 3 4⑤ X	4	20
		TOTAL SCORE (out of 500)		350

Additional comments

References

ADAC (2002), Personal correspondence.

BS 5588: Part 11 (1997), *Code of Practice for Shops, Offices, Industrial, Storage and Other Similar Buildings*, British Standards Institute, London.

BS 9999 (2008), *Code of Practice for Fire Safety in the Design, Management and Use of Buildings*, British Standards Institute, London.

BS EN 13501-1 (2000), *Reaction to Fire Testing*, British Standards Institute, London.

Budnick, E K, Kushler, B D and Watts, J M, Jr (1993), Fire risk assessment: a systematic approach for telecommunication facilities, *Annual Conference on Fire Research: Book of Abstracts*, NISTIR 5280, National Institute of Standards and Technology, Gaithersburg, MD.

Bürgi, K (1973), A method of evaluating fire risks and protective measures, Fourth International Fire Protection Seminar on Methods of Evaluating Fire Hazards of Industrial and Other Objects, Zurich.

Chemical Industry Safety and Health Council, (1977), *A Guide to Hazard and Operability Studies*, Chemical Industry Safety and Health Council, London.

Cluzel, D and Sarrat, P (1979), Méthode ERIC: Evaluation du risque incendie par le calcul. *Proceedings of Conseil International du Bâtiment Symposium on Systems Approach to Fire Safety in Buildings*, Vol I, II, CIB, Rotterdam.

Donegan, H A, Shields, T J and Silcock, G W (1989), A mathematical strategy to relate fire safety evaluation and fire safety policy formulation for buildings, *Fire Safety Science – Proceedings of the Second International Symposium*, Hemisphere Publishing Corporation, New York.

Drysdale D (2002), *An Introduction to Fire Dynamics*, 2nd edn, John Wiley and Sons, Chichester.

FRAME (2000), Fire Risk Assessment Method for Engineering, http://user.online.be.notr.034926/webengels.doc.htm, 7 September.

FSD (2000) Risk Design, http://www.fsd.se/eng/index.html, 7 September.

Gretener, M (1968), Attempt to calculate the fire risk of industrial and other objects, Third International Fire Protection Symposium, Eindhoven.

HTM 86 (1987), *Fire Risk Assessment in Hospitals*, Department of Health, HMSO, London.

HTM 86 (1994), *Fire Risk Assessment in Hospitals*, FIRECODE, NHS Estates, HMSO, London.

Kaiser, J (1980), Experiences of the Gretener method. *Fire Safety Journal*, 2, 213–222.

Mannan, S (ed.) (1980), *Lees' Loss Prevention in the Process Industry*, Butterworth, Oxford.

NFPA 101A (1995), *Alternative Approaches to Life Safety*, National Fire Protection Association, Quincy, MA.

Purt, G (1971), The evaluation of fire risk as basis for the planning of automatic fire protection systems, *Sixth International Seminar for Automatic Fire Detection*, IENT, Aachen.

Ramachandran, G (1982), A review of mathematical models for assessing fire risk, *Fire Prevention*, 149, May, 28–32.

RiskPro (2000), Simco Consulting, Wantirna, Australia.

Shields, T J and Silcock, G W (1986), An application of the hierarchical approach to fire safety, *Fire Safety Journal*, 2, 3, 235–242.

Stollard, P (1984), The development of a points scheme to assess fire safety in hospitals, *Fire Safety Journal*, 7, 2, 145–153.

TSO (2005)The Regulatory Reform (Fire Safety) Order, The Stationery Office, London

Watts, J M, Jr, (1995a), 'Software review: fire risk evaluation model', *Fire Technology*, 31, 4, 369–371.

Watts, J M, Jr (1995b), Fire risk ranking, *SFPE Handbook of Fire Protection Engineering*, 2nd Edition, Section 5, Chapter 2, National Fire Protection Association, Quincy, MA.

Wessels, E C (1980), Rating techniques for risk assessment, *Fire International*, 67, 80–89.

3 Quantitative risk assessment techniques

Due to uncertainties caused by several factors affecting fire safety in buildings, it may be realistic to treat these factors as non-deterministic random phenomena. This generally means adopting a probabilistic approach to the evaluation of fire risk and assessment of the fire protection requirements of a building (Ramachandran 1988a). In this approach, there are essentially four types of models that are used in the quantitative analysis of fire risk in which probabilities enter the calculations explicitly. These are:

- statistical methods;
- logic tree analysis;
- stochastic models; and/or
- sensitivity analysis.

These models are discussed below. Further types of analysis, or variations on the types of analysis shown, can be used as appropriate and some data can be found in the tables in this chapter.

The analysis of statistics provided by real fires is the basis of most probabilistic fire risk assessment, from the frequency of ignition to the conditional probability of failure of a fire protection system. Statistical analysis takes data that has been collected on building fires and transforms it into information that can be used to predict the likelihood of occurrence of future events and their consequences. This can take the form of the simple assessment of the average probability of an event (occurrence of fire) over a set of buildings and the damage caused and their probability distribution over a period of time or a complex regression analysis.

Statistical analysis has the advantages that it is based on actual events and that the results are usually simple to apply. It is, however, based on historical data which is then averaged. This assumes that future performance can be predicted from past experience and that an average measure can be applied to a particular building with some adjustment. In most cases these assumptions are reasonable and there is less uncertainty in undertaking a risk assessment based on historical data than to take no account of the probability of failure of the various fire precautions in a building design.

Another limitation of statistical analysis is that it is often not possible to collect sufficient data to directly predict, with confidence, the kind of high consequence, low frequency events, such as multiple fatality fires, that are of concern. Statistical data is much better for more frequent events such as ignition, and the conditional probability of success or failure of fire precautions. These individual pieces of information can be used to predict the frequency of low frequency events, by using logic trees and other techniques. Alternatively, other statistical methods have been developed to deal with such insufficient outcome data, such as extreme value theory (Rabash *et al.* 2004).

For most practical problems in fire protection, it may be sufficient to carry out a probabilistic fire risk assessment based on one or more logic trees. These provide a simple method for estimating the frequency of occurrence of an undesirable event (or events) known as an outcome. Such events include the fire reaching flashover stage or spreading beyond the room of origin and smoke causing visual obscuration on an escape route. In this approach, the sub-events leading to the outcome are identified and placed in their sequential order. This process is continued until a basic event (usually ignition) or set of basic events is identified, for which the probabilities can be estimated from statistical data (DOE (1996). Probabilities associated with sub-events are then combined in a suitable logical manner to derive the frequency of occurrence of the outcomes of concern. The calculation procedure is facilitated by the use of logic diagrams or trees, which provide a graphical representation of a sequence of sub-events.

There are normally two types of logic trees used in a probabilistic risk assessment, event trees and fault trees, and these are described in Section 3.2. To understand these methods it is first important to understand probability theory.

3.1 Theoretical basis

3.1.1 Probability theory

Sections 3.3 to 3.5 introduce statistical analysis, probability distributions and other associated techniques such as regression analysis, Bayes' theorem or Monte Carlo analysis. This section looks at set theory and Boolean algebra – the basis of logic trees which are used to quantify the probability of events.

An experiment is any process where an observation is made. Experiments comprise three parts: the conditions, performance and outcome. If, under identical conditions, the performance of an experiment produces the same outcome then the process is said to be deterministic. If, under identical conditions, the performance of an experiment can produce any one of a finite or infinite number of different outcomes then the process is said to be random.

Probability theory is used to analyse random processes. For example, if an experiment can result in N different outcomes and there is no reason to suppose that any one of them should be favoured more than any other, then if n_E of them possess the property E the probability of E is given by Johnston and Matthews (1982):

$$P(E) = \frac{n_E}{N} \tag{3.1}$$

This concept of probability was developed through assessing chances of winning at gambling and can be assigned before the experiment so there is often no need to perform the experiment at all.

If an experiment is repeated a large number of times under exactly the same conditions, the relative frequency with which an event E occurs settles down to a constant limiting value $P(E)$ which is the probability of E. The stability of relative frequency in the long run brings regularity into the uncertainty involved in random experiments. This regularity may be described by a probability model – see Table 3.1.

In real studies, interest is rarely restricted to a single possible outcome, but more usually to a group of them. Such a composite result is represented in the model by a compound event comprising the set of corresponding simple events.

For an event set ε containing simple events $(e_1, e_2, e_3, ..., e_n)$ the probability of a compound event E contained in ε is the sum of the probabilities of the simple events in E. Therefore:

$$0 \leq P(E) \leq 1 \tag{3.2}$$

If E and F are events in ε then:

1   E and F are **mutually exclusive** if they contain no simple events in common.
2   E **implies** F if E is wholly contained in F.
3   If the outcome of an experiment is a simple event in **either** E or F or both (i.e. at least one occurs), this can be expressed as $E \cup F$.
4   If the outcome of an experiment is a simple event in **both** E and F (they happen simultaneously), this can be expressed as $E \cap F$.
5   The set containing all simple events not in E is the **complement** of E and is written $\overline{E}$.

Probabilities can be added together:
If E and F are mutually exclusive events in ε then:

$$P(E \cup F) = P(E) + P(F) \tag{3.3}$$

Table 3.1 Relationship between random processes and probability modelling

Real life	Model
Random experiment	Probability model
Any possible outcome	Simple event
List of possible outcomes	Event set
Long–run relative frequency	Probability

If E and F are any two events in ε then:

$$P(E \cup F) = P(E) + P(F) - P(E \cap F) \tag{3.4}$$

Since E and $\overline{E}$ are mutually exclusive, then:

$$P(E) + P(\overline{E}) = 1 \tag{3.5}$$

Conditional probability is an important concept in the analysis of fire risk and is described as follows. The probability of an event E given that an event F is known to have happened is the conditional probability of E given F, defined as:

$$P(E/F) = \frac{P(E \cap F)}{P(F)} \tag{3.6}$$

E and F are independent if $P(E/F) = P(E)$, that is, the occurrence of E is in no way affected by the occurrence of F.

If E and F are mutually exclusive then they are totally dependent, since knowledge that F has happened tells us for certain that E has not happened, so $P(E/F) = 0$ which is not the same as the unconditional $P(E)$.

Only events that can happen simultaneously are in contention for being independent.

Probabilities can also be multiplied:

If E and F are independent events in ε then:

$$P(E \cap F) = P(E).P(F) \tag{3.7}$$

If E and F are any events in ε then:

$$P(E \cap F) = P(E).P(F/E) = P(F).P(E/F) \tag{3.8}$$

In view of the two equations above, the addition rule becomes:

$$P(E \cup F) = P(E) + P(F) - P(E).P(F/E) \tag{3.9}$$

or if E and F are independent:

$$P(E \cup F) = P(E) + P(F) - P(E).P(F) \tag{3.10}$$

Both the addition and product rules readily extend to more than two events. The application of probability theory becomes more apparent in the sections on fault and event trees.

When all the simple events of an event set ε have the same probability, for example when considering a random sample of items from a batch, the calculation of probabilities of compound events in ε reduces to counting how many simple events there are in the compound event and in ε. Therefore, knowledge of formulae for permutations and combinations is desirable.

The factorial expression $n!$ is used to mean

$$1 \times 2 \times 3 \times \ldots (n-3) \times (n-2) \times (n-1) \times n.$$

Therefore, $5! = 1 \times 2 \times 3 \times 4 \times 5 = 120$, $1! = 1$ and for convenience we define $0! = 1$ also.

Permutations are the number of ways in which r items can be selected from n distinct items, taking notice of the order of selection and is given by:

$$^{n}P_{r} = \frac{n!}{(n-r)!} \tag{3.11}$$

For example, the number of ways in which 2 items can be selected from 4 distinct items, taking notice of the order of selection, is:

$$^{4}P_{2} = \frac{4!}{2!} = 4 \times 3 = 12 \tag{3.12}$$

Combinations are when the order of selection is not important but only the final content of the r items selected. Therefore, the different permutations which contain the same r items need only be counted as 1 combination of the n items taken r at a time. The number of different combinations of n items taken r at a time is:

$$^{n}C_{r} = \frac{n!}{(n-r)!r!} \tag{3.13}$$

For example, using Equation (3.13) the number of ways 6 numbers can be drawn from 49 in a lottery is:

$$^{49}C_{6} = \frac{49!}{(49-6)!6!} = 1.4 \times 10^{7}$$

or a one in 14 million chance of success with one ticket.

3.1.2 Set theory and Boolean algebra

3.1.2.1 Set theory

Set theory is an elementary and intuitive concept that is best expressed by way of lists, that is a list of all the elements of a set. For, example:

$A = (1, 2, 3, 4, 5, 6, 7, 8, 9)$
$B = (2, 4, 6, 8, 10, 12)$
$C = (2, 4, 6, 8)$
$D = (\text{even numbers})$

Sets are normally denoted by upper case letters. The other way of expressing a set is by a Venn diagram (see Figure 3.1).

Two sets are equal if they contain the same elements. For example:

(Football, Hockey, Chess) = (Hockey, Chess, Football)

The membership of a set can be expressed for the example in Figure 3.1:

1 is a **member** of set A, but 1 is not a member of set B.

Also for the example in Figure 3.1:

C is a **subset** of B because all of the elements of C are also elements of B.

There are two special sets that need to be considered:

1   The set that contains no elements at all is call the **null set** and is denoted by 0.
2   The set that contains all potential objects under consideration is called the **universal set** and is denoted by 1.

The universal set is often only implicitly used in set theoretical calculations and manipulations and it enables the identification of:

All the elements that do not belong to set A is the **complement** of A.

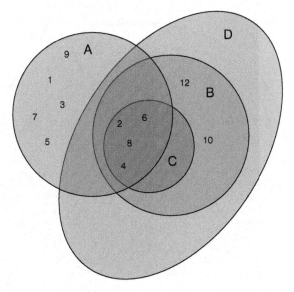

Figure 3.1 Venn diagram

Sets can also be defined in terms of other sets. For example, a set C could be defined as the set consisting of all of the elements in a set A together with all of the elements in a set B and contain no other elements:

If   A = (1, 2, 3)
and   B = (1, 4, 7)
then C = (1, 2, 3, 4, 7)

Note that elements will only appear once in the list of elements in a set. The set C is defined as the **union** of sets A and B (Figure 3.2) and is expressed as:

$$C = A + B \tag{3.14}$$

This is the notation commonly used in switching theory and is the one adopted by most fault tree analysts (there are two other notations used in set theory).

Similarly, we can define C as the set that consists of all the elements that appear in both A and B and contains no other elements:

If   A = (1, 2, 3)
and   B = (1, 4, 7)
then C = (1)

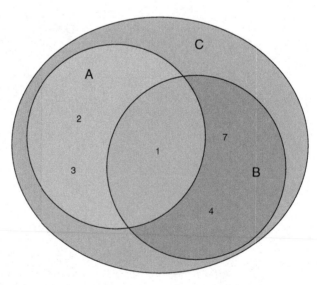

Figure 3.2 Venn diagram of union set

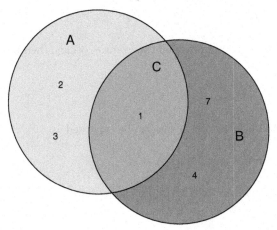

Figure 3.3 Venn diagram of intersection set

The set C is defined as the **intersection** of sets A and B (Figure 3.3) and is expressed as:

$$C = A \cdot B \qquad\qquad (3.15)$$

This notation is used in fault tree analysis because union and intersection in sets have many of the same properties as addition and multiplication of numbers.

3.1.2.2 Algebra of events

An algebra of events is used to link probability theory (associated with the outcomes of experiments/trials) with events. For example, if the trial is the throw of a die, then the event A1 can be defined:

A1 = (lands on 1)

Similarly:

A2 = (lands on 2) etc.

The universal set in this algebra is the set/event that encompasses all of the possible outcomes of the experiment in question. It is often called the event space. The null set in this algebra corresponds to an outcome that cannot occur. So, for the above die example, the sample space is the set:

U = (lands on 1, lands on 2, lands on 3, lands on 4, lands on 5, lands on 6)

and an example of the null set would be:

0 = (lands on 7)

If the experiment is performed and the event A occurs, then we can express this as:

A = 1

Conversely, if A does not occur this can be expressed as:

A = 0 (or **not** A equals one; $\bar{A} = 1$)

A probability can be associated with each potential outcome. This is expressed as:

P (A = 1)

that is, the probability that outcome A will be the result of the experiment.

Combinations of events can be described in the same way as the combination of sets. Therefore, A + B is the compound event 'the occurrence of event A or event B or both'. The truth table shown in Table 3.2 can be used to represent the outcome. This is equivalent to an OR gate in an event tree.

Similarly, A . B can be defined as the compound event 'the occurrence of both event A and event B'. The truth table for this is shown in Table 3.3.

Fault trees contain two classes of events. Firstly, those events which form termination points in the fault tree and generally will consist of compound failures; these events are called basic or **base events**. Secondly, each gate of the fault tree can be associated with a combined event. The system failure for which the fault tree was developed is called the **top event**.

Table 3.2 Truth table for an OR gate

A	B	A + B
0	0	0
1	0	1
0	1	1
1	1	1

Table 3.3 Truth table for an AND gate

A	B	A · B
0	0	0
1	0	0
0	1	0
1	1	1

For example in Figure 3.4:

G, X, Y are base events
G_1, G_2 *TOP* are combined events where:
$$G_1 = G + X$$
$$G_2 = G + Y$$
$$TOP = G_1 . G_2$$

Therefore:

- an **AND** gate corresponds to a Boolean product;
- an **OR** gate corresponds to a Boolean sum; and
- an **EXCLUSIVE OR** gate corresponds to a combination of Boolean sums, products and complements but can usually be approximated by Boolean sum.

EXAMPLE OF ITS USE IN FAULT TREE ANALYSIS

The main interest in fault tree analysis is to gain qualitative and quantitative information about the top event. To do this the Boolean expression for the top event must be manipulated into a form where:

1 all of the variables are base events;
2 the expression is in a standard format.

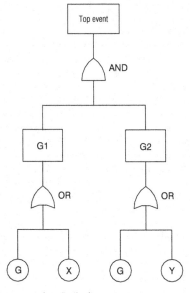

Figure 3.4 An example of a fault tree

One way of expressing this for a top event is:

$$Top = C_1 + C_2 + C_3 ... C_n \tag{3.16}$$

where C_1 is a combination event formed by the product of base events:

$$C_1 = A_1.A_2.A_3...A_n \tag{3.17}$$

where A_1 is a base event and $n \geq 1$.

These equations express the condition that if all the base events of a combination event exist simultaneously, then the combination event and so the top event will be produced. Note only one combination event is required to exist for the top event to be produced. These equations are known as the 'sum of product' form.

BOOLEAN MANIPULATION

In many ways, union and intersection operations behave in a similar way to addition and multiplication. These similarities can be seen below:

- The order the operations are performed is:

  before + unless overridden by a bracket

  For example, in $A + B.C$ the operation order is to first evaluate $B.C$ and call the result say X. Then evaluate $A + X$. However, if the expression was written $(A + B).C$ then the operation order would be to evaluate $(A + B)$ first and call the result say Y and then evaluate $Y.C$.

- Both operations are independent of the order of the variables:

$$A + B = B + A \qquad A.B = B.A \tag{3.18}$$

- Both operations are independent of order within themselves:

$$(A + B) + C = A + (B + C) \qquad (A.B).C = A.(B.C) \tag{3.19}$$

- Both operations obey the laws of multiplication:

$$(A + B).(C + D) = A.C + A.D + B.C + B.D \tag{3.20}$$

- Both operations interact with binary values:

$$A + 0 = A \qquad A.1 = A \tag{3.21}$$

There are also several areas where Boolean algebra varies from normal arithmetic:

- Complementary events:
  For any event A there exists a complementary event, $\overline{A}$ such that:

$$A + \overline{A} = 1 \quad \text{(either } A \text{ or } \overline{A} \text{ must occur)} \tag{3.22}$$

$$A \cdot \overline{A} = 0 \quad \text{(} A \text{ and } \overline{A} \text{ cannot co-exist)} \tag{3.23}$$

Complementation should occur before . and + operations.

- Idempotent laws

$$A + A = A \qquad A \cdot A = A \tag{3.24}$$

- Absorption law

$$A + A \cdot B = A \tag{3.25}$$

This law can be proved using truth tables or Venn diagrams.

If there are no complement events in an expression, the top event expression can be manipulated into the sum of product form:

1   Examine all the parts of the expression that are not bracketed and try to reduce their length using the idempotent and absorption laws.

2   Take parts of the expression of the form:

$$(A + \ldots + N) \cdot (M + \ldots + Z)$$

and expand using multiplication. Note that A, B and so on will not generally be base events and will generally be unbracketed expressions.

3   Repeat stages 1 and 2 until the expression cannot be reduced further.

This process is the way of analysing fault trees.
For example, a top event may be written:

$$
\begin{aligned}
\text{Top} \quad &= (G \cdot X) + (G \cdot Y) \\
&= G \cdot G + G \cdot Y + X \cdot G + X \cdot Y \quad \text{(expand)} \\
&= G + G \cdot Y + G \cdot X + X \cdot Y \quad (G = G \cdot G) \\
&= G + X \cdot Y \quad \text{(absorb)}
\end{aligned}
$$

Other processes of Boolean manipulation are also available. If complementary expressions are present, these have to be modified to produce the standard form of expression before Boolean reduction can commence:

1   Apply De Morgan's laws of complementary combination events until all the complementary operations act on the base events. De Morgan's laws are:

$$\bar{A} + \bar{E} = \bar{A} \cdot \bar{E} \tag{3.26}$$

$$\overline{A \cdot E} = \bar{A} + \bar{E} \tag{3.27}$$

2   Examine all the parts of the expression that are not bracketed and try to reduce their length using the following expressions:
$A + A = A = A \cdot A$
$A + A \cdot B = A$
$A \cdot \bar{A} = 0$
$A + \bar{A} = 1$
$A \cdot B \cdot C + \bar{A} \cdot B = \bar{A} \cdot B + B \cdot C$
$A \cdot B + \bar{A} \cdot B = B$

3   Expand the inner most level of brackets using multiplication.
4   Repeat stages 2 and 3 until the expression cannot be reduced further.

Again further methods are available in the literature.

MINIMAL CUT SETS

One of the most important results of the analysis is a qualitative expression of all the ways in which the top event can occur. This information is easily obtained from the sum of products form of the expression for the top event. For example:

$$Top = G + X \cdot Y$$

So if G occurs, the top event results. Equally, if X and Y occur, the top event results. Conversely, if X occurs on its own or Y occurs on its own, the top event does not result. This can be seen as an expression of redundancy or defence in depth.
    Therefore:

*   A **cut set** can be defined as a combination of base events, the simultaneous existence of which will produce the top event.
*   A **minimal cut set** can be defined as a cut set that has no other cut set as a proper subset.

So for the above example:

G
G X
G Y
G X Y
X Y

are all cut sets but only:

G
X Y

are minimal.

For fault tree analysis of a top event, Boolean expressions can be manipulated to the sum of products form. Where NOT logic is involved, the sum of product form can be obtained where:

1   each product is a minimal cut set;
2   all of the minimal cut sets are found as products.

Johnston and Matthews (1982) showed that in other expressions where NOT logic is present, other forms of analysis are necessary. Normally, fault tree analysis software is used to produce the minimal cut sets. However, knowledge of set theory and Boolean algebra is essential to the understanding and construction of fault trees.

3.1.3 *Reliability and availability of systems*

Most fire protection and detection systems are installed because they are needed to satisfy the Building Regulations or at the request of an insurance company covering the risk. The deterministic approach to fire safety engineering assumes that the installed system will work on the day. Deterministic fire safety engineering does not quantitatively address the reliability of systems. This section considers system reliability; it shows how reliabilities can be calculated and suggests values of reliability for different systems and hazards. For completeness, a brief introduction to reliability theory is given by Finucane and Pinkney (1988).

3.1.3.1 *Reliability*

Reliability is a measure of the ability of an item to perform its required function in the desired manner under all relevant conditions, and on the occasions or during the time intervals when it is required so to perform (Green and Bourne 1972). Reliability is normally expressed as a probability. For example a system that fails randomly in time but once a year on average will have a probability of failing (P_F) in any one particular month of a 1/12, i.e. $P_F = 0.0833$. Conversely the probability of success (P_S), i.e. not failing during that particular month, is $11/12 = 0.9167$ which is the same as $1 - P_P$ i.e. $P_S = 1 - P_F$ and by transposition $P_F = 1 - P_S$.

Mathematically these expressions can be expressed as:

$$P_S = e^{-t/T} \text{ and } P_F = 1 - e^{-t/T} \tag{3.28}$$

where t = the time interval during which success is required and

T = mean time between failures.

For values where $t/T = 0.1$ or less, then P_F is approximately equal to t/T.

$$P_S = 1 - P_F \tag{3.29}$$

$$P_S = 1 - t/T \tag{3.30}$$

$$P_s = \frac{T-t}{T} \tag{3.31}$$

The exponential probability distribution defined in Equation (3.28) is applicable to the major middle useful period of the life of a system. Other probability distributions such as Weibull are applicable to the early period and old age related wear out failures at the end of a system's life.

For example, if the mean time between failures is one year and the time interval during which success is required is one year, then the probability of failure P_F is not actually 12/12, i.e. 1, but

$$\begin{aligned} P_F &= 1 - e^{-t/T} \quad \text{where } t/T = 1 \\ &= 1 - 0.37 \\ &= 0.63 \end{aligned} \tag{3.32}$$

So there is a 63 per cent chance of failure in a year and the probability of success is given by:

$$\begin{aligned} P_S &= 1 - P_F \\ &= 1 - 0.63 \\ &= 0.37 \end{aligned} \tag{3.33}$$

So there is a 37 per cent chance of not failing in any one particular year.

In practice, when considering the reliability of fire protection systems, it is easier to talk in terms of unreliability or probability of failure (P_F). Taking, for example, the previously discussed case where the mean time between failures was one year, $P_F = 0.0833$ and $P_S = 0.9167$. If the mean time between failures were improved by a factor of 10, i.e. to 10 years, then P_F changes from 0.0833 to 0.00833 but P_S only changes from 0.9167 to 0.99167. For a system where failure creates a potential hazard, e.g. failure of a compartment wall or suppression system, the probability of failure P_F is a more direct measure of the risk involved.

3.1.3.2 Availability

Availability is the proportion of the total time that a system is performing in the desired manner. For protection or warning systems such as a fire alarm system, failure of the system does not, in itself, create an immediate hazard. Only if the failure exists when a fire occurs does an unprotected hazard result.

If we take the original example of a system with a mean time between failures of one year and assume that the fault is immediately alarmed, but it takes one week to repair it, on average, the system is out of action for one week per year, i.e. its unavailability is $1/52 = 0.019$ and its availability is $51/52 = 0.981$. Alternatively, assume that the fault is not alarmed, but is only revealed when a comprehensive weekly test is performed. In this case, the outage time can vary from near zero (i.e. fault occurs immediately prior to test) to nearly one week (i.e. fault occurs immediately after the test). The average outage will therefore be half a week. The unavailability from this cause will therefore be $0.5/52 = 1/104 = 0.0096$. It should be noted that this is half the probability of failure P_F for a similar one-week period. The total outage time will be the sum of the two types of outage, i.e. from immediately revealed faults and from faults only revealed at regular test intervals. As with reliability, the unavailability is a more sensitive indicator of how well a system performs.

Assume the original system with a one year mean time between failures is a fire alarm system and that the total outage is, on average, one week per fault. The unavailability will therefore be $1/52$. Assume that fires occur randomly in the protected area, again with an average mean time between fires of one year. The probability of a fire occurring within the particular week when the equipment is dead is therefore $1/52$ per fire. Since there is only one fire per year on average, only once in 52 years (mean time between hazard) is there likely to be a fire at the same time as the fire alarm system is not working. In other words, the mean time between undetected fires is the mean time between fires, divided by the fractional dead time of the fire alarm system.

$$\text{Mean time between hazards} = \frac{\text{Mean time between fires}}{\text{Unavailability of fire alarm system}}$$

$$= \frac{1 \text{ year}}{1/52} \qquad\qquad (3.34)$$

$$= 52 \text{ years}$$

3.1.3.3 Factors influencing system reliability

When considering the reliability of any system, various factors have to be taken into account. For example, the quality of the components used in the system and their suitability for the particular application; the stress imposed on these components by the designer; additional stresses imposed by the environment in which the system is installed; the tolerance of the design to variations in component performance; the test and certification procedures adopted for the system, and the time intervals between these tests. All these factors could cause a consequent, and possibly unacceptable, reduction in the system reliability (BS 5760). Evaluation of the reliability of a whole system based on the reliabilities of its components is a complex statistical problem and like many others in risk and engineering, it is not yet fully researched.

When considering reliability issues, care needs to be taken when analysing data and interpreting the results. According to UK fire statistics, in a significant proportion of fires, the sprinklers may not operate due to the fact that the fire is 'small' such that the heat generated may be insufficient for activating the sprinkler heads. Mechanical defect and the system having been turned off are main reasons for the non-operation of sprinklers. Although sprinklers operate in only 9 per cent of all fires they do so in 87 per cent (= 39/45) of the cases in which their action is required. This denotes a probability of 0.87 for sprinkler operation in 'big' or 'growing' fires. Some of the fires in which sprinklers operate are extinguished by the system itself, and some by the fire brigade.

3.1.4 Frequency of ignition

The frequency of ignition is one of the key parameters of most probabilistic risk assessments. It is usually the initiating event in most event trees and can be a base event in fault trees. A more detailed summary of the statistical derivation of frequency of ignition can be found in Section 3.3 Statistical models. The annual probability/frequency of ignition for a building increases as a 'fractional power' of the size of the building expressed in terms of total floor area.

Statistical studies (Ramachandran 1970, 1979/80, 1988a) have shown that the frequency (or probability during a period) of ignition is approximately given by:

$$F_i = aA_b^{\,b} \tag{3.35}$$

where a and b are constants for a particular type of building related to occupancy. The parameter a includes the ratio of the number of fires in a period over the number of buildings at risk (n/N) while b measures the increase in the value of F_i for an increase in A_b, denoting the total floor area of the building. For a full derivation of this power function see Section 3.3.1 and Equation (3.40).

A value of unity for b would indicate that the probability of fire starting is directly proportional to the size of the building; this would also imply that all parts of a building have the same risk of fire breaking out. This is not true, since different parts have different types and numbers of ignition sources. Hence, the probability of fire starting is not likely to increase in direct proportion to building size so that b would be less than unity. If two buildings are considered, one twice the size of the other, the probability for the larger building will be less than two times the probability for the smaller building. These theoretical arguments are confirmed by actuarial studies on frequency of insurance claims as a function of the financial value (size) of the risk insured (Ramachandran 1979/80, Benktander 1973).

Based on fire statistics and a special survey as mentioned earlier, the Home Office estimated the values of a and b for major groups of buildings; Rutstein (1979) (Table 3.4). For all manufacturing industries in the UK with A_b (m²), the values of a and b were estimated as 0.0017 and about 0.53 (respectively).

Actuarial studies (Benktander 1973) in some European countries confirm that the value of b is about 0.5 for industrial buildings. For a particular building

Table 3.4 Frequency of ignition per year

Occupancy	Probability of fire per year	
	a	b
Industrial buildings		
Food, drink and tobacco	0.0011	0.60
Chemical and allied	0.0069	0.46
Mechanical engineering and other metal goods	0.00086	0.56
Electrical engineering	0.0061	0.59
Vehicles	0.00012	0.86
Textiles	0.0075	0.35
Timber, furniture	0.00037	0.77
Paper, printing and publishing	0.000069	0.91
Other manufacturing	0.0084	0.41
All manufacturing industry	0.0017	0.53
Other occupancies		
Storage	0.00067	0.5
Shops	0.000066	1.0
Offices	0.000059	0.9
Hotels etc.	0.00008	1.0
Hospitals	0.0007	0.75
Schools	0.0002	0.75

the 'global' value of F_i given by Equation (3.35) can be adjusted. The ratio of number of fires over the number of buildings at risk provides an overall measure, unadjusted for building size, of the probability of fire starting (see Table 3.5).

Using data for the years 1968 to 1970 (North 1973) , a figure of 0.092 was estimated for all manufacturing industries in the UK for the risk of having a fire per annum, per establishment; an establishment can have more than one building. An estimate for the probability of fire starting according to building size is also given by number of fires starting per unit of floor area (see Table 3.6). It should be noted that the figures in Tables 3.4, 3.5 and 3.6 are now quite dated, but they are the best available.

3.2 Logic trees

For most practical problems in fire protection, it may be sufficient to carry out a probabilistic fire risk assessment based on one or more logic trees. These provide a simple method for estimating the probability of occurrence of an undesirable event (or events) known as an outcome. Such events include the fire reaching flashover

Table 3.5 Frequency of ignition normalised per building

Occupancy	Probability of fire starts per occupancy (fires·y⁻¹)
Industrial	4.4×10^{-2}
Storage	1.3×10^{-2}
Offices	6.2×10^{-3}
Assembly entertainment	1.2×10^{-1}
Assembly non–residential	2.0×10^{-2}
Hospitals	3.0×10^{-1}
Schools	4.0×10^{-2}
Dwellings	3.0×10^{-3}

Table 3.6 Frequency of ignition normalised by floor area

Occupancy	Probability of fire starting (starts per y⁻¹ m⁻² floor area)
Offices	1.2×10^{-5}
Storage	3.3×10^{-5}
Public assembly	9.7×10^{-5}

stage or spreading beyond the room of origin and smoke causing visual obscuration on an escape route. In this approach, the sub-events leading to the outcome are identified and placed in their sequential order. This process is continued until a basic event (usually ignition) or a set of basic events is identified for which the probabilities can be estimated from statistical data. Probabilities associated with sub-events are then continued in a suitable logical manner to derive the probability of occurrence of the outcome of concern. The calculation procedure is facilitated by the use of logic diagrams or trees which provide a graphical representation of a sequence of sub-events.

Normally two types of logic trees are used in a probabilistic risk assessment: event trees and fault trees.

3.2.1 Event tree analysis

Event trees are most useful when there is little data on the frequency of outcomes of concern (usually because they are infrequent) e.g. multiple fire deaths. Event trees can be used to predict the frequency of infrequent events by the logical connection of a series of much more frequent sub-events, for which data is available.

Event trees work forward from an initiating event (such as ignition) to generate branches defining events and paths resulting from secondary (or nodal) events to

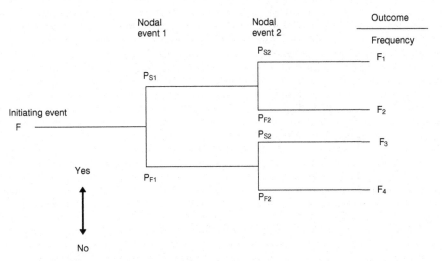

Figure 3.5 General form of an event tree

give a whole range of outcomes. Some of the outcomes may represent a very low risk event, others may represent very high risk events.

The construction of an event tree starts by defining an initiating event leading to the final outcome, following a series of branches, each denoting a possible outcome of a chain of events. Figure 3.5 is an example of an event tree representing a range of outcomes resulting from an initiating event via two nodal events. Care should be taken that the event tree reflects the actual order of events in real fires and that all the nodal events of importance have been included.

It is crucial that all the sub-events included in an event tree are **independent**. That is, they can all occur in the same event and so are not **mutually exclusive**.

The frequency associated with each branch (outcome) is given by multiplying out the initiating frequency F and the relevant conditional probabilities of success and/or failure (P_S and P_F respectively). For example:

$$F_2 = F.P_{S1}.P_{F2} \qquad (3.36)$$

Figure 3.6 shows how an event tree could be applied to the early stages of a fire.

The initiating event is ignition. The two nodal events are 'Is the fire restricted to the item first ignited?' and 'Is the fire detected less than five minutes from ignition?' The outcomes in descending order are:

1 A fire where ignition occurs, but the fire does not grow beyond the item first ignited.
2 A fire where it grows beyond the item first ignited, but is detected in less than five minutes from ignition.
3 A fire where it grows beyond the item first ignited and is not detected in less than five minutes from ignition.

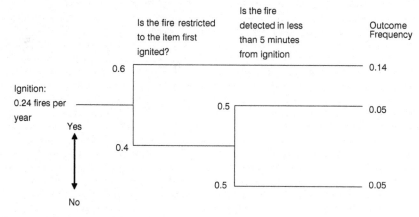

Figure 3.6 Event tree of the early stages of fire event development

The frequencies of the outcomes can be calculated as above and indicate that although ignition can be expected just under once in four years, the frequency of events where a fire would be expected to grow and not be detected is about once in twenty years. This could be used to measure the benefit of materials that are fire retardant and ignition sources that are low in number and energy. This event tree could also be used to demonstrate that an alternative mode of fire detection is equivalent to that of a code-compliant solution. Care should be taken to ensure that the conditional probability of the first nodal event does not include events that can only follow the second nodal event, e.g. first aid fire fighting.

3.2.1.1 Example of probabilistic fire risk assessment using an event tree

This subsection shows how event tree analysis can be used as part of a probabilistic fire risk assessment (Charters 1992). Fire risk assessment can be used to assess the risk/cost benefit of fire precautions for property protection. The following example is a risk assessment carried out for a major bus operator (Charters and Smith 1992, Charters 1998).

A major bus operator expressed concern about the risk to business from fires in bus garages. In particular, the operator was interested in whether or not they should install sprinkler systems in their existing bus garages or take some other action. Obviously, the cost of this would be considerable and so the bus operator commissioned a study to quantify the benefits in terms of property protection.

This risk assessment involves:

1   identifying events that could give rise to the outcome of concern;
2   estimating how often the events happen;
3   estimating what the severity of the outcome of those events would be; and
4   assessing the implications of the level of risk.

3.2.1.2 Identifying events

The events of concern are fires causing significant damage to vehicles and property in bus garages. From operating experience, fire safety judgement and full-scale fire tests, these events were narrowed down to one 'reasonable worst case' event: a seat fire at three points on a double deck bus parked amongst others.

The risk parameter chosen for the study was the cost of fires per calendar year. This could allow the bus operator to put these risks in context with historical data on other risks.

3.2.1.3 Estimating the frequency of events

To estimate how often the fire event happens, historical data was collected on how often fires occur on buses in garages. Because fires on buses are relatively infrequent, there was insufficient information to estimate how often the event occurs. Therefore, an event tree was constructed to help estimate the missing information.

An event tree is a logic diagram, which predicts the possible outcomes from an initial event (see Figure 3.7). For example, an initial event of 'seat fire in the lower saloon of a double deck bus' may have outcomes of 'damage less than £200,000' and 'damage greater than £500,000'. The likelihood of each outcome depends on other factors such as 'Is the fire noticed at an early stage?', 'Does the fire spread to neighbouring buses?' or 'Is the fire put out with fire extinguishers?'

The conditional probability of each of these other factors is estimated using historical data. Therefore, using the likelihood of an initial event and the probabilities of the other factors, an estimate can be made of how often an event occurs (Rutstein and Cooke 1978, Rutstein and Gilbert 1978, Charters 1997).

3.2.1.4 Estimating the severity of the outcome

There are several ways to estimate the severity of the outcome: from historical information, using simple analytical methods, using computer models and/or using full-scale tests. Each approach has its advantages and disadvantages. Historical data describes what the outcomes have been in the past but may not be complete or relevant. Simple analytical methods can predict the severity of outcomes cost effectively but the answer is only as good as the assumptions made. Computer models can predict the severity of outcomes more closely but can be expensive and time consuming. Full-scale testing probably gives the most accurate assessment of the severity of outcomes but it is usually even more costly and time consuming.

In this case the severity of the outcome (i.e. losses due to damaged buses/garage) depended heavily on the spread of fire from bus to bus and the effective spray density of different sprinkler systems. Therefore, a combination of full-scale testing and computer modelling was used to predict fire growth, fire spread and the effectiveness of sprinklers (Fardell and Kumar 1991). Typical data can be found in Table 3.7.

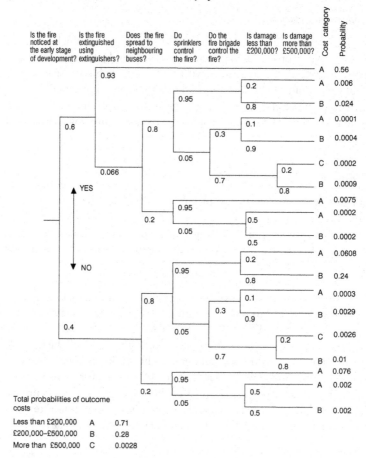

Figure 3.7 Simplified event tree for bus garage fires

3.2.1.5 Results

The risk assessment indicated that, for the event identified, a higher than ordinary hazard level of sprinkler spray density was necessary to prevent fire spread from bus to bus. The frequencies of fires in bus garages was about 0.1 per year. The fire risk was then calculated for the bus garages with and without sprinklers. The difference between the two figures is the benefit rate from reduced property losses by fitting sprinkler systems. This was approximately £2,000 per year (but this varied with the size of garage).

Historical accident data indicate that the predicted risk of damage is pessimistic; very few records of such fire damage could be found. Having quantified the benefits of sprinklers in reducing risks in bus garages, how can we tell whether they would represent a good investment in fire safety? The answer of course is cost-benefit analysis (Health and Safety Executive 1989, Ramachandran 1998b).

Table 3.7 General data on the reliability of fire protection systems

General fire data			Reference
Probability of fire occurring and not being reported to the local authority fire service (UK).	Industrial	0.5	DOE (1996)
	Commercial	0.8	Appleton (1992)
	Dwellings	0.8	Crossman and Zachary (1974) & Reynolds and Bosley (1995)
	Average	0.8	All the above
Frequency of reported fires per occupancy per year.	Industrial	4.4×10^{-2}	
	Storage	1.2×10^{-2}	
	Shops	8.4×10^{-3}	
	Offices	5.7×10^{-3}	
	Hotels etc.	3.7×10^{-2}	
	Dwellings	2.7×10^{-3}	
Probability of a reported fire causing property loss in excess of £1M (1992 prices).	Industrial	0.004	Scoones (1995)
	Other commercial	0.001	
	Educational	0.003	
Typical probability of fire spread for reported fires.	Beyond room of origin	0.1	Government Statistics Service (1992)
	To other buildings	0.2	

Fire alarm and detection systems

Improvement in probability of early detection in buildings with AFDA.	General value	0.5–0.6	
Reliability of alarm box, wiring and sounders.	General value	0.95–1.0	
Reliability of detectors. [a3]	Commercial smoke	0.90	
	Domestic smoke	0.75	
	Aspirating smoke	0.90	
	Heat	0.90	
	Flame	0.50	

Automatic fire suppression systems

Overall reduction in loss due to provision of sprinklers.	General value	50%	
Probability of successful sprinkler operation.[a1]	Maximum	0.95	
	General:	0.90	
	Property protection	0.80	
	Life safety	0.75	
	Minimum		
Probability of successful operation of other AFS systems.	General value	0.90	

Continued ...

Table.3.7 continued

General fire data			Reference
Smoke control systems (mechanical and natural)			
Probability of system operating as designed, on demand.	General value		0.90
	Passive fire systems		
Probability that fire-resisting structures will achieve at least 75% of the designated fire resistance standard.	Masonry walls		0.75
	Partition walls		0.65
	Glazing		0.40
	Suspended ceiling		0.25
Probability of fire doors being blocked open.	General value		0.30
Probability of self-closing doors failing to close correctly on demand (excluding those blocked open).	General value		0.20

3.2.1.6 Summary

A study to assess the benefits of installing sprinkler systems in bus garages indicated that there were business continuity and property protection benefits to the operator. However, the cost-benefit analysis and the operator's contingency plans meant that there was no cost-benefit for the consequence case for installing sprinklers in bus garages. As a result of the risk assessment, the operator implemented other forms of safeguard and fire precaution.

3.2.2 Fault tree analysis

Fault trees trace the root causes of a given final or 'top event' of concern by working backwards logically to base events. A fault tree is a graphical representation of logical relations between an undesirable top event and primary cause events. The construction of a fault tree starts with the definition of the top event identified at the hazard identification stage. The tree is constructed by placing various cause events in correct sequential order. This is generally done by working backwards from the top event and specifying the events' causes, faults or conditions that could lead to the occurrence of the top event, working backwards from each of these which, in effect, become secondary top events and so on. This process is continued and terminated when a final set of base (or root) events, faults or conditions are identified. A diagrammatic representation of the process would then generate the branches of a tree. Probabilities or frequencies are assigned to the root events.

The events in a fault tree are connected by logic gates, which show what combination of the constituent events could cause the particular top event. These are mainly AND gates in which all the constituent events have to occur and OR gates in which only one of the constituent events needs to occur to cause the specific top event. The probability of occurrence of the top event is calculated using Boolean algebra. Simple fault trees can be calculated directly using Boolean algebra, but more complex fault trees require that 'minimum cut sets' or 'path sets' be established using 'Boolean reduction' techniques.

Figure 3.8 shows a general fault tree and the use of the logic underlying the AND and OR gates. Computer software is available that can speed up the use of complex fault trees.

An example of a simple fault tree applied to fire detection is given in Figure 3.9. Here the top event is 'failure to detect fire within 5 minutes of ignition'. The causes of this top event can be followed through the four root causes for which data can be generated.

OR gates are usually calculated by adding the root probabilities together and subtracting their multiplied value:

$$P_{OR} = (P_A + P_B) - P_A.P_B \qquad (3.37)$$

AND gates are calculated by multiplying the root probabilities together.

The top events of fault trees can very often supply the conditional probabilities for event trees.

$$P_{AND} = P_A.P_B \qquad (3.38)$$

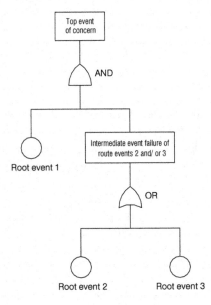

Figure 3.8 General form of a fault tree

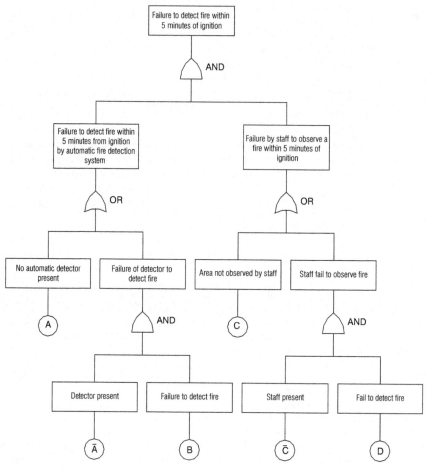

Figure 3.9 Fault tree for failure to detect a fire within five minutes of ignition

3.2.2.1 Example of probabilistic fire risk assessment applied to the detection of fires in wards by people

Traditional hospital wards were open in layout with the staff able to see most areas of the ward. This allowed staff to detect a fire by observation and so no automatic fire detection was required. A move towards patient privacy means that the vast majority of wards now comprise single-bed rooms. This means that staff no longer have visual access to the majority of the ward. Therefore, what level of automatic fire detection is required to maintain current safety levels?

The acceptance criterion is 'equivalency', i.e. the level of safety provided by the alternative fire safety solution is as good, if not better, than the standard prescriptive code solution.

A fault tree was constructed to assess the probabilistic nature of fire detection in hospital wards and is shown in Figure 3.9. Here the top event is the 'failure to detect a fire within five minutes of ignition'. The causes of this top event can be followed through the four root causes for which data can be generated.

These root causes are:

- no automatic detector present;
- automatic detector fails to detect the fire;
- no staff member present; and
- staff member fails to detect the fire.

This assumes that fires detected by patients are neglected so that the analysis errs on the side of safety.

The two cases of interest are:

1   good visual access and no automatic fire detection (AFD);
2   poor visual access but with AFD throughout.

The fault tree minimum cut set was developed using Boolean algebra and then quantified for both cases using the data in Table 3.8.

The results of the analysis for the two cases are:

1   probability of failure to detect fire within five minutes (good visual access and no AFD) = 0.3;
2   probability of failure to detect fire within five minutes (poorer visual access but with AFD throughout) = 0.1.

The results from Case 1 are consistent with those recorded generally for the probability of failure to detect a fire within 5 minutes of ignition in hospitals wards (NHS Estates 1996).

Table 3.8 Data used to quantify the probability of failure to detect a fire within five minutes of ignition

Root event	Case 1	Case 2	Comments
A	1.0	0.0	Automatic detection present/not present
Č	0.0	1.0	
B	0.1	0.1	British Standards Institute PD 7974 Part 7 (2003)
C	0.3	0.8	Charters, Barnett et al. (1997)
Č	0.7	0.2	
D	0.001	0.001	Probability that trained operators will not detect a warning signal (a low value errs on the side of safety); Cullen (1990)

Sensitivity analysis indicates that the probability of failure of AFD could be as high as 0.26 before the results reverse (a factor of 2.6 greater than the probability given by fire statistics).

Equally, the proportion of area covered by AFD could be as low as 63 per cent before the results reverse. Because there is relatively little additional cost between 63 per cent and 100 per cent AFD coverage, the principle of practicability indicates that full coverage of AFD is the appropriate solution in this case.

Therefore, the alternative fire safety solution (Case 2) provides equivalency with the standard prescriptive guidance solution (Case 1).

3.3 Statistical models

3.3.1 *Power functions*

For any type of building, the probability of fire starting or frequency of fire occurrence will increase with the number of ignition sources and hence with the size of the building expressed in terms of total floor area, A, in square metres. This probability or frequency denoted by $F(A)$, as proposed by Ramachandran (1979/80), can be estimated for any period by applying the following formula:

$$F(A) = \frac{n}{N} \frac{S_n(A)}{S_N(A)} \tag{3.39}$$

where:

n = number of fires during the period, say a year
N = number of all buildings of the type considered at risk i.e. involved and not involved in fires
$S_n(A)$ = proportion of buildings of size A involved in fires
$S_N(A)$ = proportion of buildings of size A at risk.

The parameters n and $S_n(A)$ can be estimated by analysing data on real fires in the buildings of the type considered for a period of years. Such data are provided by fire brigades in the United Kingdom. But a special survey of all buildings of the type considered at risk needs to be carried out to estimate N and $S_N(A)$. To obtain approximate estimates of N and $S_N(A)$, it may be possible to analyse some other statistics such as the distribution of manufacturing units according to employment size which may be combined with an estimate of average area occupied by each person employed in the manufacturing industry considered.

Statistical studies reviewed by Ramachandran (1970, 1979/80, 1988a) have shown that $F(A)$ is approximately given by the 'power' functions

$$F(A) = K A^{\alpha} \tag{3.40}$$

where K and α are constants for a particular type of building. The parameter K includes the ratio (n/N) in Table 3.9 while α measures the increase in the value of

Table 3.9 Annual chance of a fire outbreak for various occupancies

Hazard	Number of buildings* (N)	Number of fires annually+ (n)	P = n/N
Industry	183 377	8 075	4.4×10^{-2}
Houses	14 202 359	38 142	5.7×10^{-4}
Commercial – shops	664 817	5 574	8.4×10^{-3}
Commercial – offices	152 430	866	5.7×10^{-3}
Assembly – entertainment	12 540	1 446	1.2×10^{-1}
Assembly – non-residential	143 019	2 810	2.0×10^{-2}
Residential – clubs, hotels etc	36 609	1 352	3.7×10^{-2}
Residential – institutions	–	803	–
Storage	199 612	2 420	1.2×10^{-2}

Notes
* Source: 108th Report of the Commissioners of HM Inland Revenue
+ Fires during 1967

$F(A)$ for an increase in A. Equation (3.40) is the same as equation (3.35) but with a different notation; $K=a$, $A=A_b$ and $\alpha=b$. To estimate K and α, as mentioned in section 3.1.4, it is necessary to analyse fire statistics and data collected through a special survey of buildings at risk. This exercise was carried out by the Home Office in the United Kingdom to estimate K and α for major groups of buildings – see Table 3.4 for the values of $K=a$ and $\alpha=b$ with A in m^2]

It may be seen from the figures in Table 3.4 that the value of α (or b) is less than unity for most of the occupancy groups, e.g. 0.53 for all manufacturing industries. Actuarial studies in some European countries, based on frequency of fire insurance claims as a function of the financial value (size) of the risk insured, confirm that the value of α is about 0.5 for industrial buildings – see Benktander (1973).

A value less than unity for α indicates that the probability of fire starting does not increase in direct proportion to building size; it increases according to a fractional power. If two buildings are considered, one twice the size of the other, the probability of fire starting in the larger building will be less than two times the probability for the smaller building. If these two are industrial buildings, for example, the probability for the larger building will be 1.44 ($=2^{0.53}$) times the probability for the smaller building. The value of α will be less than unity, since different parts of a building will have different types and number of ignition sources. A value equal to unity for α would imply that all parts of a building have the same risk of fire breaking out. Actuarial studies reviewed by Benktander (1973) support the theoretical arguments mentioned above. These theoretical studies are confirmed by Ramachandran (1970, 1979/80 and 1988a).

The actuarial and statistical studies mentioned above have also revealed that, in the event of a fire occurring, the probable or expected or average area damage is approximately given by the 'power' function

$$D(A) = C A^\beta \tag{3.41}$$

where, as in Equation (3.40), A is the total floor area (size) of a building and C and β are constants for a particular risk category or type of building. These constants can be estimated by fitting the power function in Equation (3.41) if, as for fires in the United Kingdom, data are available for area damage, $D(A)$, and building size, A. Based on the survey by the Home Office, UK mentioned earlier, Rutstein (1979) estimated the values of C and β for major groups of buildings – see Table 3.10.

A fire in a large building is more likely than one in a small building to be discovered and extinguished before involving the whole building. The proportion, $D(A)/A$ destroyed in a large building can, therefore, be expected to be smaller than the proportion destroyed in a small building. Hence, this proportion or damage rate would decrease with increasing values of A; in other words, the value

Table 3.10 Probable damage in a fire – parameters of Equation (3.41)

Occupancy	Probable damage in a fire	
	C (m²)	β
Industrial buildings		
Food, drink and tobacco	2.7	0.45
Chemical and allied	11.8	0.12
Mechanical engineering and other metal goods	1.5	0.43
Electrical engineering	18.5	0.17
Vehicles	0.80	0.58
Textiles	2.6	0.39
Timber, furniture	24.2	0.21
Paper, printing and publishing	6.7	0.36
Other manufacturing	8.7	0.38
All manufacturing industry	2.25	0.45
Other occupancies		
Storage	3.5	0.52
Shops	0.95	0.50
Offices	15.0	0.00
Hotels etc.	5.4	0.22
Hospitals	5.0	0.00
Schools	2.8	0.37

of β would be less than unity as revealed by the figures in Table 3.10. This result is supported by statistical and actuarial studies.

Provision of reliable fire suppression equipment, if it operates satisfactorily when a fire occurs, would reduce the damage rate, $D(A)/A$, and the value of β. Consider, for example, the value of 0.45 for β estimated by Rutstein (1979) for industrial buildings without sprinklers. For industrial buildings with sprinklers, Rutstein's estimate of 16m^2 for average area damage in a building of total floor area 1500m^2 indicates a value of 0.27 for β. This result is based on the assumption that the value of C denoting initial conditions is 2.25m^2 whether a building is sprinklered or not.

Based on data for fire spread provided by UK fire statistics, Ramachandran (1988a) estimated an initial damage of 4.43m^2 for the parameter C and average damage of 187m^2 in an 'average' textile industry building of total floor area 8000m^2 without sprinklers. This provided a value of 0.42 for β. With an average damage of 31m^2 and C = 4.43, β was estimated to be about 0.22 for a textile industry building equipped with sprinklers for fires in which the heat produced was sufficient to activate the system. For an unsprinklered textile industry building, Rutstein (1979) estimated β = 0.39; his estimate of 2.6m^2 for C apparently included very small fires. The estimate for expected damage, $D(A)$, is very sensitive to the value used for the parameter C.

In a later study with C = 4.43, Ramachandran (1990) estimated values of 0.68 and 0.60 for β for unsprinklered and sprinklered textile industry buildings. These estimates were based on maximum area damage (worst-case) in a fire of 2000m^2 in an unsprinklered building and 1000m^2 in a sprinklered building, both buildings of an average size of 8000m^2 total floor area.

For any type of building or a risk category within a type, the product of $F(A)$ given by Equation (3.40) and $D(A)$ given by Equation (3.41) provides an estimate of fire risk expressed on an annual (yearly) basis. Table 3.11 is an example showing the estimates of annual damage (risk) for industrial buildings of different sizes with and without sprinklers.

Equations similar to (3.40) and (3.41) with the total financial value V at risk instead of total floor area A, and their product are used for determining approximate 'risk premiums' for fire insurance – see Benktander (1973). These equations can be converted to provide estimates of $F(A)$ and $D(A)$ in terms of total floor area A by using an estimate for v (= V/A) denoting value density per square metre of floor area. Area damage can also be converted to financial loss by using an approximate estimate for loss per square metre of fire damage. At 1978 prices, this rate of loss was £140 per m^2 for all manufacturing industries – see Rutstein (1979).

3.3.2 Exponential model

It is necessary to discover or detect the existence of a fire in a building soon after the start of ignition in order to commence early and safe evacuation and reduce the number of fatal or non-fatal casualties. Early detection of fire would also

Table 3.11 Industrial buildings – annual damage

Building size (A) (m²)	Probability of fire starting F(A)	Probable damage D(A)		Annual damage	
		Without sprinklers (m²)	With sprinklers (m²)	Without sprinklers (m²)	With sprinklers (m²)
1000	0.0661	50.37	14.53	3.33	0.96
2000	0.0955	68.81	17.52	6.57	1.67
3000	0.1184	82.58	19.54	9.78	2.31
4000	0.1379	94.00	21.12	12.96	2.91
5000	0.1552	103.92	22.43	16.13	3.48
8000	0.1991	128.40	25.47	25.56	5.07
10,000	0.2241	141.97	27.05	31.81	6.06
20,000	0.3236	193.93	32.62	62.76	10.56
50,000	0.5259	292.89	41.77	154.03	21.97
100,000	0.7594	400.11	50.37	303.84	38.25
500,000	1.7819	825.48	77.79	1470.92	138.61
1,000,000	2.5730	1127.66	93.80	2901.48	241.34

$F(A) = 0.0017A^{0.53}$ (annual probability); $D(A) = 2.25\,A^{0.45}$ (without sprinklers); $D(A) = 2.25A^{0.27}$ (with sprinklers)

reduce property damage by enabling the commencement of fire fighting when the fire is small in size such that the fire can be controlled and extinguished quickly. To assess the value of early detection in reducing damage to life and property, it is necessary to estimate area damage, $A(T)$, as a function of duration of burning, T. For this purpose, Ramachandran (1980) proposed the following exponential model based on statistics of real fires attended by fire brigades in the United Kingdom:

$$A(T) = A(0) \exp (\theta T) \tag{3.42}$$

where $A(0)$ is the area initially ignited and θ the fire growth parameter.

It should be emphasised that $A(T)$ in Equation (3.42) is the final (cumulative) size of a fire in terms of area damage at the time (T) of its extinguishment. $A(T)$ is not the fire size at any intermediate time T. Fire statistics do not and cannot provide information on the size of a fire at any specific time, say, when the fire brigade arrives at the scene of a fire.

The model in Equation (3.42) is based on some scientific theories and experimental results according to which heat output in a fire grows exponentially with time. It is applicable to the period after the onset of 'established burning' when a fire has a steady and sustained growth. It is not applicable to the initial and very early stage of a fire which, although small in size, can be very variable in length of time; this stage can last for hours (smouldering) or it can be over in minutes.

During the very early stage of fire development, commencing with the ignition of the first object or material, area damage at time T may be directly proportional to T^2. This model is generally recommended by fire safety engineers according to whom the heat release rate of a fire during the growth phase is proportional to T^2. This result is based on a series of fire tests and analysis of some real fires. This conclusion may be true for the initial stage of a growing fire, when only one object or material in a room is involved. It is unlikely to be realistic for the later stage of fire growth, when several objects are involved. For the later stage, the exponential model in Equation (3.42) appears to be more realistic, particularly due to the stochastic nature of fire spread from object to object – see next section. The uncertainties caused by the random nature of fire spread are not taken into account by a T-square curve.

The exponential model in Equation (3.42) is also supported by statistical analysis of large samples of real fires. Fire statistics available in the United Kingdom provide, for each fire, information on area damage, $A(T)$, and the duration of burning, T. An estimate for T can be obtained as the sum of the following five periods:

- T_1 – ignition to detection or discovery of fire;
- T_2 – detection to calling of fire brigade;
- T_3 – call to arrival of the brigade at the scene of the fire;
- T_4 – arrival to the time when the fire was brought under control by the brigade;
- T_5 – control to extinction of a fire.

The fifth period, T_5, need not be included in the sum since the growth of a fire will be practically negligible during this period.

An estimate of T_1 is given by the brigade according to the following classification:

1 Discovered at ignition ($T_1 = 0$).
2 Discovered under 5 minutes after ignition.
3 Discovered between 5 and 30 minutes after ignition.
4 Discovered more than 30 minutes after ignition.

Average values of 2, 17 and 45 minutes can be adopted for the second, third and fourth classes of T_1.

In a pilot study concerned with the economic value of automatic fire detectors, Ramachandran (1980) applied the exponential model in Equation (3.42) to data on fires in the textile industry during 1978. He estimated the values of 4.69m^2 and 0.0632 for the parameters $A(0)$ and θ respectively. Based on these parameter values, the exponential growth of fire has been depicted in Figure 3.10 to show the economic value of automatic detectors connected and not connected to the fire brigade. In a later study of textile industry fires, Ramachandran (1988a) estimated the values of θ as 0.083 if not sprinklered and 0.031 if sprinklered. These overall values were applicable to fire spread in a building. The value of 4.43m^2 for $A(0)$ related to item first ignited and corresponded to the initial stage taken as zero time and the commencement of

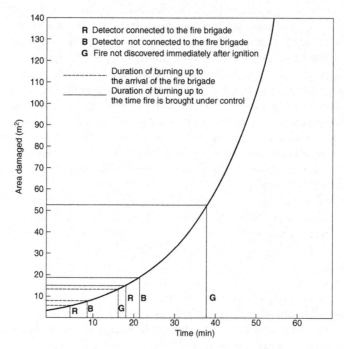

Figure 3.10 Average time (min) and area damaged (m²)

established burning. For fire spread within a room, θ was estimated to be 0.196 if not sprinklered and 0.117 if sprinklered.

Ramachandran and Chandler (1984) applied the exponential model for assessing the economic value of early detection of fires in industrial and commercial premises. In this study, the model was expanded for estimating the values of θ separately for the four periods T_1, T_2, T_3 and T_4. Fires were classified according to whether they occurred in production or storage areas. A further classification related to whether or not fires were tackled by first-aid fire-fighting before the arrival of the brigade. Public/assembly areas replaced production areas in the distributive trades group.

The following were the main conclusions in the above study:

1   Early detection would reduce the damage, especially in premises without first-aid fire-fighting or training in its use.
2   There is a clear need to achieve early fire detection in storage areas.
3   Early detection should be followed up by quick action to extinguish the fire.
4   Automatic detection systems connected to the fire brigade only marginally increase the savings from early detection as compared with systems not connected directly to the fire brigade.

The exponential model in Equation (3.42) provides an estimate of the 'doubling time' given by:

$$d = (1/\theta)\log_e^2 = (0.6931) / \theta \qquad\qquad (3.43)$$

which is the parameter generally used for characterising and comparing rates of fire growth of different objects or materials. This is the time taken by a fire to double in size and is a constant for the exponential model. For example, if it takes 5 minutes for the area damage to increase from $10m^2$ to $20m^2$, it will also take only 5 minutes for the damage to increase from $30m^2$ to $60m^2$, $50m^2$ to $100m^2$, $100m^2$ to $200m^2$ and so on. For an example considered earlier, with θ equal to 0.117 and 0.196 for sprinklered and unsprinklered rooms, the corresponding doubling times are 5.9 minutes and 3.5 minutes respectively.

Assuming an exponential model, Bengtson and Ramachandran (1994) estimated the fire growth rate, θ, for four types of occupancies – railway properties, public car parks, road tunnels and power stations. They showed clearly that the maximum value (upper confidence limit) of growth rates in individual fires is distinct from the maximum value of the average growth rate in all fires. The former maximum is considerably higher than the latter and represents more realistically the worst case scenario.

With appropriate assumptions about the ratio of vertical rate of fire spread to horizontal rate, fire growth rates and doubling times, as discussed above in terms of area damage (horizontal spread), can be converted to fire growth rates and doubling times in terms of volume destroyed – see Ramachandran (1986). As one might expect, doubling time in terms of volume involved is shorter than doubling time in terms of area.

Bengtson and Laufke (1979) used the exponential model and a combination of quadratic (T^2) and exponential models to estimate the fire area and time when sprinklers operate in different hazard categories. The authors also discussed the estimation of the time to flashover at different room volumes with and without installed fire ventilation system. Other topics discussed by them include operation time of smoke detectors, fire brigade efforts on extinguishing a fire and effects on evacuation of people. Bengtson and Hagglund (1986) described the application of an exponential fire growth curve in fire engineering problems.

3.3.3 Regression analysis

Simple regression is concerned with the estimation of the expected or average value of a dependent variable y for a given value of an independent variable x. In the context of fire protection, y can be probability or frequency of fire occurrence, area damage, financial loss or fatality rate per fire. The independent variable x can be a factor such as building size or time taken to detect or discover a fire which affects the dependent variable y. If y has a linear relationship with x which can be represented by the straight line

$$y = m x + c \tag{3.44}$$

the problem is to estimate the values of the parameters m and c.

The problem of fitting the straight line in (3.44) can be solved graphically in the first instance by plotting the pairs of observations (y_i, x_i) available for a sample of fires which occurred in the type of building or risk category analysed. The subscript i refers to the ith fire in the sample. The value of the dependent variable y_i for the ith fire corresponds to the value of the independent variable x_i for that fire. It can then be tested whether a straight line can be drawn approximately to pass through the scatter of points representing the pairs of observations (y_i, x_i). If this is possible then the parameter m is called the slope of the line and measures the ratio between the change in the value of y, say, from y_1 to y_2 and the corresponding change from x_1 to x_2 in the value of x:

$$m = (y_2 - y_1) / (x_2 - x_1) \tag{3.45}$$

The constant c which is the value of y when $x = 0$ is the intercept on the y axis and is called the y intercept.

If the graphical analysis reveals a linear relationship between y and x, the values of the parameters m and c providing the 'best' linear fit can be estimated by applying the method known as 'least squares'. Computer packages are available for this method. With the values of m and c thus estimated, Equation (3.44) can be used to estimate the expected, mean or average value of y for a given particular value of x. Computer packages also provide an estimate of the standard deviation of the 'residual error' which can be used to obtain the 'upper confidence limit' denoting the maximum value and the 'lower confidence limit' denoting the minimum value of the expected (mean) value of y.

In some cases, the straight line relationship may be applicable to the logarithm of y and logarithm of x or logarithm of y and x. For example, according to Equation (3.41) the logarithm of the expected value of area damage, $D(A)$, has a linear relationship with the logarithm of total floor area (size), A, of a building. Figure 3.11 reproduced from a previous study (Ramachandran, 1990) is an example according to which if sprinklers are installed in a building of total floor area 10,000m², area damage can be expected to reduce to 1100m² from 2300m². This figure also reveals the fact that, for an acceptable damage of 2300m² a sprinklered building of total floor area 33,000m² would be equivalent in damage to a non-sprinklered building of total floor area 10,000m². This acceptable size for a sprinklered building may be reduced to 28,000m² to take account of the probability of, say, 0.1 of the system not operating in a fire – see Ramachandran (1998b). Thus, an acceptable size of 10,000m² for an unsprinklered building can be increased to 28,000m² if the building is provided with sprinklers.

According to the exponential model in Equation (3.42), the logarithm of damage, A (T), has linear relationship with the duration of burning, T. This relationship for fire spread within a room depicted in Figure 3.12 has been reproduced from a previous study by Ramachandran (1990).

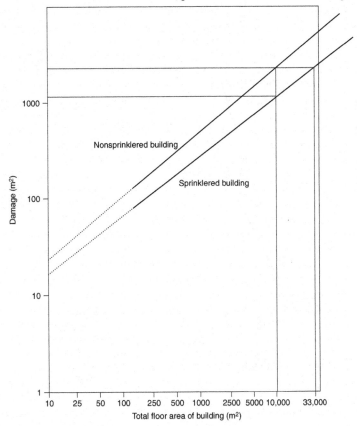

Figure 3.11 Damage and building size – textile industry, UK

According to Ramachandran (1993), the probability of one or more deaths in a fire or fatality rate per fire, P, is approximately given by the straight line:

$$P = K + \lambda D \tag{3.46}$$

where D is the time taken to discover or detect a fire and K and λ are constants for a given type of building. The relationship between P and D, shown in Figure 3.13, for single and multiple occupancy dwellings, was based on data for the period 1978–1991. Respectively for these two types of buildings, λ was estimated to be 0.0008 and 0.0006 and K to be 0.0016 and 0.0015. The fatality rate per fire for that period was about 0.012 for both types. This rate corresponded to average fire discovery times of 13 and 18 minutes for single and multiple occupancy dwellings. If it is assumed that the discovery time of an automatic detector is one minute, on average, the fatality rate would reduce to $(K + \lambda)$ if all the dwellings were fully protected by automatic detection systems. Under such protection the fatality rate per fire in single and multiple occupancy dwellings would have

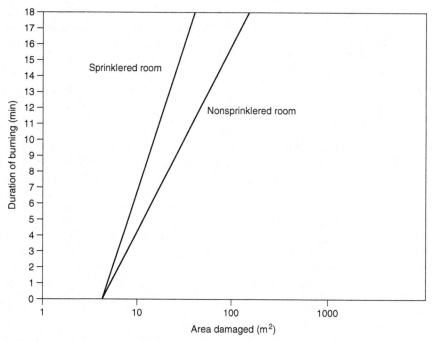

Figure 3.12 Fire growth within room of origin

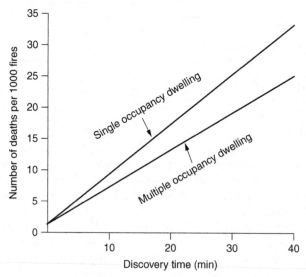

Figure 3.13 Discovery time and fatality rate (current risk level)

reduced to about 0.002, i.e. 2 deaths per 1000 fires, from 0.012, i.e. 12 deaths per 1000 fires which was the average life risk level prevailing during the period 1978–1991.

In the simple single linear regression described above, it is assumed that the value of the dependent variable, y, would be significantly affected by the magnitude of a single factor (independent variable), x. This may not be strictly true since a number of factors may jointly affect y, each factor contributing some amount towards y. For example, the area likely to be damaged in a fire may be affected by building size, building height, compartment size, ventilation, fire load, number of compartments, number of floors, fire resistance and the presence or absence of fire protection measures such as automatic detectors, sprinklers and smoke control systems. There are also other factors such as fire brigade attendance time and control time, rate of fire spread and so on. Some factors will affect property damage, some life damage (e.g. number of escape routes, widths of escape routes) and some both property and life damage. Once these factors are identified, their contribution to the damage can be estimated by performing a multiple regression analysis, with data on damage and factors for each fire for a sample of fires. Such data should be available, and if not, should be collected or estimated and their numerical values used in the analysis.

If p factors (independent variables) are considered in a multiple regression, their contributions to damage (dependent variable) quantified by the regression parameters β_j ($j = 1, 2, ..., p$) are estimated by the model

$$Z = \beta_0 + \beta_1 w_1 + \beta_2 w_2 + ... + \beta_p w_p \tag{3.47}$$

where z is the logarithm of damage, y, to base e, w_j is the numerical value (or its logarithm) of the jth factor. A preliminary simple regression analysis of data for Z and w_j may be carried out to determine whether the value of w_j to be used in a multiple regression is its numerical value or its logarithm. For a qualitative factor such as sprinklers, the value $+1$ may be assigned if the building is equipped with sprinklers or -1 if not equipped. For quantitative factors the parameter β_j measures the increase in the value of z for unit increase in the value of w_j. The constant β_0 measures the fixed effect, not depending on the factors included in the model; it is an average value for the effects of factors not included in the model. The model in Equation (3.47) assumes that damage in a fire has a log normal distribution – see Section 3.3.4.

In the application of the model in Equation (3.47), for the ith fire, z_i is the logarithm of damage and w_{ij} is the corresponding value of the jth factor. If data are available for n fires, and p factors, the n sets of ($p + 1$) values provided by z_i ($i = 1, 2, ..., n$) and w_{ij} ($i = 1, 2, ..., n$; $j = 1, 2, ..., p$) are used in a least squares multiple regression analysis to estimate the parameters β_j ($j = 0, 1, 2, ..., p$). Computer packages are available for performing this analysis. Once the parameters β_j are estimated, the expected value of the logarithm of damage can be estimated for any given set of values for the factors w_j ($j = 1, 2, ..., p$) with the aid of the Equation (3.47).

Most computer software packages on multiple regression provide an estimate of the correlation between the dependent variable z and each of the independent variables w_j ($j = 1, 2, ..., p$). An independent variable (factor) whose correlation with z is very low (close to zero) can be excluded from the analysis and the parameters β_j of the other factors re-evaluated. The contribution to damage, z, from a factor with low correlation, will be negligible.

Software packages also provide estimates of the correlation between independent variables. If two independent variables w_j and w_k are highly correlated, such a high degree of interaction will confuse the interpretation of the predicted value of z due to 'collinearity'. In such a case, only one of the two variables, w_j or w_k, may be included in the final analysis.

In both the single and multiple regression analysis, if the variable $z = \log_e y$ where y is the dependent variable such as area damage, computer packages will provide estimates of the regression parameters m and c in Equation (3.44) or regression parameters β_j ($j = 0, 1, 2, ..., p$) in Equation (3.47). These parameters and equations would provide an estimate of the expected value, μ, of z for given values of independent variable x or independent variables w_j ($j = 1, ..., p$). The median value of damage, y, on the original scale is given by $\exp(\mu)$. The probability of exceeding this median is 50 per cent assuming that the 'residual error' in the regression analysis has a normal distribution. Computer packages provide an estimate of the standard deviation σ, of the residual error. The maximum value of z is given by the upper confidence limit ($\mu + t\,\sigma$) and the corresponding maximum value of damage y by $\exp(\mu + t\,\sigma)$. For any probability level, the value of t can be obtained from a table of the standard normal distribution. For example, if $t = 1.96$, the probability of z or y exceeding the maximum is 0.025. The expected value of y is given by

$$\exp[\mu + (\sigma^2/2)]$$

Instead of area damage, the probability p_s of fire spread beyond the room of origin may be used as the dependent variable in a single or multiple regression model. In this case, the 'logit' given by

$$P_s = \tfrac{1}{2} \log [\, p_s / (1 - p_s)\,] \tag{3.48}$$

should be used in the estimation process, instead of p_s, for rendering the effects of factors approximately additive. In the logit model, the probability of area damage exceeding, say, 100m^2 or financial loss exceeding, say, £100,000 can be used for p_s.

Baldwin and Fardell (1970) applied the logit model to estimate the influence of various factors on the probability of a fire spreading beyond the room of origin. According to this study, there were significant differences in this probability between buildings used for different purposes and between some single storey and multi-storey buildings. The biggest factor affecting fire spread was the time of discovery of the fire, the chance of spread at night being twice that of the day; this was probably because of delays in the discovery of fires. The chance of spread was considerably smaller for modern buildings than for older buildings, particularly for multi-storey buildings. This was, perhaps, the result of increased building (fire) control and safety consciousness. The fire brigade attendance time had no influence on fire spread.

Shpilberg (1975) applied the logit model to quantify the relative effects of types of building construction, number of storeys, sprinkler protection, type of fire department and the subjective Factory Mutual Overall Rating on the probability of loss size. His object was to predict the probability of loss being above or below $10,000 given the particular characteristics of a group of risks. The logit transformation, Equation (3.48), was applied to the probability of loss exceeding $10,000. For purposes of illustration, Shpilberg used all fire loss claims in industrial property classified as 'machine shops' paid by Factory Mutual during 1970–1973. In particular, the overall rating adopted by Factory Mutual was found to be of great value for predicting size and degree of loss, i.e. the fraction of the value of the property that was lost. Sprinklers were also found to be a major factor in determining both expected size and degree of loss.

3.3.4 *Probability distributions*

The 'power function' in Equation (3.41) provides an approximate estimate for the expected value of area damage when a fire occurs in a building of given size. A more accurate but complex method is provided by probability distributions. The probability distribution for financial loss x or area damage d expresses mathematically the probabilities with which the loss or damage in a fire would reach various amounts. This distribution is usually referred to as the 'parent distribution' in extreme value theory (Section 3.3.5). This distribution provides the expected value, standard deviation and other statistical parameters of x or d.

The nature of the probability distribution of loss x has been investigated by Ramachandran (1974, 1975) and Shpilberg (1974) and other authors mentioned in these papers. According to these studies, fire loss distribution is skewed (non-normal) due to the fact that many fires are small and only a few fires are large. In general the transformed variable z ($= \log x$) i.e. logarithm of loss, has a probability distribution belonging to the 'exponential type'. This type, defined by Gumbel (1958) with reference to the limiting (asymptotic) behaviour of a random variable at the tail of its probability distribution, includes exponential, normal, log normal, chi square, gamma and logistic distributions. Among these distributions, normal and exponential for z have been widely recommended by actuaries, based on the analyses of data on fire insurance claim amounts. These correspond to log normal and Pareto for loss x on the original scale. Figure 3.14 is a general sketch of the probability density function, $f(z)$, of a normal distribution for z ($= \log x$). The function $f(z)$ is the derivative of the cumulative distribution function $F(z)$.

If figures for financial loss are available for all the fires which occurred in a risk category, standard statistical methods or a graphical method can be applied to identify the probability distribution which would provide the best fit for the data analysed. But in most countries these data are generally available only for

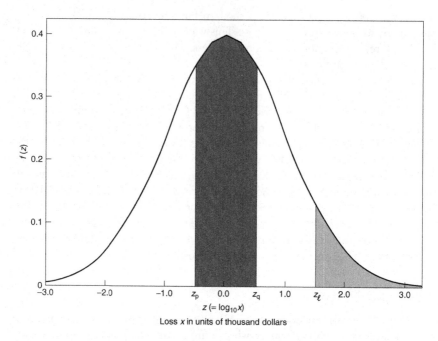

Figure 3.14 Density function (f (z)) curve of fire loss

large fires which, in the United Kingdom for example, are currently defined as fires costing £50,000 or more in property damage. The threshold level which was £10,000 until 1973 has been gradually increased over the years due to inflation and the need to keep the number of large fires to be reported by insurance companies at a manageable level. This led to the development of extreme value statistical models discussed in Section 3.3.5.

However, a probability distribution can be constructed for area damage for which, particularly in the UK, data are available for all fires. The probability of area damage being less than or equal to d is given by the cumulative distribution function $G(d)$ and the probability of damage exceeding d by $[1 - G(d)]$. Figure 3.15 is an example (textile industry) based on fire brigade data and shows the relationship between d and $[1 - G(d)]$ for a building with sprinklers and a building without sprinklers (Ramachandran, 1988a). The area damage is on a log scale since, as revealed by several statistical studies, this random variable, like financial loss, has a skewed probability distribution such as log normal. The values of the parameters of this distribution vary from one type of building to another and with the effectiveness of fire protection measures. Using Equation (3.42), the x (horizontal) axis can be converted to describe the probability distribution for duration of burning (in minutes).

It appears from Figure 3.15 that an initial damage of 3m² is likely to occur before the heat generated in a fire is sufficient to activate a sprinkler system. For both types of building, the probability of damage exceeding 3m² is 0.58. It is

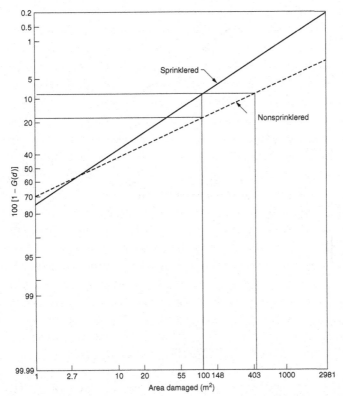

Figure 3.15 Textile industry, United Kingdom – probability distribution of area damaged
$G(d)$ = Probability of damage being less than or equal to d
$1 - G(d)$ = Probability of damage exceeding d

apparent that, in the range greater than 3m², the successful operation of sprinklers would reduce the probability of damage exceeding any given value. For example, the probability of damage in a fire exceeding 100m² is about 0.18 if the building has no sprinklers and 0.08 if the building is equipped with sprinklers. Also, for a given probability level, say, 0.08 for $[1 - G(d)]$ or 0.92 for $G(d)$, the damage would be 500m² if not sprinklered, compared with 100m² if sprinklered.

A log normal distribution was fitted to the raw data pertaining to Figure 3.15, disregarding fires with damage less than 1m² and following a method appropriate for 'censored' samples – see Ramachandran (1988a). For the range exceeding 1m², values of 0.02 and 2.46 were obtained for the mean and standard deviation of logarithm (z) of area damage in a sprinklered building. The expected (average) damage was calculated as 41.64m². For a non-sprinklered building the mean and standard deviation of z were 0.75 and 2.87 leading to an expected damage of 216.67m².

Figure 3.16 is an example based on Pareto distribution for area damage which is the same as exponential distribution for logarithm of damage. If this

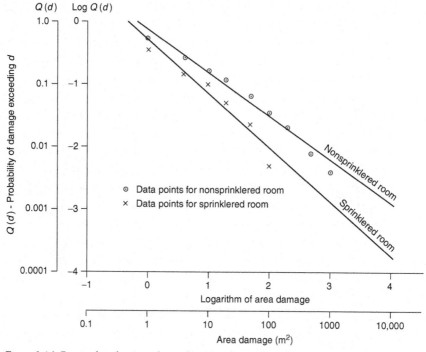

Figure 3.16 Pareto distribution of area damage – retail premises (public areas)

distribution is appropriate, as in Figure 3.16, logarithm of damage and logarithm of the survivor function $[1 - G(d)]$, denoted as $Q(d)$ in the figure, should have approximately a straight line relationship. Values for plotting the points in Figure 3.16 were obtained from the figures for the frequency distribution of damage in Table 3.12.

If occupants in a building, attempting to escape from a fire, are exposed to untenable (lethal) conditions caused by a combustion product, some of them may sustain fatal injuries. The number of deaths occurring in a fire has a discrete (discontinuous) probability distribution applicable for a random variable which takes integer values. Poisson is one such distribution, which has been widely used in the statistical literature for modelling the occurrence of rare events. According to an extended form of this distribution, the probability $p(x, t)$ of exactly x deaths occurring in a fire due to an exposure period of t minutes as discussed by Rasbash, Ramachandran et al. (2004), is given by

$$p(x, t) = \exp(-\delta t)(\delta t)^x / x! \tag{3.49}$$

where

$$x! = x(x-1)(x-2)\dots 2.1$$

Table 3.12 Retail premises – assembly areas – frequency distribution of area damage

Area damage (m²)	Number of fires in assembly areas	
	Without sprinklers	With sprinklers
less than 1	4197 (48.9)	154 (31.3)
2–4	1987 (24.7)	37 (14.7)
5–9	619 (17.1)	9 (10.7)
10–19	463 (11.5)	13 (4.9)
20–49	430 (6.2)	6 (2.2)
50–99	221 (3.5)	4 (0.5)
100–199	127 (2.0)	–
200–499	100 (0.8)	1
500–999	29 (0.4)	–
1000 and above	34	–
Total number of fires	8207	224

Note: Figures within brackets are percentages of fires exceeding upper limits of damage ranges in the first column.

If $\bar{t}$ is the average value of t in a population of fires, $\delta\bar{t} = P$ which, as defined with reference to Equation (3.46), is the fatality rate per fire or probability of one or more deaths.

As discussed with reference to Equation (3.46), data are available for a number of years for estimating the fatality rate per fire, P, for any type of occupancy. Deterministic (scientific) models may provide, for any occupancy type, an estimate of the parameter δ which is the increase in the probability of death for every extra minute of exposure to untenable conditions caused by any combustion product. It may be noted that the probability of no death is given by $x = 0$ in Equation (3.49) or by $\exp(-\delta t)$ or $\exp(-P)$ which may be approximated to $(1 - P)$ if P is small.

For any type of occupancy such as hotels, hospitals and office buildings with a large number of people at risk, figures for the frequency distribution of number of deaths for a number of years may provide sufficient data for fitting the Poisson distribution in Equation (3.49). Other distributions, e.g. negative binomial, may provide a better fit to these data. The 'global' distribution thus identified can be adjusted to take account of the number of people at risk in a particular building.

3.3.5 Extreme value distributions

Large losses fall at the 'tail' of the parent distribution of loss, discussed in the previous section. These losses constitute a very small percentage of the total number of fires in a risk category as shown in Figure 3.14 by the area with dots with values of z greater than z_i at the upper tail. Hence, these losses are not amenable to analysis by standard statistical methods. Extreme order theory, as developed by Ramachandran in a series of papers, provides a mathematical framework for making the best use of the information provided by large losses – for a review of these studies see Ramachandran (1982, 1988b). The asymptotic theory of extreme values, discussed in these studies, provides approximate results for an 'exponential type' distribution. According to this theory, the number of fires (n) occurring during a period should be large, say, more than 100. Also, preferably, at least 20 large losses should be available for analysis. Due to these requirements, in some cases, it may be necessary to consider fires occurring in a group of buildings with similar fire risks over, say, four or five years. In such a case, the loss figures should be corrected for inflation using, say, indices for retail prices.

For a more detailed summary about extreme value theory see Rasbash, Ramachandran *et al.* (2004). Basic features of this theory are as follows. The logarithms of losses in n fires, occurring in a risk category over a period of years, constitute a sample of observations generated by the parent distribution $F(z)$. If these loss figures are arranged in decreasing order of magnitude, the logarithm of the mth loss in this arrangement may be denoted by $z(m)n$ which is referred to as an extreme order statistic. For the largest value, the subscript m takes the value one (first rank). Over repeated samples (periods), $z_{(m)n}$ is a random variable with an extreme value probability distribution. Extreme order theory is concerned with the individual probability distributions generated by extreme order statistics of varying rank m and their joint distribution.

In the absence of any knowledge about the exact nature of the parent distribution, the parameters of the extreme value distribution of $z_{(m)n}$ can be estimated from observations on $z_{(m)n}$ in repeated samples. Three methods available for this purpose, as discussed by Ramachandran (1982), involve corrections due to the varying value of n, number of fires, from period to period. The estimated values of the parameters for different ranks (m) would describe the behaviour of the tail of the parent distribution as a function of n. The parameters would also provide an indication of the nature of the parent distribution. Parent distributions satisfying this behaviour can be fitted to the large losses and the errors estimated, in order to select a distribution which would provide the best fit.

Another application of the extreme value theory is concerned with the estimation of the mean (μ) and standard deviation (σ) of z, logarithm of loss, in all n fires, large and small. But this estimation has to be based on, say, r consecutive large losses, $m = 1$ to r, above a threshold level. Information on financial loss may be available only for r large fires out of n fires. To obtain the best estimates of μ and σ in all n fires from r large fires, Ramachandran has developed two methods – generalised least squares and maximum likelihood. The first method provides

Table 3.13 Average loss per fire at 1966 prices (£000s)

	Sprinklered single storey	Sprinklered multi-storey	Non-sprinklered single storey	Non-sprinklered multi-storey
Textiles	2.9	3.5	6.6	25.2
Timber and furniture	1.2	3.2	2.4	6.5
Paper, printing and publishing	5.2	5.0	7.1	16.2
Chemical and allied	3.6	4.3	4.3	8.2
Wholesale distributive trades	–	4.7	3.8	9.4
Retail distributive trades	–	1.4	0.4	2.4

'unbiased' estimates but involves complex calculations for which a computer program has been developed. The second method is quite easy to apply and would only require a pocket calculator. This method would provide 'biased' estimates but formulae have been developed to adjust the results for biases. Both the methods require an assumption, such as log normal, to be made about the parent distribution.

Assuming a log normal distribution and applying the generalised least squares method, Rogers (1977) estimated the average losses due to fires in industrial and commercial buildings with and without sprinklers. His results are reproduced in Table 3.13. Figure 3.17 is an example based on his investigation.

3.4 Stochastic models

3.4.1 Stochastic nature of fire spread

The statistical models discussed in the previous section are useful for assessing fire risk in a group or category of buildings with similar risk of fire starting and causing damage. The quantitative estimates of risk provided by these models are generally applicable to a building of 'average' characteristics belonging to the risk category considered. These estimates can be adjusted to provide an approximate assessment of fire risk in a particular building within that category or group – see Chapter 5.

However, it may be desirable to assess the risk in a particular building based mostly on the characteristics of that building. These characteristics include:

* building layout
* design
* fire load, i.e.
  * amount of combustible materials and objects; and
  * arrangement of these objects, i.e. distance between the objects
* ventilation

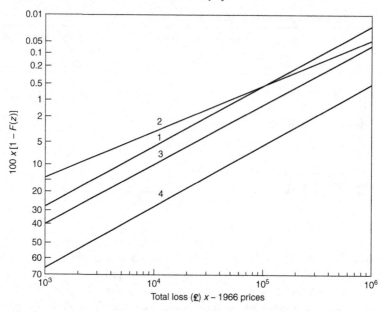

Line	Subpopulation	Parameters
1	Sprink/singlestorey	m= –0.616 s = 1.024
2	Sprink/multistorey	m= –1.419 s = 1.340
3	Nonsprink/singlestorey	m= –0.334 s = 1.062
4	Nonsprink/multistorey	m= 0.401 s = 0.992

$z = \log x$
$m = $ mean (average) value of z
$s = $ standard deviation of z

$F(z)$ (cumulative distribution for z) = Probability of loss less than or equal to z
$V(x)$ (cumulative distribution for x) = Probability of loss less than or equal to x
Survivor probability = $1 - F(z) = 1 - V(x)$ = Probability of loss exceeding x or z

Figure 3.17 The survivor probability distribution of fire loss for each class in the textile industry

- fire resistance of the structural boundaries; and
- installed fire protection measures such as sprinklers.

Fire risk in a particular building can be assessed by applying a stochastic model of fire spread. Detailed reviews of stochastic models of fire growth were carried out by Ramachandran (1995a, 2002) and Rasbash, Ramachandran *et al.* (2004). Basic features of three of these models are discussed in the next three subsections.

For the following reasons the spread of a real fire in a building is a stochastic (probabilistic) phenomenon. It is not a deterministic process, obeying in exact terms the course predicted by scientific theories and experimental results.

A fire in a room usually starts with the ignition of one of the objects. Then it spreads to other objects depending on the distances between the objects and other factors such as fire load and ventilation. This process produces a chain of ignitions that can lead to fully developed fire conditions, defined as 'flashover'. If such conditions occur, the fire can spread beyond the room, depending on the fire resistance of the barriers (walls, floor, ceiling).

There will, however, be a chance that the fire chain can break at some stage for various reasons with the fire being extinguished before spreading further – statistics of real fires support this hypothesis.

As described above, a fire passes through several stages during its development with a chance of being extinguished during any stage. It stays for a random length of time in each stage before moving to the next stage. Its movement (spread) from stage to stage is governed by 'transition probabilities' which are functions of time since the start of ignition. This hypothesis forms the basis of a stochastic model of fire spread.

Due to uncertainties caused by several factors, the spread of fire is really a stochastic phenomenon, although a fire experiences certain deterministic (physio-chemical and thermodynamic) processes during its development. A stochastic model is a non-deterministic model which provides a probabilistic prediction of spatial spread of fire as a function of time – see Ramachandran (1991).

3.4.2 Random walk

In a simple stochastic representation, the fire process involving any single object or number of objects can be regarded as a random walk. The fire takes a random step every short period either to spread with a probability λ or to be extinguished (or burn out) with a probability μ ($= 1 - \lambda$). The parameter λ denotes the success probability of the fire and μ the success probability of an extinguishing agent. The process stops when the fire is extinguished; extinguishment is an 'absorbing boundary' to the random walk.

If the fight between the fire and the extinguishing agent goes on continuously every minute, a random walk as described above will lead to an exponential model (Ramachandran, 1985) for the random variable t:

$$Q(t) = \exp(-\mu t); c = \mu - \lambda \tag{3.50}$$

where $Q(t)$ is the probability of duration of burning exceeding t minutes. The fire-fighting effort is adequate if c is positive with $\mu > \lambda$ and hence greater than $\frac{1}{2}$; it is inadequate if c is negative with μ less than λ and hence less than $\frac{1}{2}$. If $c = 0$, such that $\mu = \lambda = \frac{1}{2}$ there is an equal balance between fire-fighting efforts and the propensity of fire to spread.

The area destroyed (d) is also a random variable whose logarithm is directly proportional to t as a first approximation – see Equation (3.42). This assumption would transform Equation (3.50) to the Pareto distribution:

$$\varphi(d) = d^{-w}; d \geq 1 \tag{3.51}$$

denoting the probability of damage exceeding d (see Section 3.3.4). The use of the Pareto distribution for fire damage was proposed by Mandelbrot (1964) who derived this distribution following a random walk process.

Ramachandran (1969) found that a constant value for μ in Equation (3.50) or for w in Equation (3.51) was unrealistic for fires which were fought and extinguished at some stage. He suggested a distribution with the 'failure rate' μ in Equation (3.50) increasing for large values of t. The failure rate can be decreasing in the early stages of fire development, denoting a success for fire in spreading but it would eventually increase since fire-fighting efforts would succeed ultimately.

Consider the spread of fire as a random walk along spaces arranged in a linear sequence. Every minute, the fire takes a random step to move forwards to an adjacent space with a probability of λ or stay in its present space with a probability of μ $(= 1 - \lambda)$. It does not move backwards to the space from which it has moved. In a simple model, λ and μ are regarded as constants, although, realistically, they are functions of time since the commencement of burning.

Under the above assumptions, if the random walk process starts in the first space with probability one, the probabilities of the fire being in the first two spaces after one minute are μ and λ respectively. After two minutes, the fire is in one of the first three spaces. At that time, the fire may be in the first space with the probability of μ^2 or it may be in the third space with the probability of λ^2, since it should have moved to the second phase after one minute with the probability of λ. Hence, after two minutes, the probability of the fire remaining in the second space is $2\lambda\mu$ $(= 1 - \lambda^2 - \mu^2)$.

Generalising, after t minutes, the probability of the fire being in the rth space is the rth term in the binomial expansion of $(\mu + \lambda)^t$. After t minutes, the probabilities of fire burning in the first and $(t + 1)$th spaces are μ^t and λ^t. If μ is a small quantity, $(1 - \mu)$ or λ is an approximation for $\exp(-\mu)$ in Equation (3.50) such that λ^t is an approximation for $\exp(-\mu t)$. Hence the probability of duration of burning exceeding t minutes is λ^t approximately. The probability of the fire being extinguished during the $(t + 1)$th minute is $\lambda^t.\mu$.

3.4.3 Markov model

Stages of fire growth can generally be defined as states or spatial modules (Watts, 1986), phases (Morishita, 1977), or realms (Berlin, 1980). Fire spreads, moves or makes a transition from state to state. Mathematically, if a fire is in state a_i at the nth minute, it can be in state a_j at the $(n + 1)$th minute according to the transition probability $\lambda_{ij}(n)$. The probability of remaining in state a_i at the nth minute without making a move to another state is denoted by $\lambda_{ii}(n)$. In a Markov process with stationary transition probabilities, the value of $\lambda_{ij}(n)$ is a constant λ_{ij} not varying with the time denoted by n. With m states the transition probabilities per minute are most conveniently handled in the matrix form:

$$P = \begin{matrix} \lambda_{11} & \lambda_{12} & \lambda_{13} & \cdots & \lambda_{1m} \\ \lambda_{21} & \lambda_{22} & \lambda_{23} & \cdots & \lambda_{2m} \\ & \cdots & & & \\ \lambda_{m1} & \lambda_{m2} & \lambda_{m3} & \cdots & \lambda_{mm} \end{matrix}$$

where $\sum_{j=1}^{m} \lambda_{ij} = 1; \quad i = 1, 2, \ldots\ldots\ldots m$

The probabilities of the fire being in different states at time n can be expressed as the vector

$$P_n = (q_1 n \; q_2 n \; \cdots \; q_{mn})$$

where q_{in} is the probability of the fire burning in state a_i at time n. Since a fire can be in one of the m states, for any time n, $\sum_{i=1}^{m} q_{in} = 1$. The vector P_{n+1}, given by the matrix product $P_n.P$, expresses the probabilities of fire burning in different states one transition (minute) later. If the fire starts in state a_1, the first element in the vector P_0 for the initial time denoted by q_{10} is unity and the rest of the other elements in this vector are zero. With this initial condition, the elements in the vectors P_n for different times n can be obtained by performing the matrix multiplication $P_n.P$ repeatedly starting with $P_0.P$.

As an example, consider a Markov model of fire growth in a room in which state a_i represents i objects burning. Suppose, with 3 objects, i.e. $m = 3$, and no extinguishment, the process terminates with the occurrence of flashover at state a_3 when all the three objects are burning. There is no recession in growth and hence there is no transition to a lower state from a higher state. The problem now is to estimate the transition probabilities per minute in the matrix P. This is possible by applying scientific and statistical models to experimental data on heat output or release rate for the three objects and other information such as distances between the objects.

With the assumptions mentioned above, let the transition matrix be

$$P = \begin{matrix} 0.5 & 0.3 & 0.2 \\ 0 & 0.8 & 0.2 \\ 0 & 0 & 1.0 \end{matrix}$$

The process starts with the ignition of one of the objects with the other two objects not yet ignited. We may express the probability distribution of the system at the initial time 0 as the vector

$$P_0 = (1 \; 0 \; 0)$$

By performing the matrix multiplication $P_0.P$, the probability distribution of the system after one minute is:

$$P_1 = (0.5 \ 0.3 \ 0.2)$$

By performing the matrix multiplication successively, it may be seen that

$$
\begin{aligned}
P_2 &= P_1 P &&= (0.25 &&0.39 &&0.36) \\
P_3 &= P_2 P &&= (0.125 &&0.387 &&0.488) \\
P_{10} &= P_9.P &&= (0.0010 &&0.1064 &&0.8926) \\
P_{15} &= P_{14}.P &&= (0.0001 &&0.0352 &&0.9647)
\end{aligned}
$$

and so on. Hence after 15 minutes, the probability of flashover (state a_3), with all the three objects burning, increases to such a high figure as 0.9647.

Following the model described above, Berlin (1980) estimated λ_{ij} for six realms for residential occupancies:

- no-fire state
- sustained burning
- vigorous burning
- interactive burning
- remote burning; and
- full room involvement.

The realms were defined by critical events characterised by heat release rate, flame height and upper room gas temperature. Estimation of λ_{ij} for different i and j was based on data from over a hundred full-scale fire tests. The first realm, no fire, was an 'absorbing state' since all fires eventually terminate in this state.

Berlin applied his model to a smouldering fire in a couch with cotton cushions. He estimated the probabilities of maximum extent of flame development as 0.33, 0.07, 0.02 and 0.58 for a fire reaching the second, third, fourth and fifth realms respectively but not growing beyond these realms. Berlin also discussed other fire effects such as probability of self-termination and distribution of fire intensity. The fire growth model of Beck (1987) was based on the six realms defined by Berlin.

The state transition model (STM) is a particular (simple) version of a Markov model with stationary (constant) transition probabilities. An event tree such as in Figure 3.18 constitutes a simple STM in which fire in a room is described as developing through four successive stages or states E_1 to E_4. A fire can 'jump' to E_4 from E_1 or E_2 without passing through E_2 and E_3 but such 'jumps' have not been considered in this simple STM. The probability E_i is the probability of confinement of fire (extinguishment) in the ith state if the fire has spread beyond the previous states. The probabilities E_1, E_2, E_3 and E_4 add up to unity. These probabilities are limiting probabilities of a fire being extinguished ultimately in the four states. Estimates of these probabilities for most of the building types and some risk categories can be obtained by collecting and analysing data such as those compiled by the fire brigades in the UK.

Based on estimates for E_i ($i = 1$ to 4), the probabilities λ_i and μ_i ($i = 1$ to 3) can be estimated by applying the formulae given at the foot of Figure 3.18; $\lambda_i + \mu_i = 1$. The parameters λ_i and μ_i are respectively 'conditional' probabilities of fire spreading beyond and being extinguished in the ith state given that the

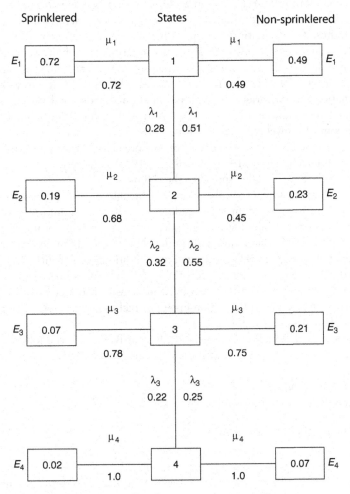

Figure 3.18 Probability tree for textile industry

E_1 = Probability of confinement to item first ignited.
E_2 = Probability of spreading beyond item first ignited but confinement to contents of room of fire origin.
E_3 = Probability of spreading beyond item first ignited and other contents, but confinement to room of fire origin and involvement of structure.
E_4 = Probability of spreading beyond room of fire origin, but confinement to building.

$E_1 = \mu_1$; $E_2 = \lambda_1 \cdot \mu_2$; $E_3 = \lambda_1 \cdot \lambda_2 \cdot \mu_3$; $E_4 = \lambda_1 \cdot \lambda_2 \cdot \lambda_3 \cdot \mu_4$; $\mu_4 = 1$

fire has spread to this state. These parameters are values to which the transition probabilities of spread and extinguishment would tend ultimately over a period of time; they are not probabilities per minute. The parameters λ_i and μ_i can be expressed on a per minute basis by estimating the duration for which their values in the event tree are applicable. It may be noted that $\mu_4 = 1$ and $\lambda_4 = 0$ since fire spread beyond the building of origin is not considered in this model.

Ramachandran (1985) applied an STM similar to Figure 3.18 in the overall framework but expanded it to include sub-realms (periods), each of a fixed duration of 5 minutes, for evaluating the spreading and extinguishment probabilities λ_i and μ_i as functions of time. He also considered the probability w_i of burning (remaining) in the ith state without extinguishment or spreading; in this case, $\lambda_i + \mu_i + w_i = 1$. Statistics on real fires provided by the fire brigades were used for estimating the transition probabilities mentioned above and the probability distributions of duration of burning in each stage. Four materials ignited first in the bedroom of a dwelling were used as examples. The states defined by Aoki (1978) were based on the spatial extent of fire spread and his analysis was similar to that of Ramachandran (1985). Morishita (1977) considered eight phases of spatial spread which included spread to the ceiling.

The STM approach can also be adopted for evaluating the probability of fire spreading from room to room in a building – see Ramachandran (1995). Each room or corridor in a building has an independent probability of fire spreading beyond its structural boundaries. This probability for a room or compartment is the product of probability of flashover and the conditional probability of structural (thermal) failure, given flashover. Using these probabilities, estimated for different rooms and corridors, fire spread in a building can be considered as a discrete propagation process of burning among points which abstractly represent rooms, spaces or elements of a building.

For example, three adjoining rooms R_1, R_2 and R_3 provide the following four states with the fire commencing with the ignition of objects in R_1:

- only R_1 is burning
- R_1 and R_2 (not R_3) are burning
- R_1 and R_3 (not R_2) are burning; and,
- all the three rooms are burning.

There is no transition from the first to the fourth state or from second to the third state or from third to the second state. There is also no transition from the second or third or fourth state to the first state, i.e. no recession of fire growth.

A transition from the second to the fourth state involves the spread of fire to R_3 from R_1 or R_2. The probability for this transition is, therefore, the sum of probabilities for spread from R_1 to R_3 and R_2 to R_3. Likewise, the probability of transition from the third to the fourth state is the sum of probabilities for spread from R_1 to R_2 and R_3 to R_2. A fire can burn in the same state without transition to another state. The process terminates when the fourth state is reached. With the assumptions mentioned above, a transition matrix can be formed specifying the probability of fire spread from one room to another. For estimating the average time for transition to the fourth state, denoting the burning of all the three rooms, Morishita (1985) proposed a method based on partitioning of the transition matrix.

An event tree such as the one in Figure 3.18 can provide a 'global' estimate of the probability of flashover for a room or compartment belonging to any occupancy type and area of fire origin separately for the two cases – unsprinklered and sprinklered. This method and other methods of estimating the probability of flashover are discussed in Section 5.3. Given flashover, the probability of compartment failure depends on the severity likely to be attained when a fire occurs in the compartment and the fire resistance of the structural boundaries of the compartment. Estimation of this probability is discussed in Sections 3.7.3 Safety factors and 3.7.4 Beta method.

The major weakness of the Markov model discussed above is the assumption that the transition probabilities are constants, remaining unchanged regardless of the number of transitions representing the passage of time. The length of time a fire burns in a given state affects future fire spread. For example, the probability of a wall burn-through increases with fire severity which is a function of time. The time spent by fire in a particular state may also depend on how that state was reached, i.e. whether the fire was growing or receding. Some fires may grow quickly and some may grow slowly depending on high or low heat release. In a Markov model with constant transition probabilities, no distinction is made between a growing fire and a dying fire. An application of a Markov model with time-varying transition probabilities would require large amounts of data and involve complex computation procedures.

3.4.4 Network models

As explained in the previous section, the STM can provide for each room in a building a cumulative probability, p_s, at time t_s, when the structural boundaries of the room are breached. The duration t_s is the sum of t_f representing the time to the occurrence of flashover and t_b representing the time for which the structural barriers of the room can withstand fire severity attained during the post-flashover stage. The probability p_s is the product of probability p_f of flashover and probability p_b of structural failure given flashover. The pairs of values (p_s and t_s) for different rooms can then be used in a stochastic model in which a building is represented by a network with rooms or compartments as nodes and the links between these nodes as possible paths for fire spread.

Consider, as an example, the simple layout of Figure 3.19(a) relating to four rooms and the corresponding graph shown in Figure 3.19(b) which also shows the probability (p_{ij}) of fire spread between each pair of rooms (i, j). The probability p_{ij} refers to p_s as defined above, whereas Dusing et al. (1979) and Elms and Buchanan (1981) considered only the barrier failure probabilities p_b ignoring the probability of flashover p_f. The specific problem considered by these authors was to compute

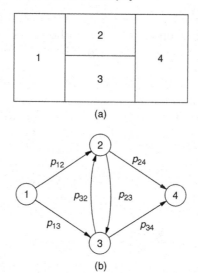

Figure 3.19a Room layout; *b* Corresponding graph

the probability of fire spread from room 1 to room 4 which might follow any of the four paths

$(1) \rightarrow (2) \rightarrow (4)$
$(1) \rightarrow (3) \rightarrow (4)$
$(1) \rightarrow (2) \rightarrow (3) \rightarrow (4)$ and
$(1) \rightarrow (3) \rightarrow (2) \rightarrow (4)$

Using the event space method, Elms and Buchanan considered first all possible 'events' or combinations of fire spreading or not spreading along various links. If a_{ij} represents spread of fire along link ij, and $\bar{a}_{ij}$ represents fire not spreading along the link, then one event might be

$$[a_{12}, \bar{a}_{13}, a_{23}, \bar{a}_{32}, \bar{a}_{24}, a_{34}]$$

There will be $2^6 = 64$ events which will all be exclusive as any pair of events will contain at least one link for which fire spreads in one event and does not spread in the other. The probability of each event occurring is the product of the probabilities of its elements assuming that the elements are independent. Thus, for the example given above, the event probability will be

$$P_{12}(1 - P_{13})P_{23} (1 - P_{32})(1 - P_{24})P_{34}$$

and the overall probability is the sum of all 64 event probabilities.

Representing the completed event space as a tree with 64 branches, Elms and Buchanan adopted a procedure known as 'depth-first search of a graph' for identifying or searching possible paths of links leading to node (room) 4 from node 1. The calculation was carried out for each pair of rooms and the results assembled in a 'fire spread matrix' with unit values for the diagonal elements. The core of this model is a probabilistic network analysis to compute the probability of fire spreading to any compartment within a building. The dimension of time was not explicitly considered in this model, although it was implicit in many of the functions used. In a similar network model proposed by Platt (1989), the probability of spread is dependent on time.

Ling and Williamson (1986) proposed a model in which a floor plan is first transformed into a network. Each link in their network represents a possible route of fire spread and those links between nodes corresponding to spaces separated by walls with doors are possible exit paths. The space network is then transformed into a probabilistic fire spread network as in the example in Figure 3.20 with four rooms, Rm 1 to Rm 4 and two corridor segments c_1 and c_2. With Rm 1 and Rm 1´ with a 'prime' denoting the pre-flashover and post-flashover stages the first link is represented by

$$Rm\ 1 \rightarrow Rm\ 1´$$
$$(p_f, t_f)$$

where p_f represents the probability of flashover and t_f represents the time to flashover.

In Figure 3.20, three different types of links are identified. The first corresponds to the fire growth in a compartment, the second to the fire breaching a barrier element, and the third to fire spread along the corridors. To each link i, a pair of numbers (p_i, t_i) is assigned with p_i representing the distributed probability that a fire will go through link i and t_i representing the time distribution that it will take for such a fire to go through link i. The section of the corridor, c_1, opposite Rm 1 is treated as a separate fire compartment and is assigned a (p_f, t_f) for the link from c_1 to $c_1´$. The number pair (p_s, t_s) represents the probability and time for the

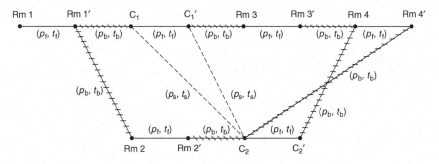

Figure 3.20 Probabilistic network of fire spread of Room 1 to C_2

Fire growth within compartment; ————
Fire breaches barrier elements ⼗⼖⼖⼗⼀
Fire spread along corridor   – – – –

pre-flashover spread of fire along the corridor from c_1 to c_2. Once full involvement occurs in the section c_1 of the corridor outside Rm 1 (i.e. node c_1' is reached) the fire spread in the corridor is influenced more by the ventilation in the corridor and by the contribution of Rm 1 than by the materials properties of the corridor itself. Thus there is a separate link, c_1' to c_2 which has its own (p_s, t_s). The number pair (p_b, t_b) represents the probability of failure of the barrier element with t_b representing the endurance of the barrier element.

Once one has constructed the probabilistic network, the next step is to 'solve' it by obtaining a listing of possible paths of fire spread with quantitative probabilities and times associated with each path. For this purpose, Ling and Williamson adopted a method based on 'emergency equivalent network' developed by Mirchandani (1976) to compute the expected shortest distance through a network. (The word 'shortest' has been used instead of 'fastest' to be consistent with the literature.) This new 'equivalent' network would yield the same probability of connectivity and the same expected shortest time as the original probabilistic network. In this method, each link has a Bernoulli probability of success and the link delay time is deterministic.

It must be noted that there are multiple links between nodes in the equivalent fire spread network. For example, the door between Rm 1 and the corridor could be either open or closed at the time that fire flashed over in Rm 1. Ling and Williamson assumed, as an example, that there is a 50 per cent chance of the door being open and that an open door has zero fire resistance. Furthermore, they assumed that the door, if closed, would have a 5-minute fire rating. With further assumptions, they constructed the equivalent fire spread network (Figure 3.21) with 12 possible paths for the example in Figure 3.20 to find the expected shortest time for the fire in Rm 1 to spread to the portion of corridor c_2. A similar network was constructed for the case with self-closing 20-minute fire rated doors which had ten possible paths. For the two equivalent networks, all the possible paths are listed in Tables 3.14 and 3.15 with increasing time and with all the component links identified. Each of the paths describe a fire scenario. For instance, the scenario for path 1 in Table 3.14 would be: the fire flashes over, escapes from Rm

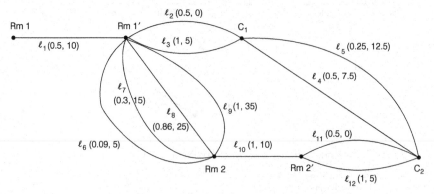

Figure 3.21 Equivalent fire spread network with five-minute unrated doors

1 through an open door into the corridor c_1 and spreads along the corridor to c_2. The probability for that scenario is 0.13. The time of 17.5 mins is the sum of 10 mins for flashover and 7.5 mins for fire spread from c_1 to c_2.

Ling and Williamson derived a formula for calculating, from the figures in Tables 3.14 and 3.15, the probability of connectivity which is 0.5 for both the networks. This probability is a direct result of the assumed probability of 0.50 for flashover in the room of fire origin and the occurrence of unity probabilities in the remaining links which make up certain paths through the network. According to another formula, the expected shortest time is 29.6 minutes for Figure 3.21 which increases to 47.1 minutes due to the presence of the 20-minute fire rated door. The equivalent fire spread network thus facilitates an evaluation of design changes and affords ready comparison of different strategies to effect such changes.

Following Ling and Williamson (1986), Connolly and Charters (1997) applied a network model to evaluate the effectiveness of passive fire protection measures in contributing to fire safety in hospitals. The particular objective considered by

Table 3.14 Pathways for fire spread equivalent network assuming five-minute unrated corridor doors

Paths	Component links	Probability p_i	Time t_i (minutes)
1	1–2–4	1/8 = 0.13	17.5
2	1–2–5	1/16 = 0.06	22.5
3	1–3–4	1/4 = 0.25	22.5
4	1–6–10–11	1/44 = 0.02	25.0
5	1–3–5	1/8 = 0.13	27.5
7	1–6–10–12	1/22 = 0.05	30.0
8	1–7–10–12	3/40 = 0.08	35.0
9	1–8–10–11	3/14 = 0.21	40.0
10	1–8–10–12	3/7 = 0.43	50.0
11	1–9–10–11	1/4 = 0.25	55.0
12	1–9–10–12	1/2 = 0.50	60.0

Table 3.15 Pathways for the fire spread equivalent network assuming self–closing 20-minute rated corridor doors

Paths	Component links	Probability p_i	Time t_i (minutes)
1	1–2–3	1/4 = 0.25	37.5
2	1–2–4	1/8 = 0.13	42.5
3	1–5–9–10	1/22 = 0.05	45.0
4	1–6–9–10	3/20 = 0.15	55.0
5	1–7–9–10	3/7 = 0.43	65.0
6	1–8–9–10	1/2 = 0.50	75.0

Table 3.16 Passive fire protection measures in hospitals

Fire safety measure	$t_{shortest}$ (mins)	
	$t_f = 5$ mins	Safety factor
Datum		
30-minute wall, 5-minute doors, 50/50 chance of being open	27.7	1.00
(i) Lock store room door	30.0	1.08
(ii) Lock store room door, s/c on ward door	30.4	1.10
(iii) Store room door 30 min FR, s/c on store room door	33.7	1.22
(iv) Firecode 30 min FR doors both with s/c	34.8	1.25

Note: s/c = self-closer assumed to ensure the door is closed 80% of time; FR = fire-resisting

them was to prevent a fire starting in a store room from spreading to the ward. They assumed a minimal level of fire protection for the datum case analysed. The fire resistance of the wall separating the store from the ward was 30 minutes. The doors to the store room and to the ward had a fire resistance of only 5 minutes. It was assumed further that there was a 50 per cent chance of the doors being open in which event their fire resistance would be zero. The probability of flashover, p_f, was set equal to the probability of fire spreading beyond the item first ignited. The time, t_f, for the occurrence of flashover was not specified.

For the datum level of fire protection described above, Connolly and Charters (1997) calculated the expected shortest time, t_m, for fire to spread from the store to the ward as

$$t_m = 1.22t_f + 21.6 \text{ minutes}$$

This result suggested that fire spread was likely to take at least 21.6 minutes for the nature of fire protection defined in the datum case. Fire protection measures considered by the authors included:

- locking of store room door; and
- fitting self-closing device on ward door and upgrading the fire resistance of doors to 30 minutes.

A summary of their findings, reproduced in Table 3.16, is applicable to situations where time to flashover is small, say, 5 minutes.

3.5 Monte Carlo simulation

3.5.1 Introduction

Fire safety engineers are required to deal with complex fire scenarios which include human reactions and behaviour in addition to physical and chemical fire processes evolved by a variety of burning materials. Physical models representing such scenarios involve intractable mathematical relationships which cannot be solved analytically. Also, sufficient and realistic experimental or statistical data are unlikely to be available for estimating all the parameters of a physical model. For such complex models, only numerical solutions can be obtained by applying a simulation procedure step by step.

Simulation involves the construction of a working mathematical model, representing a dynamic system in which the processes or interactions bear a close resemblance or relationship to those of the specific or actual system being simulated or studied. The model should include realistic input parameters capable of generating outputs that are similar or analogous to those of the system represented. Then, by varying the numerical values of the input parameters, it would be possible to predict the time-varying behaviour of the system and determine how the system will respond to changes in structure or in its environment. Such simulation experiments can be performed on a computer by developing an appropriate software package.

Simulation models can be either discrete or continuous. As time progresses, the state of a building changes continuously as a small fire develops into a big fire. The physical and chemical processes involved in such a fire growth lend themselves to a continuous simulation model. On the other hand, discrete simulations are more appropriate for determining 'design times' concerned with fire detection and fighting and building evacuation. These times define critical events occurring discretely during a sequence of clear-cut stages. In a continuous model, changes in the variables are directly based on changes in time. Phillips (1995) discussed in detail the various aspects of computer simulation for fire protection engineering together with some examples.

3.5.2 Monte Carlo simulation

Monte Carlo analysis is a simulation technique applicable to problems involving stochastic or probabilistic parameters. For example, some input parameters such as compartment size and ventilation factor may be of deterministic nature such that, for each of these parameters, a range of possible values can be used in simulation experiments. On the other hand, some input parameters may be random variables, taking values according to probability distributions during the course of fire development. Examples of such variables are:

- rate of flame spread and fire growth
- temperature of the fire
- smoke concentration
- ambient air temperature
- wind speed and wind direction
- number of doors open
- number of windows open; and
- the response of occupants to fire alarms.

Consider, as an example, a stochastic parameter x_i whose value at time T during the course of fire development is $x_i(T)$. The exact value of $x_i(T)$ may not be known but it might be possible to estimate its mean $\mu_i(T)$ and standard deviation $\sigma_i(T)$ and the form of its probability distribution. Suppose this distribution is normal, such that the standardised counterpart t of $x_i(T)$ has a standard normal distribution. Then, with $t_i = 1.96$, the probability that the value of the stochastic parameter x_i at time T is less than or equal to the value given by the following equation is 0.975:

$$x_i(T) = \mu_i(T) + \sigma_i(T) \cdot t_i \qquad (3.52)$$

The probability of the value of the stochastic parameter exceeding the value given by the above equation is 0.025. This particular value of $x_i(T)$ can be regarded as the probable maximum, while the value corresponding to $t_i = -1.96$ in the above equation would be the probable minimum. The probability of the value of the stochastic parameter being less than this minimum is 0.025. Instead of the maximum or minimum value, a series of random values of $x_i(T)$ can be generated by 'spinning the Monte Carlo wheel' in the computer and selecting, at random, values of the standard normal variable t_i. Virtually every computer is equipped with a subroutine that can generate random numbers. This process will provide a random sample for estimating the time-varying relationship between the input parameter x_i and an output variable y_j. The output variable may be a quantity such as area damage representing property damage or number of fatal or non-fatal casualties representing life loss. Methods have been developed for generating random values for most of the well-known probability distributions such as Normal as well as any empirical distribution.

The probability distribution of an output variable y_j can now be estimated with the aid of random sample values of several input variables x_i generated by Monte Carlo simulation. Some input variables may be of deterministic nature and some of stochastic or probabilistic type. It would be possible to regress the output y_j on the input variables using a multiple linear regression analysis technique. In this analysis, as discussed in Section 3.3.3, it may be necessary to use the logarithm of y_j and the logarithms of some of the input variables or other appropriate transformations of the variables to reduce to a linear form the relationship between the output and input variables. There are, however, computer packages available to identify the non-linear relationship and perform a non-linear multiple regression analysis. The multiple regression equation then would provide an estimate of the expected value of the output y_j for a given set of random or extreme (maximum or minimum) values of the input variables x_i at any time T during the period of fire development.

Monte Carlo simulation can be used to generate sample values for constructing the probability distribution of an input variable which might not be known due to lack of data or whose mathematical structure is too difficult to be derived theoretically. This method would provide the mean, standard deviation and other parameters of the variable which can be used to confirm or reject theoretical results.

The object of Monte Carlo simulation is to take account of uncertainties governing the input and output variables involved in the fire safety system and estimate the effects of input variables on the output variables. Suppose that, at a given time, the output variables y_j ($j = 1, 2, ..., N$) are dependent on the input variables x_i ($i = 1, 2, ..., n$) according to a set of functions:

$$y_j = f_j (x_1, x_2, ..., x_n) \tag{3.53}$$

Then, in the neighbourhood of $x_1, x_2, ..., x_n$, y_j can be evaluated approximately by expanding the function in Equation (3.53) in a Taylor series and then omitting all terms after the second. This method would provide the variance-covariance matrices for the input and output variables – see Phillips (1995).

Suppose the following linear hypothesis is valid:

$$y_j = a_0 + a_1 x_1 + a_2 x_2 + ... + a_n x_n \tag{3.54}$$

If x_i ($i = 1, 2, ..., n$) are independent random variables with mean $\bar{x}_i$ and variance σ_i^2, the mean and variance of y_j as given by Taylor series expansion are

$$\bar{y}_j = a_0 + a_1 \bar{x}_1 + a_2 \bar{x} + \cdots a_n \bar{x}_n \tag{3.55}$$

$$\sigma_j^2 = a_1^2 \sigma_1^2 + a_2^2 \sigma_2^2 + \cdots a_n^2 \sigma_n^2 \tag{3.56}$$

For the input variable x_1, consider as an example, the rate of heat output $\dot{Q}$ that may increase with time T according to a T^2 or exponential function. This function will provide an estimate of $\dot{Q}$ at time T which may be regarded as the expected or mean value $\mu_q(T)$ of $\dot{Q}$. But $\dot{Q}$ is a random variable, since it is affected by ventilation and other factors. Hence, as discussed earlier:

$$\dot{Q}(T) = \mu_q(T) + \sigma_q(T) \cdot t \tag{3.57}$$

where $\sigma_q(T)$ is the standard deviation of $\dot{Q}(T)$ and the random variable t may be assumed to have a standard normal distribution. Experimental data would provide an estimate of $\sigma_q(T)$ for any material or object. Random variables $\dot{Q}(T)$ can then be generated by simulating random values of t.

The mass loss rate of fuel, $\dot{m}$, is another input variable, whose mean value and standard deviation can be estimated directly from experimental data or by considering the relationship

$$\dot{Q} = \dot{m} \Delta H \tag{3.58}$$

where ΔH is the effective heat of combustion of the fuel, usually assumed to have the value 18,800 kilojoules per kilogram. $\dot{Q}$ is measured in kilowatts and $\dot{m}$ in kilograms per second.

The parameters $\dot{m}$ and $\dot{Q}$ are directly correlated with the rate at which the floor area of a compartment is destroyed per unit of time – see Ramachandran (1995b) and Section 5.2. Area damage is an output variable which is also affected by other input variables such as fire load, compartment dimensions, ventilation factor and delays in detecting and commencing fire fighting.

3.5.3 Simulation models – examples

Based on physical equations of motion, Evers and Waterhouse (1978) developed a computer model for simulating and predicting the movement of smoke from a fire in a building. The building was considered as a series of spaces or nodes, each at a specific pressure with air movement between them, from areas of high to areas of low pressure. The inflow and outflow of air from each node, through paths such as windows, doors and ventilation openings, were determined in order to analyse the smoke flow and examine the way in which the smoke concentration increases with time at each node. The principal stochastic inputs to the calculations were the number of doors and windows left open, burnt down or broken, the ambient wind and temperature conditions and the location and severity of the fire. The values of these stochastic variables were sampled at random from specified statistical distributions. The model was applied to two types of buildings – the main building at the Fire Research Station, Borehamwood and law courts in London.

Coward (1975) used a simulation method to estimate the statistical distribution of fire severity from data obtained from surveys of office rooms. Fire severity was defined as the time taken for an equivalent exposure in a standard furnace test. A physical model was used to investigate how fire severity varies with the fire load, the area of ventilation and the dimensions of a compartment. Values of these factors were sampled at random from their frequency distributions ascertained from the survey data. Each distribution used in the simulation was built up by a pseudo-random number series, independent of all other series. The program generated a model population of 20,000 office rooms. From the data provided by such a large sample, fire severity, S (in units of time (minutes)), was estimated to have the exponential distribution

$$p = \exp(-0.04S) \tag{3.59}$$

where p is the probability that any room will have a severity greater than S. Accordingly, the average severity for an office room was estimated as 25 (= 1/0.04) minutes.

The mathematical structure of the Comparison of Risk Indices by Simulation Procedures (CRISP), developed by the Fire Research Station (see Phillips, 1992) is based on systems of simultaneous differential equations:

$$\frac{dx_i}{dT} = f(x_1, \ldots, x_i, \ldots, x_n) \tag{3.60}$$

In these equations, the rate at which a state variable x_i on the right-hand side changes with time T is expressed by the differential coefficient on the left-hand side. When the differential equations are solved, they yield a solution which represents the evolution of the system over time.

In CRISP, the history of a 'reference fire' is analysed into a sequence of five stages:

- initiation;
- accelerating growth;
- decelerating growth;
- full fire equilibrium; and
- extinction.

These stages are modelled with the aid of physical parameters such as combustion, heat and temperature. Alarm and suppression come into action during the last stage. Physiological effects and human behaviour are also taken into consideration. Monte Carlo tests are conducted by carrying out a sequence of runs of the model using sample values of parameters drawn from their probability distributions. From the model output, it may be possible to estimate probability densities representing fire conditions or number of casualties.

The Building Fire Simulation Model (BFSM) developed by the National Fire Protection Association, USA (see Fahy, 1985) allows the user to examine the interrelationships among fire development, spread of combustion products and people movement in residential occupancies. Data from full-scale fire tests are used in the model. Fire growth is defined in terms of six discrete stages called 'realms'. The realms are:

- the non-fire state;
- sustained burning;
- vigorous burning;
- interactive burning;
- remote burning; and
- full room involvement.

These realms are based on measurable criteria such as heat release rate and air temperature. The levels of combustion products estimated by the model are based on the realm the fire is in and the time it is in that realm. The ability of people to escape depends on these estimated levels of combustion products throughout the building.

Fire Risk Assessment by Simulation (Fire sim), developed in Norway, combines simulation techniques with statistics to calculate the expected annual fire risk of industrial plants (see Hansen-Tangen and Baunan, 1983). A large number of fires are simulated to estimate the percentage damage of the total value. The expected frequency of fires for the plant is estimated from statistics.

Sasaki and Jin (1979) carried out simulations of urban fires by applying probabilistic percolation theory. From fire statistics, probability of fire spread was obtained as a function of distance between buildings. The model provided an estimate of number of burnt buildings per fire incident in Tokyo.

3.5.4 Merits and demerits of simulation

Simulation is the best tool for obtaining solutions to problems lacking sufficient data for applying statistical, probabilistic, deterministic and other types of mathematical models. It may be expensive and time-consuming to develop a computer program/package for a simulation model but, once developed, the package can provide rapid and inexpensive useful information about the time-varying behaviour of the system it represents. It is a cheaper alternative method than costly and time-consuming exercises involving statistical data collection surveys, small or large scale experiments on fire characteristics of different materials or evacuation drills. A computer program based on a simulation model permits all the critical factors, processes and events that characterise a real system to be internally represented so that alternative fire and human behaviour patterns may be identified by varying the numerical values of the input parameters. Simulation is the only practical tool to deal with uncertainties and probabilities governing the behavioural patterns of a complex system consisting of several interactive factors varying with time.

Simulation does not produce a general solution. It does not identify all the possible behaviour patterns. Instead, simulation gives one time history of system operation corresponding to initial conditions and coefficients of the model parameters, whose numerical values are selected or specified. Several sets of numerical values for initial conditions and coefficients will have to be used as inputs for predicting the time-varying history of the system. Hence, simulation exercises of a complex system can be time-consuming and expensive. The results of simulation exercises in fire safety engineering should be checked and validated against data provided by real fires.

Simulation is not the model to be used for preliminary evaluations and comparisons. Analytical models, e.g. stochastic, are cheaper to use than simulation, particularly if good-quality data are available in sufficient quantities for evaluating the major component parts of a system. It is, however, necessary to base analytical models on assumptions which are realistic and relevant according to scientific theories supported by experimental results.

3.6 Consequence analysis

3.6.1 Introduction

Consequence analysis is intended to assess how severe a fire event or set of fire events are likely to be.

The ideal consequence analysis should:

- be physically similar to the system or building being assessed; and
- cover the full range of possible fire events.

The implication of these objectives is that the ideal consequence analysis is an infinite number of full-size experiments with real people and assets which experience all possible fire events over an infinite period of time. Clearly this is not practical or ethical, so like many other forms of quantitative risk assessment we seek to find the information that we need to assess event consequences in a number of other ways. These other methods include:

- historical data;
- disasters and near misses;
- experiments and fire tests;
- modelling.

3.6.2 Historical data

Historical data is where information about previous fire events has been collected. Typically, historical data can be national for a certain building type or for a specific organisation or business. The fire report, filled in by fire brigades, is an example of the latter and this collects data on the consequences of each fire event in terms of number of injuries and fatalities and area damaged by burning, heat and smoke.

Figure 3.22 shows an example of a consequence probability distribution.

The advantage of historical data is that it is a measure of the fire consequences from real fire events in real buildings. If these buildings are similar to the one being assessed, this can be very useful information.

The disadvantages of historical data include:

- under-reporting or bias in reporting or analysis;
- lack of physical similarity or homogeneity in the set being studied;
- short sample duration compared with the events of concern.

To a degree, under-reporting and bias can be dealt with by ensuring that the risk analysis is undertaken with a consistent data set. For example, if a study shows that small fire events are under-reported, the data may still be useful in assessing the risks for large fire events. Bias is a little harder to deal with, but an assessment of the nature and extent of the bias may lead to use of data, whose pessimism (i.e. it errs on the side of safety or a robust decision) there is a strong case for.

Lack of physical similarity or homogeneity in the set from which the data is collected can lead to consequences from one set of circumstances being used to assess a physically quite different set of circumstances. It is worth noting that, in insurance risk assessments, if the set is a portfolio of the insured, homogeneity is less important in assessing companies' risk exposure and hence in setting premiums and excesses.

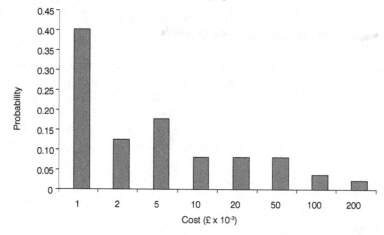

Figure 3.22 Typical consequence probability distribution

Short sample duration can mean that there has been insufficient time to collect enough data to cover the consequences of the events of concern. For example, 1,000 buildings studied for 10 years represents 10^4 building years of experience. This may provide a reasonable basis for consequences whose events have a frequency of 10^{-3} per building year or greater, but is clearly not adequate to address consequences for events whose frequency is of the order of 10^{-6} per building year.

For these reasons most quantitative fire risk assessments include other forms of consequence analysis.

An example of the application of historical data is the assessment of consequences from small fires. In many fires consequences can occur when the fire is too small to cause untenability in the assessment area. In these cases, statistical analysis is used to predict the consequences of small fires, e.g. people who die as a result of intimate contact with the fire through accident or attempts to extinguish the fire. Figure 3.23 shows a typical correlation of the area damaged by fire/heat/ smoke and the probability that occupants will be injured.

3.6.3 Disasters and near misses

Disasters like the King's Cross (see Figure 3.24) and Piper Alpha fires can provide significant insight into the various failure modes and potential severity of fire events. They can provide insight into the way that the fire behaved, what the main mechanisms of event escalation were, what the main hazards/damage mechanisms were and how people responded to the event.

However, it is essential to bear in mind that they are only a single event for a specific facility and therefore are of relatively limited use when assessing the infinite number of possible fire events that could occur in other different facilities. If we address any one disaster too literally, there is a danger that we are constantly addressing the last disaster rather than providing a broad understanding of the potential consequences of fires in a building.

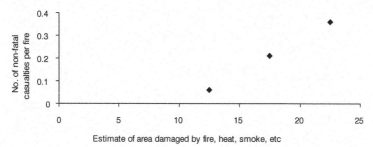

No. of non-fatal casualties per fire vs estimate of area damaged by fire, heat, smoke, etc
(shopping malls, indoor markets, etc)

Figure 3.23 Correlation of the number of non-fatal injuries per fire and the area damaged
by fire, heat and smoke

Figure 3.24 Ticket hall after the fire at King's Cross underground station

3.6.4 Experiments and fire tests

Experiments and fire tests can provide insight into the physical behaviour of
materials, items or spaces in a building.

Fire tests are applied to a wide range of building materials and contents and
so can provide insight into the likely behaviour of materials when exposed to
a specific physical hazard (for example a radiant heat flux or time/temperature
curve). Fire tests are primarily intended for the classification of building products
and so may not fully represent the physical processes (many are small or reduced
scale to minimise economic burdens on manufacturers) or situation of interest,
although assessments can be made where the situation of interest varies only
slightly from the standard test. Figure 3.25 shows the fire testing of a shutter.

Figure 3.25 Fire testing of a shutter

Experiments, if designed to address a specific item, can provide much more relevant information to a specific building and/or type of fire event. Fire experiments are more likely to be full or large scale and so more representative of the physical processes. For example, Figure 3.26 shows a fire experiment to assess the likely fire growth and smoke spread from an engine compartment fire in a bus. Experiments, if ethically sound, may also involve people in evacuation exercises and desktop event simulations. The disadvantage of experiments is that they are usually large and expensive and so, like disasters, they only usually represent a small sample of events or a limited part of the event progression. Therefore, the use of test and experimental data usually require some form of modelling.

3.6.5 Modelling

3.6.5.1 Introduction

Modelling is an approach for predicting various consequences of fires in buildings. There are three main approaches:

1   simple calculations such as those in CIBSE TM19 (1995);
2   zone (or control volume) models;
3   field (or computational fluid dynamic) models.

Figure 3.26 Example of a full-scale fire experiment

Simple calculation procedures are available in many fire engineering textbooks and handbooks and an example in this section is the addressing of fire tenability. This section will mostly discribe control volume modelling before discussing more complex computational models.

Control volume models are often known as 'zone' modelling. The approach works by dividing the building fire 'system' in to a series of control volumes (or zones). Each control volume represents a part of the system that is homogeneous in nature, that is, it is assumed to have the same properties throughout (e.g. temperature, velocity, density, species concentration etc). Conservation equations are applied to each control volume to predict how 'source terms', such as the fire or plume, and processes between control volumes, such as radiative and convective heat transfer, affect the control volume's properties. Figure 3.27 shows how the building fire domain can be divided up into different control volumes (Drysdale 1999).

Control volume modelling has the capacity to predict various aspects of fires in buildings and is an approach used in many fields of engineering. Indeed, control volume modelling has been developed and applied to fire in compartments since the 1970s (Quintiere 2003). However, any radical departure by the fire system from the conceptual basis of the control volume model can seriously affect the accuracy and validity of the approach.

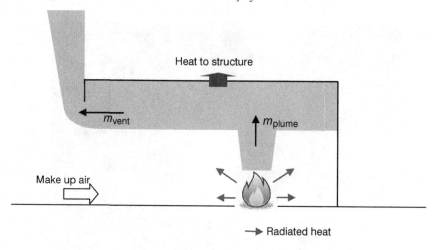

Figure 3.27 Typical schematic of building fire control volume model

3.6.5.2 *Application of control volumes to fires*

Control volume modelling of fires is built on three types of model: conservation equations, source terms and mass and heat transfer sub-models. These models are described in brief below; an extensive review of control volume equations as applied to compartment fire can be found in the literature e.g. Quintiere (1989).

The kinds of assumptions that are typical of control volume modelling include:

- all properties in the control volume are homogeneous;
- the gas is treated as an ideal gas (usually as air);
- combustion is treated as a source term of heat and mass;
- mass transport times within a control volume are instant;
- heat transfer to building contents such as vehicle is neglected;
- the section of the building is constant;
- the pressure in the building is assumed to be constant;
- frictional effects at boundaries are not explicitly treated.

When applying control volume equations to building fires, consideration should be given to the unique nature of some fire phenomena in buildings, for example:

- An assumption that hot layer properties are homogeneous along the length of the spaces will only be tenable for relatively short spaces.
- Ambient and forced ventilation flows in buildings may affect air entrainment in plumes.
- The relative velocities of hot and cold layers may mean that shear mixing effects at the interface may not be negligible.

CONSERVATION EQUATIONS

The main basis of control volume modelling is the conservation of fundamental properties. The concept of conservation is applied to mass, energy and momentum. However, momentum is not normally explicitly applied since the information needed to calculate velocities and pressures is based on assumptions at the boundaries.

The conservation of mass for a control volume states that the rate of change of mass in the volume plus the sum of the net mass flow rates out to the volume is zero:

$$A\frac{d(\rho z)}{dt} + \sum m_o = 0 \tag{3.61}$$

where:　A is the area of the control volume
　　　　ρ is the density of the gas in the control volume
　　　　z is the height of the control volume, and
　　　　m is a net mass flow rate out of the control volume.

Figure 3.28 shows how this differential equation for the conservation of mass can be applied to a control volume (Rylands *et al.* 1998).

The mass contained within the control volume at time t is given by the following equation (Rylands *et al.* 1998):

$$M(t) = M(t - \Delta t) + \Delta t(m_{in} - m_{out} + m_{in\_hot} + m_{in\_cool}) \tag{3.62}$$

where:　M is the mass in the control volume
　　　　Δt is a small interval of time
　　　　m_{in}, m_{out}, m_{in_hot} and m_{in_cool} are the mass flow rates into and out of the control volume due to convection and shear mixing effects.

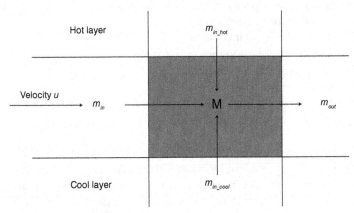

Figure 3.28 Schematic of mass flow between control volumes

Similar equations can be used for the conservation of species (that is mass concentrations such as carbon monoxide) and energy (Quintiere 2003).

The mass concentration of a species Y can be conserved by modifying Equation (3.61):

$$\rho z A \frac{dY}{dt} + \sum m_o \left(Y - Y_{cv} \right) = \omega \tag{3.63}$$

where: Y is the mass concentration of a species in the flow
   Y_{cv} is the mass concentration in the control volume, and
   ω is the mass production of that species due to the fire.

The mass concentration of a species depends strongly on plume entrainment (see 'Mass and heat transfer sub-models' below) and the mass production rate of that species by the fire. The mass production rate of a species will vary significantly depending on the type of fuel and the air/fuel ratio near the fire. For example, the mass production rate of carbon dioxide depends on the rate of heat release, the amount of carbon in the fuel and the air/fuel ratio. Limited knowledge of the detailed chemical reactions in real fires' combustion means that a simplified one-step chemical model and studies of stoichiometry can be used to estimate species production. However, uncertainties around the application of stoichiometry mean that most control volume models use a mass conversion factor of the mass flow rate of burnt fuel. For carbon monoxide these mass conversion factors can vary tremendously depending on the type of fuel and whether the fire is fuel bed or ventilation controlled. Therefore, results of large-scale fire tests are used to derive mass conversion factors of CO.

The conservation of energy for a control volume combines Equation (3.61) and the equation of state, $p = \rho R T$, such that

$$\rho C_p z A \frac{dT}{dt} - z A \frac{dp}{dt} + C_p \sum m \left(T - T_{cv} \right) = m_f \chi \Delta H - Q_{net\_loss} \tag{3.64}$$

where: C_p is the specific heat capacity of the gas in the control volume
   p is the pressure in the control volume
   T is the temperature of gas in the flow
   T_{cv} is the temperature of gas in the control volume
   m_f is the rate at which fuel is volatilised
   χ is the combustion efficiency
   ΔH is the heat of combustion (taken as positive), and
   Q_{net_loss} is the net rate at which heat is lost to the boundary.

This model assumes that there is sufficient oxygen to react with the fuel with a combustion efficiency factor to adjust for incomplete combustion. There is also difficulty in dealing with combustion in vitiated layers. Often in control volume models, the rate of heat release is a user input from which the mass flow rate of fuel is derived.

If the rate of temperature changes in the control volume is low, the first term can be neglected. This leads to a simpler quasi-steady analysis for growing fires. In well-ventilated conditions (some transport buildings are well ventilated in terms of fire dynamics), the second term can also be neglected. This leaves an enthalpy term for flows into and out of the control volume and two source terms. For flows out to the control volume the lumped parameter assumption means that $T = T_{cv}$.

SOURCE TERMS

The main source terms in fire modelling are the rate heat release and the mass flow rate of fuel. Where the fire source is known and well controlled, such as a gas burner, precise values can be used. In most fire safety situations the rate of heat release and mass flow rate of fuel is the result of a spreading fire over a variety of material and surfaces. In these circumstances, an empirical model to give the rate of heat release at time t can be used (CIBSE TM19 1995):

$$q_f = \alpha(t - t_o)^2 \tag{3.65}$$

where: q_f is the rate of heat release of the fire
α is the fire growth coefficient
t_o is the time between ignition and fire growth (incubation period).

Similar source term models exist for the mass flow rate of fuel, for example:

$$m_f = \frac{q_f}{\Delta H \chi} \tag{3.66}$$

Given the conversion rate of fuel to carbon monoxide (a function of the materials and how well ventilated the fire is) the mass concentration of CO can be estimated.

These empirical models provide approximations and so consideration should be given so that values for the coefficient of fire growth, combustion efficiency and mass conversion of rates are appropriate.

MASS AND HEAT TRANSFER SUB-MODELS

Mass and heat transfer models are an essential feature of control volume models. These models may include:

- entrainment in plumes;
- flows through openings;
- mixing between layers;
- convective heat transfer to surfaces;
- radiative heat transfer;
- conductive heat transfer;
- other effects.

Entrainment in plumes has been shown to have a critical effect on the tenability and development of fire hazards. There is a range of models available to predict flame heights and mass entrainment in plumes (CIBSE TM19 1995, Heskestad 2003). It is essential that the model used is appropriate to the physical situation of the real fire. For example, entrainment in an axisymmetric plume may be appropriate for a fire in the floor in the middle of a building section, but may not be appropriate for a spill plume from a window or a plume attached to one of the building walls. The effect on plume properties of any ambient ventilation and for a situation where flames reach the hot layer should also be taken into account. Figure 3.29 shows a schematic of a building fire plume.

Flows through openings are crucial to the modelling of fires in compartments, such as doors and windows of vehicles, flows into and out of building ventilation system openings and entrance/exits. Detailed models for flows through openings and building ventilation systems can be found in the literature, e.g. Emmons (2003).

Mixing between layers can occur in one of three ways:

1   a cold flow injected into a hot layer;
2   shear mixing associated with lateral layer flows;
3   mixing due to building wall flows.

Cold flows injected into a hot layer can be resolved by computational fluid dynamics or physical modelling research. Shear mixing of layers has been studied to a certain extent and some correlations are available for counter flow between layers and back-layering against forced ventilation. Mixing due to wall flows has been studied in compartment fires. None of these three phenomena is as critical as the primary buoyant mixing in the plume (Emmons 1991, Vantelon 1991).

Convective heat transfer to surfaces is one of the main processes of heat loss between the hot layer control volume and the building lining. Convective heat transfer to ceilings has been studied extensively. Convective flows will vary along the building walls and ceiling depending on their position with respect to the fire. Where convective heat transfer is dependent on local boundary layer temperatures (rather than hot layer control volume temperature) an adiabatic wall temperature approach can be used. Convective heat transfer for ceilings and walls in building fires has not been developed and so most control volume models use natural convection correlations (Atreya 2003, Evans 2003).

Radiative heat transfer theory is generally sufficient for control volume models of building fires. Grey body radiation from uniform temperature hot gas layers can be predicted, although emissivity values require careful consideration. For radiation from flames, empirical data is used because complex temperature distributions for radiation from flames and the role of soot are not well understood (Tien *et al.* 2002, McCaffrey 2003).

Conductive heat transfer through the building linings should be balanced with the radiative and convective heat transfer from the hot gas layer control volumes. This entails a numerical (or graphical) solution to a set of partial differential

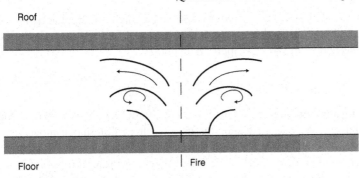

Figure 3.29 Schematic of a plume in natural ventilation

equations. Most control volume models consider conductive heat transfer in one dimension only and that the building linings are thermally thick or conductive heat transfer is to an infinite ambient heat sink (Carslaw and Jaegar 1959, Rockett and Milke 2003).

Figure 3.30 shows a schematic of the kinds of heat transfer processes that may be relevant to fires in buildings.

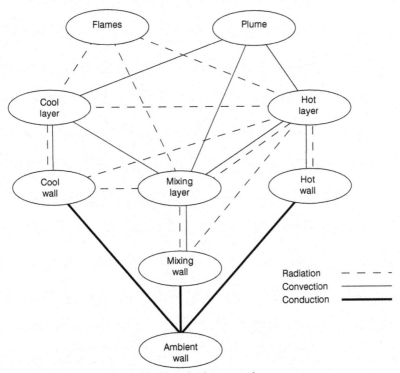

Figure 3.30 Typical schematic of building fire heat transfer processes

From the above it can be seen that other effects can be incorporated into control volume building fire models easily, as long as there is an adequate description of the phenomenon and there is appropriate data.

3.6.5.3 *Application of control volume models in fire safety*

Recent fires in buildings have illustrated how, for fires in buildings, the products of combustion are confined and this may result in fast smoke movement and rapid threat to life. Figure 3.26 shows the typical control volumes associated with a 2-layer building fire model. A review of building fire control volume models indicated a number that are being used to predict building fire smoke movement (Quintiere 2003).

There are also likely to be significant differences between the source term and heat and mass transfer sub-models used in each of the models and these should be reviewed in selecting a model. For example, some models incorporate modified plume sub-models to address flame impingement of the hot layer or building roof.

Figure 3.31 shows typical results from a transient, 3-layer, building fire control volume model. It shows a typical vertical temperature profile against experimental data.

Control volume models are being increasingly used to inform design decisions for building fire safety and the next two subsections indicate two typical applications.

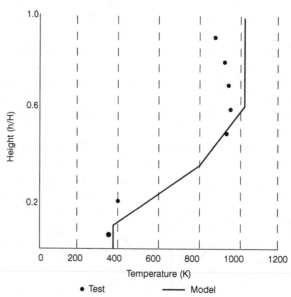

Figure 3.31 Typical results for a three-layer control volume model compared with experimental data

3.6.5.4 Evacuation

Evacuation analysis predicts how quickly occupants are likely to exit an assessment area. This uses simple models for time to detection, pre-movement time and movement time. The total time to escape is given by

$$t_e = t_a + t_p + t_m \tag{3.67}$$

where: t_e is the total time to escape (s)
   t_a is the time from ignition to alarm (s)
   t_p is the pre-movement time for occupants (s)
   t_m is the time for occupants to move to a place of relative safety (s).

The time for occupants to move to a place of safety depends on the number of occupants, the travel distance, the exit width and the speed of movement and flow rate of occupants. For each evacuation scenario, the time to move will be limited by travel distance or exit width.
   The time for the last occupant to move this distance is given by

$$t_{m(d)} = \frac{d_t}{u_m} \tag{3.68}$$

where u_m is the velocity of movement (taken to be 1 m/s).
   The time for the last occupant to flow through the exits is given by

$$t_{m(w)} = \frac{N_p}{\sum w_e . n_p} \tag{3.69}$$

where: w_e is the exit width (m)
   N_p is the number of people (people)
   n_p is the flow rate of people through an exit (people/m/s).

The movement time, t_m used in the analysis is then taken to be the larger of the travel distance and exit width limited times (Equations (3.68) and (3.69)).
   The number of non-fatal injuries (for $t_e > t_u$) is given by:

$$N_i = (t_e - t_u) . \frac{N_p}{t_m} \tag{3.70}$$

where: t_u is the time for conditions to become untenable
   From the value of N_i, the number of fatalities can be calculated using factors based on fire statistics.

3.6.5.5 Fire tenability

By analysing the response of people escaping a building fire to the hazards that the fire creates, it is possible to gain a much better understanding of the way that a building and its fire precautions will perform (Charters 1992, Charters *et al* 1994). Control volumes models are particularly well suited to providing information on fire hazards so that tenability can be assessed.

The main hazards from building fires are (Purser 2003):

- heat (hyperthermia);
- toxic gases (hypoxia); and
- thermal radiation.

The level of visibility in a building fire, although not directly hazardous, may reduce occupants' way-finding capability and so increase their exposure to other fire hazards.

Tenability to heat and toxic gases is strongly dose-related and so fractional effective dose techniques are used. For example, the heat fractional effective dose per second can be calculated by

$$F_{heat} = \frac{1}{\left(60e^{(5.2-0.027T)}\right)} \tag{3.71}$$

where: T is the temperature of the smoke (°C).

This relationship indicates that a temperature of 100°C could be tolerated for 12 minutes and 60°C for 35 minutes.

For toxicity (CO) the fractional effective dose per second is given by

$$F_{toxicity} = \frac{C}{90} \tag{3.72}$$

where: C is the percentage concentration of CO.

This relationship indicates that 0.5 per cent CO could be tolerated for 3 minutes. A maximum concentration criterion of 1 per cent CO is also applied and consideration should be given to selecting a CO conversion factor that will be conservative.

For thermal radiation, a criterion of 2.5 kW/m² is taken. Below this value thermal radiation can be tolerated over extended periods of time and above the level tolerability is measured in a few tens of seconds.

The visibility criterion may be either 5m or 10m depending on the nature of the building and its ease of way-finding. For example, for a small space in a building, a 5m visibility criterion may be appropriate. Similarly, for a large space in a building with exit doors clearly marked, a 10m visibility criterion may be appropriate.

Therefore, the hazard/time output for each control volume from the smoke movement model can be integrated with a semi-infinite stream of people moving away from the fire. As each individual moves down the building, they move from one control volume to the next. Their dose can then be calculated based on the level of hazard in each control volume as they passed through it.

3.6.6 *Complex computational analysis*

There is a range of computational methods available for analysing fluid or people flow in similar ways. The advantages of these methods are that they model the flow processes in much greater detail than control volume/hydraulic evacuation models. The disadvantage is that they are relatively expensive to undertake and so tend to be used for a limited number of fire risk assessments and for a small number of cases. Figures 3.32 and 3.33 show examples of computational fluid dynamics analysis and discrete evacuation analysis.

3.7 Dealing with uncertainty

This section describes the importance of pessimistic assumptions, sensitivity analysis and/or uncertainty analysis in testing the robustness of a decision based on an estimate of risk.

3.7.1 *Sensitivity analysis*

Sensitivity analysis can be used to draw useful conclusions in the first instance or to assess the robustness of a decision based on probabilistic risk assessment.

Probabilistic risk assessment, like all fire engineering analysis, uses analysis techniques and data to answer questions regarding fire safety. The analysis techniques and data may have simplifying assumptions and limitations that mean that they may not replicate the details of actual events. However, if meaningful conclusions are to be drawn from an analysis, they should be sufficiently representative that the correct fire safety decision is taken.

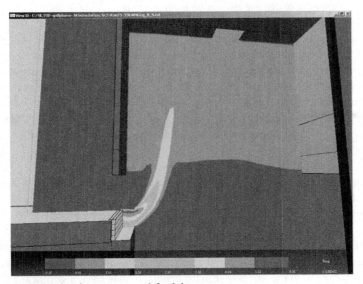

Figure 3.32 Example of computational fluid dynamics

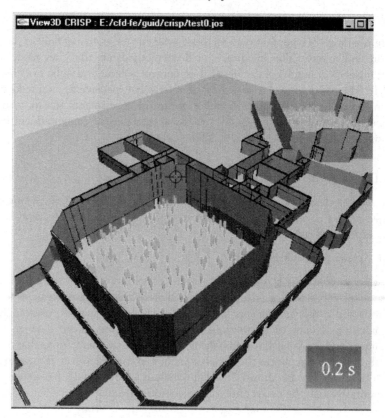

Figure 3.33 Example of discrete evacuation analysis

If the results of the probabilistic risk analysis are well within the acceptance criteria, then sensitivity analysis may not be needed. If, however, the results of the probabilistic risk analysis are close to the acceptance criteria, then variations in the variables may have a significant effect on the conclusions from the analysis and sensitivity analysis should be used to assess this.

The first step of sensitivity analysis is to identify the variable(s) that are likely to have the greatest impact on the results of the analysis. The variables can be identified as those where:

- a small change may be magnified due to its role in an equation or analysis; and/or
- its value may be subject to the greatest variability or uncertainty.

For example, a variable that has a value to the 1/3 power in an equation may not have a large impact on the final results of the analysis. Variations in another variable, which is to the 4th power in an equation, may have a significant impact on the results of the analysis. If a variable is the only one in an equation or it is used several times in the analysis, then it too may have a significant impact on the results of the analysis.

The variables identified as potentially having a significant impact on the results of the analysis can then be investigated in one of three ways:

- a single variable with an alternative value;
- a single variable over a range of values;
- a multiple point assessment of multiple variables.

A sensitivity analysis of a single variable with an alternative value is the simplest approach. The analysis is repeated with an alternative (usually more onerous) value to assess whether the conclusions of the analysis are robust. If the assessment criteria are still satisfied, then the conclusions of the analysis are further reinforced.

Sensitivity analysis based on a single alternative value of a variable is often not very conclusive. Therefore, sensitivity analysis of a variable over a range of values is used. The analysis is repeated and a graph produced showing the variation of the results of the analysis against values of the variable. This provides much greater insight into the relationship between the variable and the output of the analysis. If the results of the analyses lie across the acceptance criteria then a critical value of the variable can be identified and an assessment made of its implications.

Advanced methods of sensitivity analysis are available that allow more than one variable to be varied at a time (see the Section 3.5 on Monte Carlo simulation). The results of this analysis can then be presented in a table or, after applying regression analysis, as a mathematical expression.

Estimates of risk should assess the implications of assumptions and bias in the risk analysis. For life safety against absolute risk criteria, it is often possible to deal with assumptions and bias by making them pessimistic so that they err on the side of safety, i.e. such assumptions that tend to increase the predicted frequency and/ or consequence of an event. For comparative life risk analysis, the same approach can be used as long as the assumptions or bias do not clearly favour one option over another. For financial objectives, there is no natural 'side of safety' to err on, since all options have costs and financial benefits this type of analysis should normally be of best estimates rather than values of realistic or credible pessimism.

Sensitivity analysis is often used to test how robust a decision is to variations in values that may have been subject to assumptions and/or bias.

There is guidance in British Standards Institute (2003) PD7974 Part 7 on data sources and quality. Again the robustness of a decision to variations in data values can be assessed using sensitivity analysis. The following suggest some key questions that should be considered with respect to sources and quality of data.

Applicability:

- What is the set of cases that the data is drawn from?
- What case is the data measuring?
- How similar is my system to the cases considered?

- If the data is from another country, will variations in statutory controls or design practices skew the data?

Quality:

- How old is the data (10 years may be considered a cut-off age for high-quality data, but this will depend on the system and the extent and rate of its change)?
- Is collaborative data available?
- Is the data from statistical studies or based on engineering judgement?

Results:

- Do the answers look realistic?
- How sensitive are the results to questionable data?

3.7.2 Uncertainty analysis

The kinds of uncertainty that are built into the estimated levels of fire risk arise from:

- the choice of tangible and credible failure cases from the wide range of events that could be envisaged;
- the frequency of ignition for each case;
- the probability of failure or effectiveness of mitigation measures;
- the fire's location, growth rate and peak heat release rate;
- the level of hazard generated by the fire;
- the validity of the smoke movement and human behaviour;
- uncertainties in the values used in the models for entrainment, vent flow, number and distribution of people and their pre-movement times etc.;
- the accuracy of toxicology models in translating hazard levels and distributions into lethal effects.

There are also uncertainties arising from the hazard identification process. These are:

- failure to identify all the significant failure events;
- failure to include significant events that were identified.

These are problems of incompleteness rather than uncertainty and as such cannot be addressed easily using uncertainty analysis.

Uncertainty analysis uses Monte Carlo analysis (see Section 3.5) or Latin hypercube sampling to process probability distributions that represent the uncertainty in a parameter or parameters. For application to fire risk assessment, the output parameter would be the level of risk.

The method for estimating uncertainties in risk assessment is:

1   Identify the parameters that contribute most to the variations in risk levels, with a few key parameters usually being chosen (Charters 1999), e.g.
    a.   the fire growth rate
    b.   the time to detect a fire
    c.   the number of people present
    d.   their pre-movement time.
2   A probability distribution is derived for each parameter.
3   Monte Carlo analysis randomly selects a value of each parameter and runs the model and this is repeated a large number of times.
4   The output parameter results are then used to create a probability distribution for which statistical properties like the mean, standard deviation and 95th percentile can be determined.

Monte Carlo analysis depends on a large sample for its accuracy, whereas Latin hypercube sampling selects from discrete segments of the distribution and so its results are less sensitive to the number of calculations undertaken.

Where there is good data on the variability of a parameter, uncertainty analysis can provide useful insight into the robustness of a decision based on an estimate. Where there is uncertainty surrounding a point value due to lack of data, this may be amplified in the 'generation' of a probability distribution, it is not clear whether uncertainty analysis increases or decreases the robustness of the decision. In these cases, sensitivity analysis provides much clarity and a clear audit trail in terms of robustness.

3.7.3 Safety factors

3.7.3.1 Introduction

For many fire safety engineering components or subsystems, the performance may be formulated in terms of two random variables, X and Y. The variable X represents stress and Y the strength. Taking the compartment in a building as an example, X is the severity of fire to which the structural boundaries of the compartment are exposed and Y the fire resistance of the boundaries. Both fire severity and fire resistance are usually expressed in units of time. Another example is concerned with building evacuation, in which X is the time taken by a combustion product to produce an untenable condition on an escape route and Y the time since the start of ignition taken by an occupant to get through the escape route.

In the first example mentioned above, the compartment would 'fail' with consequential damage to life and property if X exceeds Y, particularly during the post-flashover stage. In the second example, 'egress failure' would occur with fatal or non-fatal casualties if Y exceeds X. The objective of fire safety design is to reduce the probability of failure to an acceptably small level. For estimating this probability, two methods are generally adopted. The first method discussed in this section involves partial safety factors and is semi-probabilistic. The second method, discussed in the next section, is probabilistic and involves probability distributions of X and Y; it is also known as the Beta method.

3.7.3.2 Characteristic values

The first step in this analysis is to select appropriate values for X and Y which are typical or characteristic values representing the two random variables. These values can be, for example, the mean or average values μ_x and μ_y of X and Y or other statistical parameters such as median (50th percentile) or mode (the most probable value with the highest relative frequency). A value corresponding to some other percentile, e.g. 80th or 90th or 95th can also be selected as a characteristic value for X or Y.

Consider a design problem in which failure would occur if $X > Y$ or success if $Y \geq X$. For example, thermal failure of the compartment would occur if severity S exceeds resistance R and success if $R \geq S$. It is usual to provide a structural element with minimum fire resistance, R_p, which is greater than the maximum severity, S_q, likely to be encountered during the post-flashover stage. R_p and S_q can be regarded as the characteristic values R_k and S_k of R and S.

Suppose μ_r and σ_r are the mean and standard deviation of fire resistance R and μ_s and σ_s the mean and standard deviation of fire severity S. If the values of these parameters are known, we can write

$$R_p = \mu_r - t_r \, \sigma_r \tag{3.73}$$

$$S_q = \mu_s + t_s \, \sigma_s \tag{3.74}$$

If V_R and V_S are the coefficients of variations given by

$$V_R = \sigma_r / \mu_r; \; V_S = \sigma_s / \mu_s \tag{3.75}$$

we have

$$R_p = \mu_r \, (1 - V_R \, t_r) \tag{3.76}$$

$$S_q = \mu_s \, (1 + V_S \, t_s) \tag{3.77}$$

According to the Chebyshev inequality (La Valle, 1970), whatever may be the probability distribution of S, the probability of fire severity exceeding S_q given by Equation (3.74) is less than or equal to $(1/t_s^2)$. For instance, $t_s = 2$ guarantees a safety margin of at least 75 per cent $(= 1 - \dfrac{1}{2^2})$. The probability of severity exceeding S_q in this case is at most 0.25. The values of t_s and S_q may be selected according to any specified safety margin. It may be seen, for example, that $t_s = 3.16$ would provide a safety margin of at least 90 per cent. The probability of severity exceeding S_q in this case is at most 0.10. In the case of minimum fire resistance, if $t_r = 3.16$, the probability of resistance being less than R_p given by Equation (3.73) would be at most 0.10 and the probability of resistance exceeding R_p would be at least 0.90.

Suppose the probability distributions of R and S are also known, in addition to their means and standard deviation. If these are normal, for example, the values of t_r and t_s for any specified probability levels can be obtained from tables of the standard Normal distribution. For example, $t_s = 1.96$ corresponds to the fractile value of 0.975 of the probability distribution of fire severity. In this case, the probability of severity exceeding the value of S_q given by Equation (3.74) would be 0.025. If $t_s = 2.33$ which corresponds to the fractile value 0.99, the probability of severity exceeding S_q is 0.01. The probability of fire resistance being less than the value of R_p, given by Equation (3.73), would be 0.025 if $t_r = 1.96$ and 0.01 if $t_r = 2.33$.

The mean, maximum or any other value representing the characteristic value S_k of fire severity likely to be attained in a compartment, can be estimated with the aid of an analytical model such as

$$T_e = c \cdot w \cdot q \tag{3.78}$$

where c is a constant depending on the thermal properties of the compartment boundaries, w the ventilation factor and q the fire load density. The ventilation factor is given by

$$w = \frac{A_f}{(A_T \, A_V \, \sqrt{h})^{\frac{1}{2}}} \tag{3.79}$$

where A_f is the floor area of the compartment, A_T the area of the bounding surfaces of the compartment including the area of ventilation openings (A_V) and h the weighted mean ventilation height. With area in m², h in m and q in MJ/m², fire severity T_e is expressed in minutes. Formulae (3.78) and (3.79) relating to equivalent time of fire exposure have been recommended by the CIB (1986).

In Equation (3.78) the parameters c and w may be regarded as constants for any compartment with known or given structural (thermal) characteristics, dimensions and area and height of ventilation openings. Fire load density q may be considered as a random variable such that the mean severity μ_s is estimated by inserting the average value μ_q of fire load density:

$$\mu_s = c \cdot w \cdot \mu_q \tag{3.80}$$

The standard deviation σ_s of fire severity is given by:

$$\sigma_s = c \cdot w \cdot \sigma_q \tag{3.81}$$

where σ_q is the standard deviation of fire load density. Then, from Equations (3.75), (3.80) and (3.81), it may be seen that the coefficient of variation V_S of severity is equal to that of fire load density given by σ_q / μ_q.

The fire resistance required for a structural element of a compartment may be based on the criterion that the minimum fire resistance R_p given by

Equation (3.73), exceeds the maximum severity S_q given by Equation (3.74). A standard fire resistance test would indicate whether the structural element meets this criterion or not. However, the fire resistance would be a random variable in a real fire – see Ramachandran (1990). The variability depends on materials used. For example, fire resistance of a gypsum board wall would have a greater variability than the resistance of a concrete block wall. The resistance of a steel wall would depend on the thickness of insulation, total mass of insulation and steel, average perimeter of protective material and a factor representing the insulation heat transmittance value for the material.

Fire resistance of a compartment composed of different structural elements would not be the same as the fire resistance of any of these elements. Fire resistance of a compartment is affected by weakness caused by penetrations, doors or other openings in barriers. Sufficient data are not available for estimating realistically the mean μ_r and standard deviation σ_r of the fire resistance of a compartment in an actual fire. The values of these parameters can only be assumed according to data provided by standard fire resistance tests and other experiments. These tests and experiments can provide some indication of the standard deviation σ_r or coefficient of variation V_R as defined in Equation (3.75). For the sake of simplicity, fire resistance may be assumed to have the same probability distribution as that of fire severity, e.g. Normal.

The mean fire resistance μ_r required for a compartment is an output to be estimated according to the input values μ_s and σ_s of fire severity. The output μ_r should satisfy the design criterion that the minimum fire resistance R_p, as given by Equation (3.73), exceeds the maximum severity S_q as given by Equation (3.74). R_p and S_q include safety margins provided by the standard deviations σ_r and σ_s and the parameters t_r and t_s.

As defined in Equations (3.78) and (3.79), fire severity is the product of several factors. Based on data from fire tests, fire resistance in some cases is also expressed as the product of some factors, e.g. thin wall steel members (Homer, 1979). In all such cases, it may be considered necessary to take account of uncertainties governing all the factors. Generally, if a variable y is a product of several variables $x_1, x_2, x_3, \ldots$ which are mutually independent, the mean of y is approximately given by the product

$$\bar{y} = \bar{x}_1 \cdot \bar{x}_2 \cdot \bar{x}_3 \ldots \tag{3.82}$$

where $\bar{y}, \bar{x}_1, \bar{x}_2, \bar{x}_3, \ldots$ are the means of the variables. The coefficient of variation of y is approximately given by:

$$V_y^2 = V_1^2 + V_2^2 + V_3^2 + \ldots \ldots \tag{3.83}$$

where $V_1, V_2, V_3, \ldots$ are the coefficients of variation of $x_1, x_2, x_3, \ldots$.

As derived by Hahn and Shapiro (1967), the results in Equations (3.82) and (3.83) are based on an application of truncated Taylor series expansion of the function

$$y = f(x_1, x_2, \ldots) \tag{3.84}$$

The second and higher derivatives of the function are neglected in the above-mentioned expansion. Ramachandran (1998a) discussed the above results in a detailed report on the various aspects of probabilistic evaluation of structural fire protection.

For the second example relating to building evacuation, the design criterion is that the total evacuation time H, Y as defined previously, should not exceed the time F, X as defined earlier, taken by a combustion product, e.g. smoke, to travel from the place of fire origin and produce an untenable condition, e.g. visual obscuration on an escape route. The total time H is the sum of three periods. In sequential order, the first period D is the time taken to detect or discover the existence of a fire after it started. The second period B is known as 'recognition time' or 'gathering phase' in human behaviour studies. This period is the elapsed time from discovery of fire to the commencement of evacuation. The third period E, known as 'design evacuation time', is the time taken by an occupant to reach the entrance to an escape route, e.g. protected staircase, after leaving his/her place of occupation.

The time period D depends on the presence or absence of automatic fire detection systems or suppression systems such as sprinklers. A characteristic value for D can be estimated from fire statistics or detector tests, together with its standard deviation. Human behaviour studies suggest a characteristic value of 2 minutes for B. But this parameter for any occupancy type can be estimated by carrying out evacuation exercises. For any type of building, the characteristic value of E and its standard deviation can be estimated from fire drills or computer models of evacuation. A value of 2.5 minutes for E has been recommended in British Standard BS 5588. The actual value of E would depend on building type and the physical (disabled etc.) and mental conditions of the occupants, apart from other factors such as widths of staircases and exits. Deterministic models and associated computer packages can be used to estimate the characteristic value and standard deviation of F for any type of building. By reducing the rate of growth of fire and smoke, sprinklers would increase the value of F if they fail to extinguish a fire. Sprinklers also have a high probability of extinguishing a fire in which case F will have an infinite or high value.

The mean value μ_h of total evacuation time H is the sum of the mean values of D, B and E. The standard deviation σ_h of H is given by:

$$\sigma_h^2 = \sigma_d^2 + \sigma_b^2 + \sigma_e^2 \tag{3.85}$$

where σ_d, σ_b and σ_e are the standard deviations of D, B and E. For any escape route and place of fire origin, the mean value μ_f is the sum of the means of the F values for different combustion products. By considering different places of fire origin, escape routes and combustion products, the overall mean value of F can be estimated for any building or any floor of the building. An estimate of this mean is given by the sum of mean values of F for all the factors mentioned above. Following Equation (3.85), the square of the standard deviation of the overall value of F is the sum of squares of the standard deviations of the factors. The formula in Equation (3.83) is applicable for estimating approximately the coefficient of variation of the overall value of H or F.

Ramachandran (1993) discussed in detail the model described above for building evacuation and has derived equations similar to (3.73) to (3.77).

3.7.3.3 Design values

In practical fire safety engineering, it is necessary to determine design values which include partial safety factors α_x and α_y to account for uncertainties in the estimation of characteristic values for the random variables X and Y. The sources of uncertainties are mainly parameters included in, or excluded from, analytical models, data used, hypotheses and assumptions. The corrections for uncertainties should be in the direction of greater safety after assigning values greater than unity for the partial safety factors α_x and α_y.

Consider first the fire protection given by the fire resistance of the structural boundaries of a compartment. With the partial safety factor α_r greater than unity, the design value R_d for fire resistance can be estimated by

$$R_d = R_k / \alpha_r \qquad (3.86)$$

where R_k is the characteristic value. R_d will be less than R_k according to Equation (3.86). This design condition will also be satisfied if the minimum value R_p in Equation (3.73) is considered as the design value and the mean value μ_r as the characteristic value. In this case, from Equations (3.76) and (3.86) α_r is the reciprocal of $(1 - V_R t_r)$.

The formula for the design value S_d for fire severity is

$$S_d = \alpha_s . S_k \qquad (3.87)$$

where S_k is the characteristic value and α_s greater than unity is the partial safety factor. Accordingly, S_d will be greater than S_k. This design condition will also be satisfied if the maximum value S_q in Equation (3.74) is considered as the design value and the mean value μ_s as the characteristic value. In this case, from Equations (3.77) and (3.87), α_s is equal to $(1 + V_s t_s)$.

For example, if the estimate of R_k is correct to 15 per cent, then

$$\alpha_r = 1.176, R_d = 0.85 R_k$$

It may also be seen that, if $V_R = 0.2$ and a value of 1.96 is adopted for t_r, $\alpha_r = 1.64$.

Likewise, if the estimate of S_k is known within 25 per cent, then

$$\alpha_s = 1.25, S_d = 1.25 S_k$$

Also, if $V_S = 0.2$ and $t_s = 1.96$, then $\alpha_s = 1.39$.
Since the design requirement is $R_d \geq S_d$, from Equations (3.86) and (3.87)

$$R_k \geq \alpha_r \cdot \alpha_s \cdot S_k \tag{3.88}$$

Equation (3.88) provides a method for adjusting the characteristic value S_k of fire severity to take account of uncertainties with the aid of partial safety factors α_r and α_s. For adopting to reliability requirements differing from the average or normal requirements, additional adjustment factors can be included on the right-hand side of Equation (3.88) as additional (multiplicative) partial safety factors. The adjustments for a particular building or type of building should reflect the increase or decrease in fire risk from the average risk, compartment size, effectiveness of sprinklers (if installed), efficiency of fire brigade and other such factors affecting fire severity.

For the evacuation model, the design value H_d for the total evacuation time H is given by

$$H_d = H_k \cdot \alpha_h \tag{3.89}$$

where H_k is the characteristic value and α_h the partial safety factor greater than unity. The maximum total evacuation time

$$H_q = \mu_h (1 + V_h t_h)$$

can be considered as the design value and the mean μ_h as the characteristic value. In this case,

$$\alpha_h = (1 + V_h t_h)$$

where V_h is the coefficient of variation of H and μ_h a parameter similar to t_r in Equation (3.73) or t_s in Equation (3.74).

The design value for the combustion product time F is given by

$$F_d = F_k / \alpha_f \tag{3.90}$$

where F_k is the characteristic value and α_f the partial safety factor greater than unity. The minimum value of F given by

$$F_p = \mu_f (1 - V_f t_f)$$

can be considered as the design value and the mean value μ_f as the characteristic value. In this case, α_f is the reciprocal of $(1 - V_f t_f)$. The parameter V_f is the coefficient of variation of F and t_f is a constant similar to t_h.

Since the design criterion for successful evacuation is $H_d \leq F_d$,

$$H_k \leq F_k / \alpha_f \cdot \alpha_h \text{ or } F_k \geq H_k \cdot \alpha_f \cdot \alpha_h \tag{3.91}$$

The purpose of including the partial safety factors α_f and α_h in the design process is to ensure that the maximum or any other design value for the total evacuation time H does not exceed the minimum or any other design value for the combustion product time F.

In the semi-probabilistic approach discussed in this section, the choices for the values of the partial safety factors are usually based on the expert judgement of the fire safety engineer and the quality of information available to him/her for estimating the values of the parameters. Instead of adopting such empirical and intuitive methods, the partial safety factors can be derived from the probability distributions of the variables involved. This method, based on the 'design point', was described in detail by Ramachandran (1998a).

3.7.4 Beta method

3.7.4.1 Probabilistic design criterion

In a probabilistic procedure, the deterministic design criterion, $Y \geq X$, discussed in Section 3.7.3.1 is modified to:

$$P(Y \geq X) \geq 1 - P_g \tag{3.92}$$

where $P(Y \geq X)$ denotes the probability of strength Y being greater than or equal to stress X; this is equivalent to the probability of success. P_g is a (small) target probability (risk) acceptable to a property owner or society at large. The value of P_g depends on consequences in terms of damage to life and property if failure occurs. The probability of failure should be less than P_g:

$$P(Y < X) < P_g \tag{3.93}$$

If Y is fire resistance R and X is fire severity S the probabilistic design criterion for compartment success is

$$P(R \geq S) \geq 1 - P_g \tag{3.94}$$

Probability of compartment failure should be less than P_g:

$$P(R < S) < P_g \tag{3.95}$$

For building evacuation, X is the time F taken by a combustion product to produce an untenable condition on an escape route and Y is the total evacuation time H. In this case, Equation (3.92) is modified to

$$P(H \leq F) \geq 1 - P_g \tag{3.96}$$

for egress success. Probability of egress failure should be less than P_g:

$$P(H > F) < P_g \qquad (3.97)$$

Probabilistic methods are concerned with the evaluation of P_g and $(1 - P_g)$ for different combinations of X and Y. The evaluation procedure takes account of uncertainties through the probability distributions of X and Y.

3.7.4.2 Univariate approach

In this approach, discussed by Ramachandran (1998a), the stress variable X only is considered as a random variable with the strength variable Y treated as a constant. This is the approach traditionally adopted by fire safety engineers for determining the fire resistance required for a structural element. The cumulative probability distribution function of severity S is denoted by $F_s(x)$, the probability of severity being less than or equal to x. If the fire resistance R of a structural element is set equal to x, the probability of severity exceeding R is $[1 - F_s(R)]$ which is the probability of failure of the element.

Consider first the exponential probability distribution for fire severity S:

$$F_s(x) = 1 - \exp(-\lambda_s x) \qquad (3.98)$$

According to a property of this distribution, λ_s is the reciprocal of the mean value μ_s of fire severity. Baldwin (1975) estimated $\mu_s = 25$ minutes for office buildings such that $\lambda_s = 0.04$. It may be seen from Equation (3.98) that if $R = x = 25$ minutes, the probability of failure, would be

$$1 - F_s(R) = \exp(-\lambda_s R)$$
$$= \exp(-1) = 0.37$$

which is not a small quantity. But the probability of failure would reduce to 0.09 if $R = 60$ minutes and to 0.03 if $R = 90$ minutes and so on.

If fire severity S has a normal distribution with mean μ_s and standard deviation σ_s, the standardised random variable t given by

$$t = (S - \mu_s) / \sigma_s \qquad (3.99)$$

has a standard normal distribution with mean zero and standard deviation unity. From Equation (3.99)

$$S = \mu_s + t\sigma_s \qquad (3.100)$$

If the fire resistance R of a structural element is set equal to S with $t = 0$ such that $R = \mu_s$, the probability of success or failure of the element in a fire is 0.5. But if $R = S$ with $t = 1.96$ in Equation (3.100), the probability of success given by the cumulative distribution function of t is 0.975; the probability of failure is 0.025. For $t > 1.96$, the probability of failure would be less than 0.025. For $t = 2.33$, the probability of success would be 0.99 with 0.01 for the probability of failure. Probabilities of success and failure for different values of t can be obtained from a table of standard Normal distribution. Using this table, the fire resistance required to meet any target level for the probability of failure can be determined by using, in Equation (3.100), the value of t corresponding to this level.

3.7.4.3 Bivariate approach

In this approach, more commonly known as the Beta method, both the stress and strength variables are considered as random variables affected by uncertainties – see Ramachandran (1998a) and Magnusson (1974). The difference $(Y - X)$ is the 'safety margin' which is also referred to as the 'state function'. The expected value of the random variable

$$Z = Y - X \tag{3.101}$$

is given by

$$\mu_z = \mu_y - \mu_x \tag{3.102}$$

where μ_y and μ_x are the mean values of Y and X. The standard deviation of z is given by

$$\sigma_z = (\sigma_y^2 + \sigma_x^2)^{\frac{1}{2}} \tag{3.103}$$

where σ_y and σ_x are the standard deviations of Y and X. The 'safety index' β is given by

$$\beta = \mu_z / \sigma_z \tag{3.104}$$

Consider first the determination of fire resistance required for a structural element to satisfy a specified level for the probability of failure. If the mean and standard deviation of fire resistance R are μ_r and σ_r and the mean and standard deviation of fire severity S are μ_s and σ_s, the mean and standard deviation of the state function $z = R - S$ are

$$\mu_z = \mu_r - \mu_s \tag{3.105}$$

$$\sigma_z = (\sigma_r^2 + \sigma_s^2)^{\frac{1}{2}} \tag{3.106}$$

The 'safety index' β is given by

$$\beta = \mu_z / \sigma_z = (\mu_r - \mu_s) / (\sigma_r^2 + \sigma_s^2)^{\frac{1}{2}} \tag{3.107}$$

The fire resistance required may be set according to μ_r given by

$$\mu_r = \mu_s + \beta(\sigma_r^2 + \sigma_s^2)^{\frac{1}{2}} \tag{3.108}$$

If R and S have normal distributions, the parameter β has a standard normal distribution. In this case, the value of β corresponding to any target level for probability of failure can be obtained from a table of standard Normal distribution. This value can then be inserted in Equation (3.108) to provide the fire resistance μ_r required for the structural element. As discussed in Section 3.7.4.2 in terms of the variable t, the probability of structural failure would be 0.5 if $\beta = 0$ and $\mu_r = \mu_s$, less than 0.5 if β is positive and $\mu_r > \mu_s$ and greater than 0.5 if β is negative and $\mu_r < \mu_s$. The probability of failure would be 0.025 if $\beta = 1.96$, 0.01 if $\beta = 2.33$ and 0.001 if $\beta = 3.09$. For a selection of values of β, probabilities of structural success and failure are given in Table 3.17.

Consider now the determination of the total evacuation time H that will satisfy a specified level for the probability of egress failure – see Ramachandran (1993). The state function in this case may be written as $z = F - H$ such that the mean and standard deviation of z are

$$\mu_z = \mu_f - \mu_h \tag{3.109}$$

$$\sigma_z = (\sigma_f^2 + \sigma_h^2)^{\frac{1}{2}} \tag{3.110}$$

The parameters μ_f and σ_f are the mean and standard deviation of F and μ_h and σ_h are the mean and standard deviation of H. The safety index β is given by

$$\beta = \mu_z / \sigma_z \tag{3.111}$$

$$= (\mu_f - \mu_h) / (\sigma_f^2 + \sigma_h^2)^{\frac{1}{2}}$$

The total evacuation time required may be set according to μ_h in the following equation:

$$\mu_h = \mu_f - \beta(\sigma_f^2 + \sigma_h^2)^{\frac{1}{2}} \tag{3.112}$$

If F and H have normal distributions, as discussed earlier, β has a standard normal distribution. The probability of egress failure would be 0.5 if $\beta = 0$ and $\mu_h = \mu_f$, less than 0.5 if β is positive and $\mu_h < \mu_f$ and greater than 0.5 if β is negative and $\mu_h > \mu_f$. The probability of failure would be 0.025 if $\beta = 1.96$, 0.01 if $\beta = 2.33$ and 0.001 if $\beta = 3.09$. The figures in Table 3.17 can be used in conjunction with Equation (3.112) to determine the total evacuation time according to a specified level for the probability of egress failure. It should be noted that β has a positive sign attached to it in Equation (3.108) but a negative sign in Equation (3.112).

To satisfy the condition specified in Equation (3.112), it may be necessary to install automatic detectors and/or sprinklers if the building is not already equipped with these devices. These devices would reduce the detection time D and hence reduce μ_h. Sprinklers would also increase the combustion product time μ_r. The total evacuation time μ_h can also be reduced by providing additional or wider staircases which will reduce the design evacuation time E.

If egress failure occurs, there is a probability, K, that one or more deaths may occur. This probability can be estimated by analysing fire statistics. According to an analysis of these statistics for the period 1978 to 1988 by Ramachandran (1993), the average detection time for single and multiple occupancy dwellings was 10 minutes. With $B = 2$ minutes and $E = 3$ minutes, the average total evacuation time, μ_h, was 15 minutes. The mean value of combustion product time, μ_r, for causing death, was assumed to be 15 minutes such that the probability of egress failure was estimated to be 0.5. With a fatality rate per fire of 0.013, the value of K was estimated as 0.026 (= 0.013/0.5).

3.7.4.4 Safety factor

Corresponding to the safety index β, a safety factor θ may be defined as the ratio between the mean values of the stress and strength variables. In the case of structural failure

$$\theta = \mu_r / \mu_s \tag{3.113}$$

such that, from Equation (3.107)

$$\beta = (\theta - 1) / (V_R^2 \, \theta^2 + V_s^2)^{\frac{1}{2}} \tag{3.114}$$

where V_R and V_S are coefficients of variation given by

$$V_R = \sigma_r / \mu_r \, ; V_S = \sigma_s / \mu_s$$

For facilitating calculations, Equation (3.114) may be inverted to give

$$\theta = \frac{1 + \beta(V_R^2 + V_s^2 - \beta^2 \, V_R^2 \, V_s^2)^{\frac{1}{2}}}{1 - \beta^2 \, V_R^2} \tag{3.115}$$

Equation (3.115) has a solution only if $V_R < 1/\beta$.
If it is is assumed that $V_R = V_s = r$, then Equation (3.115) reduces to

$$\theta = \frac{1 + \beta r(2 - \beta^2 \, r^2)^{\frac{1}{2}}}{1 - \beta^2 \, r^2} \tag{3.116}$$

For $r = 0.15$, the values of θ corresponding to those of β given in Table 3.17 for different failure probabilities.

In the safety factor approach, the mean value of fire resistance, μ_r, should be set equal to or greater than the value given by $\theta\mu_s$. Suppose, for example, the probability of structural failure should be less than 0.005. In this case, from Table 3.17 for $P_g = 0.005$, $\beta = 2.5758$ and $\theta = 1.7934$ if $r = 0.15$. Hence, to achieve

Table 3.17 Probabilities of success and failure (standard normal distribution)

Probability of success $(1-P_g)$	Probability of failure (P_g)	β	θ $(r = 0.15)$
0.0001	0.9999	−3.7190	0.3993
0.0005	0.9995	−3.2905	0.4573
0.0010	0.9990	−3.0902	0.4848
0.0020	0.9980	−2.8782	0.5145
0.0025	0.9975	−2.8070	0.5245
0.0050	0.9950	−2.5758	0.5576
0.0100	0.9900	−2.3263	0.5941
0.0250	0.9750	−1.9600	0.6494
0.0500	0.9500	−1.6449	0.6990
0.1000	0.9000	−1.2816	0.7587
0.2000	0.8000	−0.8416	0.8355
0.3000	0.7000	−0.5244	0.8945
0.4000	0.6000	−0.2533	0.9477
0.5000	0.5000	0.0000	1.0000
0.6000	0.4000	0.2533	1.0552
0.7000	0.3000	0.5244	1.1180
0.8000	0.2000	0.8416	1.1969
0.9000	0.1000	1.2816	1.3181
0.9500	0.0500	1.6449	1.4307
0.9750	0.0250	1.9600	1.5398
0.9900	0.0100	2.3263	1.6832
0.9950	0.0050	2.5758	1.7934
0.9975	0.0025	2.8070	1.9064
0.9980	0.0020	2.8782	1.9437
0.9990	0.0010	3.0902	2.0626
0.9995	0.0005	3.2905	2.1869
0.9999	0.0001	3.7190	2.5043

the desired target, the mean fire resistance, μ_r, should be set equal to or greater than $1.79\mu_s$.

For the evacuation model, the safety factor is given by

$$\theta = \mu_f / \mu_h \tag{3.117}$$

such that, following Equation (3.114)

$$\beta = (\theta - 1) / (V_f^2 \theta^2 + V_h^2)^{\frac{1}{2}}$$

where V_f and V_h are the coefficients of variation of F and H given by
$$V_f = \sigma_f / \mu_f ; V_h = \sigma_h / \mu_h$$

Also,

$$\theta = \frac{1 + \beta(V_f^2 + V_h^2 - \beta^2 V_f^2 V_h^2)^{\frac{1}{2}}}{1 - \beta^2 V_f^2} \qquad (3.118)$$

Equation (3.116) is applicable if $V_f = V_h = r$.

If $r = 0.15$, from Table 3.17, $\beta = 2.3263$ and $\theta = 1.6832$, for a target maximum value of 0.01 for the probability of egress failure. To achieve this target, the mean value, μ_h, of total evacuation time should not exceed $0.59\mu_f$. This result follows from Equation (3.117) according to which μ_f should be greater than $\theta\mu_h$ or μ_h should be less than (μ_f / θ). Under such a protection for life safety, the fatality rate per fire would be less than 0.00026 (= 0.01 × 0.026) if the probability, K, of one or more deaths occurring given egress failure is 0.026 as estimated earlier. The value of K would vary from one type of occupancy to another.

Thus, the fatality rate per fire in single and multiple occupancy dwellings can be reduced to 0.00026 from the level of 0.013 estimated from the data for 1978–1988 if the total evacuation time, μ_h, is reduced to 9 minutes (= 0.59 × 15) from the level of 15 minutes. With $B = 2$ minutes and $E = 3$ minutes, the detection or discovery time, D, should be reduced to 4 minutes from the level of 10 minutes.

3.7.4.5 Log normal safety index

For structural fire resistance, Esteva and Rosenblueth (1971) put forward a design format based on the state variable

$$y = \log_e (R/S) \qquad (3.119)$$

which is applicable if resistance R and severity S have log normal probability distributions. Approximate values of the mean $\bar{y}$ and standard deviation σ_y of y are given by

$$\bar{y} = \log_e(\mu_r / \mu_s)$$
$$\sigma_y = (V_R^2 + V_S^2)^{\frac{1}{2}}$$

where, as defined earlier, μ_r and V_R are the mean and coefficient of variation of R and μ_s and V_s are the mean and coefficient of variation of S.

The safety index corresponding to the state variable y in Equation (3.119) is

$$\beta_{ER} = \overline{y} / \sigma_y \tag{3.120}$$

$$= \log_e(\mu_r / \mu_s) / (V_R^2 + V_S^R)^{\frac{1}{2}}$$

The fire resistance required may be determined according to μ_r given by

$$\log_e \mu_r = \log_e \mu_s + \beta_{ER}(V_R^2 + V_s^2)^{\frac{1}{2}} \tag{3.121}$$

The safety factor θ_{ER} is given by

$$\theta_{ER} = (\mu_r / \mu_s) = \exp\left[\beta_{ER}(V_R^2 + V_s^2)^{\frac{1}{2}}\right] \tag{3.122}$$

The mean fire resistance μ_r should be set equal to or greater than $\beta_{ER} \cdot \mu_s$.

Values of β_{ER} for different probabilities of structural failure are the same as those in Table 3.17. $\theta_{ER} = 1$ if $\beta_{ER} = 0$, less than 1 if β_{ER} is negative and greater than 1 if β_{ER} is positive. If $V_R = V_s = r$:

$$\theta_{ER} = \exp\left[\beta_{ER} r\sqrt{2}\right] \tag{3.123}$$

Calculations based on Equation (3.123) show that, for any target probability of failure less than 0.3, the value of θ given by Equation (3.115) is marginally greater than the corresponding value of θ_{ER}. Hence, in this range of failure probability which is of interest in structural fire safety design, an assumption of normal distributions for R and S would provide a slightly greater safety margin than an assumption of log normal distributions.

It is a somewhat complex statistical problem to construct an appropriate safety index if both R and S have exponential probability distributions or they have different distributions. The safety index proposed in Equation (3.107) or (3.120) would be sufficient for all practical purposes. Ramachandran (1998a) discussed in detail other problems such as 'design point', full probabilistic approach, extreme value technique and determination of tolerable failure probability.

References

Aoki, Y (1978), *Studies on Probabilistic Spread of Fire*, Research Paper No. 80, Building Research Institute, Tokyo.

Atreya A (2003), Convection heat transfer, Section 1 Chapter 3, *SFPE Handbook of Fire Protection Engineering*, 3rd edn, SFPE/NFPA, Gaithersburg, MD.

Baldwin, R (1975), *Economics of Structural Fire Protection*, Current Paper CP 45/75., Building Research Establishment, Watford.

Baldwin, R and Fardell, L G (1970), *Statistical Analysis of Fire Spread in Buildings*. Fire Research Note 848, Fire Research Station, Borehamwood.

Beck, V R (1987), A cost-effective decision-making model for fire safety and protection, *Fire Safety Journal*, 12, 121–138.

Bengtson, S and Hagglund, B (1986), The use of a zone model in fire engineering application, *Fire Safety Science: Proceedings of the First International Symposium*, Hemisphere Publishing Corporation, New York.

Bengtson, S and Laufke, H (1979/80), Methods of estimation of fire frequencies, personal safety and fire damage, *Fire Safety Journal*, 2, 167–180.

Bengtson, S and Ramachandran, G (1994), Fire growth rates in underground facilities, *Fire Safety Science: Proceedings of the Fourth International Symposium*, National Institute of Standards and Technology, Gaithersburg, MD.

Benktander, G (1973), Claims frequency and risk premium rate as a function of the size of the risk, *ASTIN Bulletin*, 7, 119–136.

Berlin, G N (1980), Managing the variability of fire behaviour, *Fire Technology*, 16, 287–302.

British Standards Institute (2003), PD 7974 Code on the application of fire safety engineering principles to the design of buildings, Part 7: Probabilistic risk assessment, British Standards Institute, London.

BS 5760 Reliability of systems, equipment and components, British Standards Institute, London.

Carslaw H S and Jaeger J C (1959), *Conduction of Heat in Solids*, 2nd edn, Oxford University Press, Oxford.

Charters D (1999), What does quantified fire risk assessment have to do to become an integral part to design decision-making, International Symposium on Uncertainty, Risk and Reliability, SFPE, Bethesda, MD.

Charters D A (1992a), Fire risk assessment of rail buildings, *Proceedings of the 1st International Conference on Safety in Road and Rail Buildings*, ITC, Basel.

Charters, D A (1992b), *Fire Safety Assessment of Bus Transportation*, C437/037, Institution of Mechanical Engineers, London.

Charters D A (1997), Risk assessment – the reliability and performance of systems, FIREX, NEC, Birmingham.

Charters, D A (1998), Fire safety at any price?, *Fire Prevention*, 313, October, 12–15.

Charters D A and Smith, F M (1992), *The Effects of Materials on Fire Hazards and Fire Risk Assessment*, C438/017, Institution of Mechanical Engineers, London.

Charters D A, Gray W A and McIntosh A C (1994), A computer model to assess fire hazards in buildings, *Fire Technology*, 30, 1, 135–155.

Charters D A, Barnett J, *et al.* (1997), Assessment of the probabilities that staff and/or patients will detect fires in hospitals, *Proceedings of the Fifth International Symposium of Fire Safety Science*, Society of Fire Protection Engineers, Melbourne.

CIB W14 (1986), Design guide: structural fire safety, *Fire Safety Journal*, 10, 2, 77–154.

CIBSE TM19 (1995), *Relationships for Smoke Control Calculations*, Chartered Institution of Building Service Engineers, London.

Connolly, R J and Charters, D A (1997), The use of probabilistic networks to evaluate passive fire protection measures in hospitals, *Proceedings of the Fifth International Symposium on Fire Safety Science*, Society of Fire Protection Engineers, Melbourne.

Coward, S K D (1975). *A Simulation Method for Estimating the Distribution of Fire Severities in Office Rooms*. Current Paper CP 31/75. Building Research Establishment, Watford.

Crossman R and Zachary W (1974), Occupant response to domestic fire incidents, US National Fire Protection Association Annual Conference.

Cullen W (1990), *The Public Inquiry into the Piper Alpha Disaster*, HMSO, London.

DOE (1996), *Data for the Application of Probabilistic Risk Assessments to the Evaluation of Building Fire Safety – Appendix II*, Department of Environment, London.

Drysdale, D, (1999) *An Introduction to Fire Dynamics*, 2nd edn, Wiley, Chichester.

Dusing J W A, Buchanan A H and Elms D G (1979), *Fire Spread Analysis of Multi-compartment Buildings*, Research Report 79/12, Department of Civil Engineering, University of Canterbury, New Zealand.

Elms D G and Buchanan A H (1981), *Fire Spread Analysis of Buildings*, Research Report R35, Building Research Association of New Zealand, Auckland.

Emmons H (1991), The ceiling jet in fires, *Proceedings of the 3rd International Symposium of Fire Safety Science*, Elsevier, London.

Emmons H (2003), Vent flows, Section 2 Chapter 5, *SFPE Handbook of Fire Protection Engineering*, 3rd edn, SFPE/NFPA, Gaithersburg, MD

Esteva L and Rosenblueth E (1971), Use of reliability theory in building codes, Conference on Application of Statistics and Probability to Soil and Structural Engineering, Hong Kong, September.

Evans D (2003),Ceiling jet flows, Chapter 2 Section 4, *SFPE Handbook of Fire Protection Engineering*, 3rd edn, SFPE/NFPA, Gaithersburg, MD.

Evers E and Waterhouse A (1978), *A Computer Model for Analysing Smoke Movement in Buildings*, Current Paper CP 69/78, Building Research Establishment, Watford.

Fahy R F (1985), Building fire simulation model – an overview, *Fire Safety Journal*, 9, 189–203.

Fardell P J and Kumar S (1991), *Fires in Buses: A Study of Life Threat and the Efficacy of Bus Garage Sprinkler Protection for London Buses Ltd. Part A Life Threat Experimental Studies and Predictive Computer Modelling*, LPC Draft Report.

Finucane M and Pinkney D (1988), *Reliability of Fire Protection and Detection Systems*, SRD R431, UKAEA, Harwell.

Government Statistics Service (1992), *Fire Statistics United Kingdom 1990*, HMSO, London.

Green A E and Bourne A J (1972), *Reliability Technology*, Wiley, London.

Gumbel, E J (1958), *Statistics of Extremes*, Columbia University Press, New York.

Hahn, G J and Shapiro, S S (1967), *Statistical Models in Engineering*, Wiley, New York.

Hansen-Tangen E and Baunan T (1983), Fire risk assessment by simulation – Fire sim, *Fire Safety Journal*, 5, 205–212.

Health and Safety Executive (1989), *Quantified Risk Assessment: Its Input to Decision Making*, HMSO, London.

Heskestad G (2003), 'Fire plumes', Section 2, Chapter 2, *SFPE Handbook of Fire Protection Engineering*, 3rd edn, SFPE/NFPA, Gaithersburg, MD.

Homer R D (1979), The protection of cold-form structural elements against fire, *Proceedings of International Conference on Thin-Wall Structures*, Wiley, New York.

Johnston B D and Matthews R H (1982), *Noncoherent Structure Theory: A Review and its Use in Fault Tree Analysis*, SRD R 245, UKAEA, Harwell.

La Valle I H (1970), *An Introduction to Probability, Decision and Inference*, Holt, Rinehart and Winston, New York.

Ling W T C and Williamson R B (1986), The modelling of fire spread through probabilistic networks, *Fire Safety Journal*, 9, 287–300.

Magnusson S E (1974), *Probabilistic Analysis of Fire Exposed Steel Structures*, Bulletin 27, Lund Institute of Technology, Lund, Sweden.

Mandelbrot, B (1964), Random walks, fire damage amount and other Paretian risk phenomena, *Operations Research*, 12, 582–585.

McCaffrey B (2003), Flame height, Section 2 Chapter 1, *SFPE Handbook of Fire Protection Engineering*, 3rd edn, SFPE/NFPA, Gaithersburg, MD

Mirchandani, P B (1976). *Computations and Operations Research*, 3, Pergamon Press, Oxford.

Morishita, Y (1977), *Establishment of Evaluating Method for Fire Safety Performance Report*, Research Project on Total Evaluating System on Housing Performances, Building Research Institute, Tokyo.

Morishita, Y (1985), A stochastic model of fire spread, *Fire Science and Technology*, 5, 1, 1–10.

NHS Estates (1996), *Fire Practice Note 9 – NHS Healthcare Fire Statistics 1994/5*, HMSO, London.

North, M A (1973), *The Estimation of Fire Risk of Various Occupancies*, Fire Research Note 989, Fire Research Station, Borehamwood.

Phillips, W G B (1992), *The Development of a Fire Risk Assessment Model*, Information Paper IP 8/92, Building Research Establishment, Watford.

Phillips, W G B (1995), Computer simulation for fire protection engineering. *SFPE Handbook of Fire Protection Engineering*, 2nd edn, Section 5, Chapter 1, National Fire Protection Association, Quincy, MA.

Platt, D G (1989), *Modelling Fire Spread: A Time-based Probability Approach*, Research Report 89/7, Department of Civil Engineering, University of Canterbury, New Zealand.

Purser, D A (2003), Toxicity assessment of combustion products, *SFPE Handbook of Fire Protection Engineering*, SFPE/NFPA 3rd Edition.

Quintiere J G (1989), 'Fundamentals of enclosure fire zone models', *Journal of Fire Protection Engineering*, Volume 1 (3).

Quintiere J G (2003), 'Compartment fire modelling', Section 3 Chapter 5, *SFPE Handbook of Fire Protection Engineering*, 3rd edn, SFPE/NFPA, Gaithersburg, MD.

Ramachandran, G (1969), The poisson process and fire loss distribution, *Bulletin of the International Statistical Institute*, 43, 2, 234–236.

Ramachandran, G (1970), *Fire Loss Indexes*, Fire Research Note 839, Fire Research Station, Borehamwood.

Ramachandran, G (1974). Extreme value theory and large fire losses. *ASTIN Bulletin*, 7, 3, 293–310.

Ramachandran, G (1975), Extreme order statistics in large samples from exponential type distributions and their application to fire loss, *Statistical Distributions in Scientific Work*, vol 2, D Reidel Publishing, Dordrecht.

Ramachandran, G (1979/80), Statistical methods in risk evaluation, *Fire Safety Journal*, 2, 125–145.

Ramachandran, G (1980), *Economic Value of Automatic Fire Detectors*, Information Paper IP27/80, Building Research Establishment, Fire Research Station, Borehamwood.

Ramachandran, G (1982), Properties of extreme order statistics and their application to fire protection and insurance problems, *Fire Safety Journal*, 5, 59–76.

Ramachandran, G (1985), Stochastic modelling of fire growth, *Fire Safety: Science and Engineering*, ASTM STP 882, American Society for Testing and Materials, Philadelphia, PA.

Ramachandran, G (1986), Exponential model of fire growth, *Fire Safety Science. Proceedings of the First International Symposium*, Hemisphere Publishing Corporation, New York.

Ramachandran, G (1988a), Probabilistic approach to fire risk evaluation, *Fire Technology*, 24, 3, 204–226.

Ramachandran, G (1988b), Extreme value theory, Section 4, Chapter 4, *SFPE Handbook of Fire Protection Engineering*, 1st edn, National Fire Protection Association, Quincy, MA (reproduced in 2nd edn, 1995, Section 5, Chapter 3).

Ramachandran, G (1990), Probability-based fire safety code, *Journal of Fire Protection Engineering*, 2, 3, 75–91.

Ramachandran, G (1991), Non-deterministic modelling of fire spread, *Journal of Fire Protection Engineering*, 3, 2, 37–48.

Ramachandran, G (1993), Probabilistic evaluation of design evacuation time. *Proceedings of the CIB W14 International Symposium on Fire Safety Engineering*, University of Ulster, Belfast.

Ramachandran, G (1995a), Stochastic models of fire growth, Section 3, Chapter 15, *SFPE Handbook of Fire Protection Engineering*, 2nd edn, National Fire Protection Association, Quincy, MA.

Ramachandran, G (1995b), Heat output and fire area, *Proceedings of the International Conference on Fire Research and Engineering*, CRC Press, Boca Raton, FL.

Ramachandran, G (1998a). *Probabilistic Evaluation of Structural Fire Protection – A Simplified Guide*. Fire Note 8, Fire Research Station, Building Research Establishment, Borehamwood.

Ramachandran, G (1998b). *The Economics of Fire Protection*. E & F N Spon, London.

Ramachandran, G (2002), Stochastic models of fire growth, Section 3, Chapter 15, *SFPE Handbook of Fire Protection Engineering*, 3rd edn, National Fire Protection Association, Quincy, MA.

Ramachandran, G and Chandler, S E (1984), *Economic Value of Early Detection of Fires in Industrial and Commercial Premises*, Information Paper IP13/84. Building Research Establishment, Fire Research Station, Borehamwood.

Rasbash, D, Ramachandran, G, Kandola, B, Watts, J M and Law, M, (2004), *Evaluation of Fire Safety*, Wiley, Chichester.

Reynolds C and Bosley K (1995), *Domestic First Aid Firefighting*, Research Report 65, Fire Research and Development Group, London.

Rockett J A and Milke J A (2003), 'Conduction of heat in solids', *SFPE Handbook of Fire Protection Engineering*, 3rd edn, National Fire Protection Association, Quincy, MA.

Rogers, F E (1977), *Fire Losses and the Effect of Sprinkler Protection of Buildings in a Variety of Industries and Trades*, Current Paper CP 9/77, Building Research Establishment, Fire Research Station, Borehamwood.

Rutstein, R (1979), The estimation of the fire hazard in different occupancies, *Fire Surveyor*, 8, 2, 21–25.

Rylands S, David P, McIntosh A C and Charters D A (1998), Predicting fire and smoke movement in buildings using zone modelling, *Proceedings of the 3rd International Conference on Safety in Road and Rail Buildings*, International Technical Conferences, Nice.

Sasaki, H and Jin, T (1979), *Probability of Fire Spread in Urban Fires and their Simulations*, Report No 47, Fire Research Institute of Japan, Tokyo.

Scoones K, (1995), FPA large fire analysis', *Fire Prevention Journal*, January/February, Fire Protection Association (Loss Prevention Council), Moreton in Marsh.

Shpilberg, D C (1974), *Risk Insurance and Fire Protection: A System Approach. Part 1: Modelling the Probability Distribution of Fire Loss Amount*, Technical Report 22431, Factory Mutual Research Corporation, Norwood, MA.

Shpilberg, D C (1975), Statistical decomposition analysis and claim distribution for industrial fire losses, Twelfth ASTIN Colloquium, Portugal, September–October 1975.

Tien C L, Lee K Y and Stretton A J (2003), 'Radiation heat transfer', Section 1 Chapter 4, *SFPE Handbook of Fire Protection Engineering*, 3rd edn, National Fire Protection Association, Quincy, MA.

Vantelon *et al.* (1991), Investigation of fire induced smoke movement in tunnels and stations: An application to the Paris Metro, *Proceedings of the 3rd International Symposium of Fire Safety Science*, Elsevier, London.

Watts, J M, Jr (1986), Dealing with uncertainty: some applications in fire protection engineering, *Fire Safety Journal*, 11, 127–134.

4   Acceptance criteria

We all take risks all the time. Whether it is crossing the road, driving to work or watching television. The risk may vary from being knocked down, to being in a car accident or suffering ill health due to lack of exercise. The same can be said of fire safety in buildings. As long as we occupy buildings where there is a chance that ignition sources and combustible materials may be present together, there will be a risk of death and injury due to fire. We need not be too fatalistic, however, this simply identifies the need to manage the risk. It also indicates that although we should work towards reducing risk, the ultimate goal of zero risk is not a realistic expectation.

They say that there are two certainties in life: death and taxation. It follows that whilst we live there is a risk of death and the only way of not dying is not to live in the first place. This may be self-evident, but is very important when we start to consider specific risks that we consider them in the context of other risks.

There is a practical benefit to looking at risks in context. Society may decide that it would like to dedicate more resources to addressing one risk than another. For example, for healthcare, it may typically cost about £20k to save a life, whereas for fire safety in buildings, it may cost more than say £1million to save a life. Therefore, society (or its representatives) could decide to put more resources into healthcare than into building fire safety

Equally, society may decide that the risk of death by fire is much worse than the risk of not being treated for a potentially fatal condition, in which case the opposite conclusion is valid. This is called 'risk acceptability' and there are several ways in which a risk may be accepted:

1   Ignorance. If the existence of a risk is not known, then it may be accepted for many years. Smoking, asbestos and the fire safety of certain types of sandwich panel fall into this category. However, once these risks were known, there was a duty to address them.
2   Negligible. If the risks are so low that they can be considered negligible, they can be accepted. What is negligible is difficult to define, but for health and safety, risks that are well below 1 death in a million years for a member of the public are generally considered negligible.

Risks that are above those that are negligible can be considered tolerable or intolerable:

1   Tolerable. Risks may be considered tolerable where the benefits of the activity causing the risk are perceived to outweigh the risks. The convenience of travelling by air, rail or car could be seen in this way. Even though the benefits are seen to outweigh the risks, there is still a need to reduce the risks until the costs of further risk reduction far outweigh reduction in risk. This principle is known as reducing the risks until they are 'as low as reasonably practicable' (ALARP) and has been a legal precedent in health and safety for many years.
2   Intolerable. There are, however, levels of risk that could not be tolerated under any circumstances and these can be considered intolerable. This could be the risk of death of a member of the public above say one in a thousand per person year.

For fire safety in buildings, the annual fire statistics indicate that there is a finite level of fire risk in buildings. This may also indicate that if a building complies with the appropriate fire safety standards, then its level of fire risk is broadly tolerable (or possibly acceptable). It could also be said that applying the fire standards to a non-standard building could result in intolerable levels of fire risk. However, no explicit numerical criteria for fire risk in buildings have been set in the UK.

For the fire safety engineering of a non-standard building, this means that the level of risk should be designed to be the same or lower than that for an equivalent standard building. Because there are no quantitative risk criteria, this means comparison of a non-standard building with a compliant building. Because the regulations are not generally framed in terms of risk, this results in an assessment of physical hazards and the balancing of qualitative arguments about risk.

Types of acceptance criteria

The acceptability of fire risks is often seen to be contentious, with engineers, approval bodies, fire authorities, psychologists, sociologists, politicians and many others contributing to the debate. Neither is the acceptability of risks a fixed phenomenon. Through history it can be seen that risks that were once accepted or tolerated become unacceptable from time to time. However, without some form of acceptance criteria it is not possible to judge whether a building is acceptably safe.

There are three general approaches to risk acceptability:

1   Risk comparison
2   Absolute risk criteria
3   Economic basis.

Approach 3 is also typically a form of risk comparison. This can be seen in Table 4.1 which contains a summary of the approaches as applied to fire safety.

Qualitative and semi-quantitative risk assessment methods

Since these methods do not explicitly predict a level of fire risk, risk acceptance is generally by comparison of the results of the analysis with an accepted standard. For example, the acceptability of a risk identified in a qualitative fire risk assessment in an office is generally made with reference to what would normally be expected in such an office in the relevant guidance document (see Section 2.1).

For the points schemes of Section 2.2, the risk acceptability is represented by a threshold score or range of scores which lead to risk reduction or no action. For matrix methods, risk may be acceptable by comparison to the rating of a compliant design or by the low level of risk implicit in the score and these can be expressed in matrices of risk acceptance.

Risk acceptance criteria for quantitative methods is altogether more complicated, but can provide a more informed and evidence-driven test of how risks should be treated.

4.1 Comparative criteria

The simplest and most common basis for assessing whether a building is acceptably safe is by comparison with other risks via historical data. Extensive lists of risks have been drawn up for many types of activity. Table 4.2 contains levels of fire risk in different building types in England and Wales. The concept is that risks that are significantly lower than those attached to common everyday activities accepted by society are safe.

For fire risks in buildings, it can often be difficult to establish the level of risk in absolute terms. However, it may be relatively straightforward to demonstrate that the design provides a level of risk equivalent to that in a building which conforms to more prescriptive codes (life safety or financial). Since the study is purely comparative, many errors contained in assumptions or data regarding ignition frequencies or reliability of systems will be cancelling in nature and have

Table 4.1 Typical types of acceptance criteria

| | | Fire safety objectives | |
		Life safety	Financial
Analysis method	Comparative	Level of risk equivalent to code-compliant solution, e.g. Approved Document B (AD B).	Comparison of design alternatives, (cost/benefit analysis).
	Absolute	Number of casualties per occupant year.	Acceptable average loss per year.

Table 4.2 Number of deaths per building year and number of deaths per occupant year

Occupancy	No buildings	No occupants	Average per year 1995–1999			Death/ building/year	Death/ occupant/year[c]
			No deaths	No injuries	No fires		
Further education	1051	845617[a]	0.0	17	535	<2.4E-04	< 3.0E-07
Schools	34731	10503100[a]	0.0	51	1669	<7.2E-06	< 2.4E-08
Licensed premises	101081	–	2.8	262	3317	2.7E-05	–
Public recreation buildings	45049	–	1.3	48	2581	2.8E-05	–
Shops	354475	–	3.3	284	5671	9.2E-06	–
Hotels	28371	389174[a]	2.5	116	1021	8.8E-05	6.4E-06
Hostels	9829	–	0.5	60	1338	5.1E-05	–
Hospitals	3486	–	3.3	113	3063	9.3E-04	–
Care homes	29080	–	4.5	130	1616	1.5E-04	–
Offices	209627	4107000[b]	0.3	219	1988	1.2E-06	7.3E-08
Factories	170972	–	4.3	286	5299	2.5E-05	–
All above occupancies	987752	15844891	22.5	1584	28096	2.3E-05	6.5E-06

Notes
a Number of occupants equals the sum of the number of employees and other occupants.
b Number of occupants equals the number of employees only.
c It may be more appropriate to use the number of deaths per occupant for large or complex buildings.

no significant influence on the decision. This can be confirmed by sensitivity analysis.

Before it can be demonstrated that a solution offers an equivalent level of risk as a prescriptive code, the intent of that code needs to be clearly understood. It is important to understand the intentions of each recommendation, as particular provisions may have more than one objective. Alternative risk management solutions can be developed to address the specific underlying objectives. The fire risk assessor should demonstrate that the solution proposed will be at least as effective as the conventional approach.

The limitations of this approach can be summarised as follows:

- It does not distinguish between risks that are accepted and those that are tolerated (for example the risk from potato poisoning is accepted since there has only been one recorded death, but the risk of choking is tolerated because of the benefits of eating and drinking).
- It takes no account of changes that may make historical risk levels out of date (the technology and regulation of air travel have changed significantly over the last 50 years, therefore basing acceptance criteria on the data from the last 50 years may be misleading).
- It does not address risk reduction following a major incident.
- It neglects other factors that affect the acceptability of the risk, e.g. involuntary and societal risks.

Risks are more likely to be accepted if the activity is voluntary, i.e. the person exposed to the risk perceives that they can control their level of exposure. For example, driving a car can be perceived as a voluntary activity, whereas being a passenger on a train can be seen as an involuntary activity.

There are several ways of expressing risk criteria and one of the most common contains three concepts:

1   There is an upper limit which cannot be tolerated on any grounds and activities whose risk levels cannot be lowered should be stopped.
2   There is a lower limit where the level of risk appears to be trivial and the use of resources to reduce it would be wasteful.
3   All activities whose risk lies between the upper and lower limits should only be carried out if the benefits justify the risks.

The third concept leads to the principle that the risks between the upper and lower limits should be as low as reasonably practicable (ALARP) which is enshrined in the UK Health and Safety at Work Act 1974 Part 1 under the general duties of persons concerned with premises:

It shall be the general duty of each person who has, to any extent control of premises ... to take such measures ... as so far as is reasonably practicable ... that they are safe and without risks to health.

As low as practicable can be defined as the point where the cost of further risk reduction far outweighs the resultant reduction in risk.

This can mean that the levels of risk and cost of several potential safeguards or sets of safeguards are evaluated against a baseline solution (i.e. a benchmark). The level of risk reduction relative to the baseline (usually an existing or code-compliant case) can then be evaluated. Care should be taken that the event scenarios quantified cover an appropriately broad range of events. For example, if a very severe fire growth and peak heat release rate is the only event quantified in the evaluation of say, a new rail tunnel design, then the alternatives are very unlikely to show any risk/cost benefit unless they change the event being modelled in some way.

The cost of the alternative solution can then be evaluated, in terms of 'cost per life saved', i.e. the cost of saving a life using the alternative solution. This can provide a very effective measure of the risk/cost effectiveness of a range of alternative solutions. The cost per life saved can be then be compared with other alternatives or against benchmarks such as safeguards costing more than £10 million pounds for a life saved are not risk/cost-effective. The cost of saving a life through a clinical intervention is generally of the order of £20,000. This is conceptually and ethically quite different to the 'value of a life' described in the absolute criteria section below (see Figure 4.1).

Comparative criteria in fire engineering use the level(s) of risk predicted for a code-compliant building as a benchmark for a non-standard building to assess equivalency (i.e. the level of risk is the same or lower). The main advantage of comparative criteria is that assumptions made during the analysis cancel each other out and so decisions made on the output should be much more reliable. The main disadvantage is that the code-compliant building may not provide an acceptable level of risk.

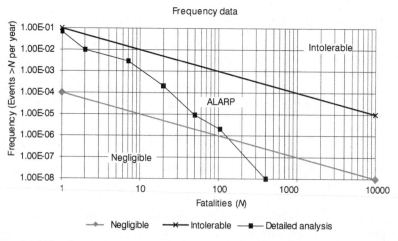

Figure 4.1 Life safety acceptance criteria shown on an F–n Curve.

4.2 Absolute criteria

Absolute acceptance criteria for life safety are derived from society's perception of the risk. The perceived level of risk is often very different from the estimated level of the risk. Extensive research by psychologists and social scientists indicates that there is a range of factors, values and beliefs that influence the public's perception of risk. Theses factors include whether the risk:

- is known or unknown in nature;
- is dreaded or not;
- is voluntary or involuntary;
- affects an individual or many people at the same time.

Risks that are unknown or known and dreaded or not can be plotted on a matrix. An example of a risk that is known and neutral in terms of dread is that of motor vehicles. An example of a risk that is known and dreaded is that of crime and an example of a risk that is unknown and dreaded is that of nuclear power. Generally speaking, risks that are unknown and/or dreaded are less acceptable to society than those that are known and not dreaded. Our perception of these risks can change with information, events and the way that they are portrayed in the media.

The degree to which a risk is voluntary also has a significant impact on the level of risk that may be tolerated. A voluntary risk may be one that an individual chooses to be exposed to and for which they feel they have a large degree of control. An example of this may be horse riding for leisure.

An involuntary risk is one where there is little alternative choice and the degree of risk is controlled by others. An example of this is air travel where the destination is on the other side of the globe. Generally, society is more willing to accept higher levels of risk in activities that are seen as voluntary rather than those that are seen as involuntary. Often the degree to which an activity is perceived to be voluntary is different to the actual situation.

For example, initially car driving may be seen as voluntary 'risk taking' but in reality the level of risk depends on many other factors including other road users, the skill of road designers, equipment maintainers, those who are near roads, the closeness and efficiency of emergency and medical services etc.

The difference between an individual following a certain pattern of activity or many people at the same time has a significant impact on tolerance. The former are known as individual risks and the latter as societal risks. Society is generally much more concerned about societal risks than it is about individual risks. Again, the different way that society (and the media) treats 3,000 largely individual road deaths each year compared with individual rail events with between 10 and 20 deaths is very clear.

Absolute life risk criteria fall into two categories: individual and societal, where:

- Individual risk is the frequency at which an individual may be expected to sustain a given level of harm from the realisation of specified hazards. This is usually related to a specific pattern of life. For fire safety this may be the individual risk of someone who works in an industrial or office building or of a shopper who visits a retail development once a week.

• Societal risk is the relationship between frequency of occurrence and the number of people in a given population suffering from a specified level of harm from the realisation of specified hazards. This is important because multi-fatality disasters are particularly repugnant to society. This may be expressed as the frequency with which ten or more people may die from fires. This is normally significantly lower than an individual level of risk.

4.2.1 Individual risk levels

The levels of risk to individual members of the public from the activities on major industrial sites (Health and Safety Executive 1988, 1989) are:

1   Maximum tolerable risk to individual member of the public (death per year) is 10^{-4}, i.e. one death can be expected from the operation of this site once every 10,000 years (or from 10,000 such sites, once per year). That is one of the implications of this type of risk criterion which is fine if the level of activity is constant and well known. This issue is more relevant for societal risks where acceptance criteria may be used to gain acceptance for one plant, but are the criteria still valid 100 plant designs later?

2   General acceptable risk to individual member of the public (death per year) is 10^{-6}.

The nature of the above risks is that they are largely involuntary, although there is an overall benefit to society from the activity.

Therefore, the generally acceptable levels of individual risk for a member of the public at home from a fire is 10^{-5} death per year. The generally acceptable levels of individual risk of the public elsewhere from a fire is 10^{-6} death per year.

4.2.2 Societal risk levels

Many studies have been carried out on the quantification of risk in the nuclear power industry. Accidents, particularly in the reactor, could cause not only early deaths but delayed deaths, sickness and genetic damage to people outside the plant. One of the earliest studies was the one carried out by Rasmussen (1975) for the US Nuclear Regulatory Commission, known as WASH 1400. Rasmussen quantified the main hazards of the pressurised water reactors in use. He also collected information on fatalities caused by a range of man-made and natural disasters in order to compare potential nuclear hazard and hazards for which there was everyday experience. He expressed the annual expectation of fatalities as the probability of a number equalling or exceeding a certain value, N. Later, Fryer and Griffiths (1979) referred to this relationship as $f(N)$ line.

Rasbash (1984) reviewed various studies on the $f(N)$ relationship to suggest criteria for acceptability for use with quantitative approaches to fire safety. According to him, two points were not taken into account in the conclusion of Rasmussen (1975), that societal risk due to the installation of 100 nuclear reactors would be small compared with the societal risk to which the nation (USA) had become accustomed from well-known man-made hazards including those of fires and explosions. Firstly, the people who would be killed after a nuclear accident would, for most of the part, be living quite near the nuclear reactor. The societal risk is, therefore, suffered disproportionately within the population. Secondly, the large majority of the people, for other hazards considered by Rasmussen, were obtaining a substantial benefit from the risk activity, they for the most part being major beneficiaries of the risk activity. For nuclear power stations, those at risk were most likely to be obtaining only marginal benefit. For a proper comparison to be made, Rasbash suggested that the number of people killed by fires and explosions should be those who were not using the building or being employed at the risk activity concerned.

Reviewing studies on the quantification of the hazard of major industrial processes, Rasbash estimated that the societal risk was about 1×10^{-4} downwards for an incident that would cause about 100 deaths outside the plant. Much of the hazard to the general public caused by process industries is due to fire and explosion associated particularly with the release of flammable liquefied gas or a large amount of flammable liquid in a vaporised or dispersed state. Rasbash put forward the view that risks associated with such hazards, both to local individuals and to local communities, should not be out of line with the fire and explosion hazard normally experienced by these people. In addition, the amount of risk that these individuals or communities should be expected to tolerate should be related to the benefit they obtain from the risk activity.

Rashbash's suggestions for societal risk were based upon experience of large kill-size fires and explosions in the UK, as shown in Figure 4.2. The full lines in this figure were based on actual fires in the UK and the extrapolation on experience in rich countries worldwide. Analysing this graph, Rasbash suggested that the total societal risk for all non-dwelling fires with N or more fatalities should be proportional to the size of the community that is threatened. On this basis and including other factors, he put forward target $f(N)$ lines as in Figure 4.3 for threatened communities of different sizes expressing the boundary of acceptability. This figure shows that a target that would not be acceptable to a community of, say, 1000, would be acceptable to a community of 10,000 and so on.

The approach discussed above has given rise to a criticism that a hazard could be made acceptable to a community if more people were crammed into the area at risk. For example, if a fire hazard associated with a plant were such that it could produce a multiple-fatality fire disaster which could kill 100 or more people 3 times in 10 million years (X in Figure 4.3), it would be unacceptable to a community of 10,000 but acceptable for 100,000. This criticism would be justified if cramming ten times the number of people into the threatened area did not affect the expected number of fatalities that might occur in an incident.

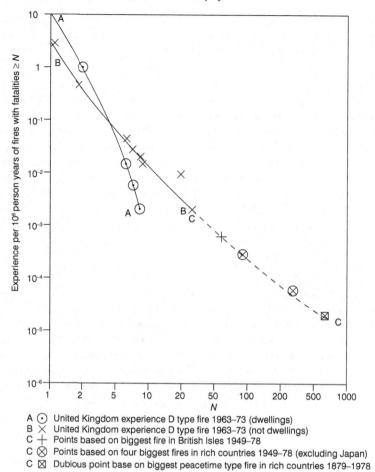

A ⊙ United Kingdom experience D type fire 1963–73 (dwellings)
B ✕ United Kingdom experience D type fire 1963–73 (not dwellings)
C ➕ Points based on biggest fire in British Isles 1949–78
C ⊗ Points based on four biggest fires in rich countries 1949–78 (excluding Japan)
C ⊠ Dubious point base on biggest peacetime type fire in rich countries 1879–1978

Figure 4.2 Experience per 10^6 person years of fires with N or more fatalities

In fact, the number of fatalities in a societal incident engendered by a nearby process plant is directly proportional to the density of population. Cramming ten times as many people into the area that would normally carry 10,000 would increase the number of expected fatalities from greater than 100 to greater than 1000 and would push the relevant risk point from X to Y. It will be seen that Y indicates a risk even less acceptable to a community of 100,000 than X is for a community of 10,000. In fact, to make the hazard more acceptable, it is necessary to reduce the density of population by a factor of 10 and shift the calculated risk point from X to Z. Thus risk which is not acceptable in an urban area could be acceptable in a rural area.

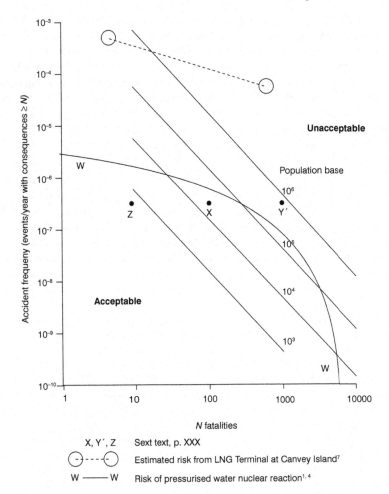

Figure 4.3 Targets for acceptability for societal risk from process industries. (Marginal benifit to those at risk)

4.2.3 Multiple-death fires

Fires can cause not only direct damage to human life in terms of fatal and non-fatal casualties but also indirect losses, e.g. distress and financial loss to the families of the victims and to society at large. The aggregate disutility or consequences due to fire deaths would be generally low for single-death fires and high for multiple-death incidents (Ramachandran, 1998). The disutility associated with a single fire with, say, 10 deaths is greater than the total disutility caused by 10 fires each with a single death. Catastrophes with several deaths have serious social, economic and political consequences.

Hence, it is necessary to analyse the characteristics of multiple-death fires and estimate the probability of occurrence of such fires, particularly in large buildings with several people at risk. Such an analysis would provide target values for accepting the occurrence of multiple-death fires. Consider, for example, Figure 4.4, produced by Rasbash (1984/1985). Information given in this figure would provide target values, firstly, for the whole country and, thereafter, for different kinds of premises. For this purpose, an $f(N)$ distribution curve such as Figure 4.5, produced by Rasbash, should be drawn to represent a risk no greater than the risk prevalent at the present time. The lines AA, BB and CC in this figure represent the total target risk respectively for populations of 10^6, 10^7 and 5.6×10^7, the last line being equal to the population of the UK at that time. There is a need to update $f(N)$ curves such as those in Figures 4.4 and 4.5 using more recent data on the frequency distribution of fire deaths which are available for many countries.

Multiple-fatality fires generally occur in buildings other than dwellings to which a figure such as Figure 4.5 is applicable. The figure provides target values for the acceptability of the total risk for different population sizes. It is necessary to determine how this total risk should be distributed amongst different types of non-dwellings at risk. For this evaluation, it is necessary to have information on

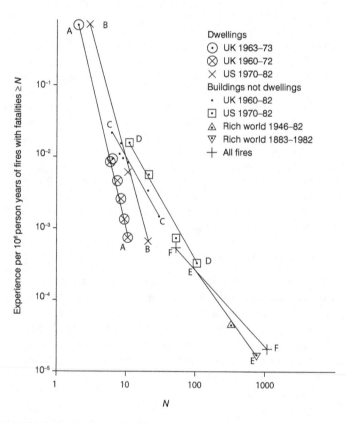

Figure 4.4 Multiple fatality fires in buildings

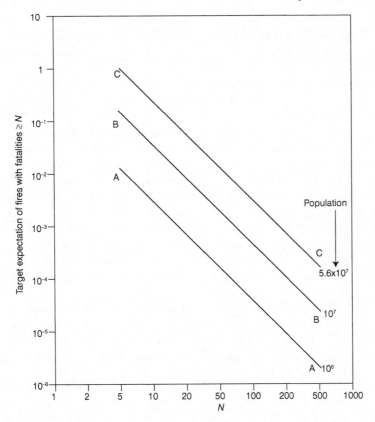

Figure 4.5 Proposed target for acceptibility from multiple fatality fires in all buildings other than dwellings for different populations

the distribution, numbers and size of various types of such buildings. Information of this kind is scarce.

Rasbash (1984) attempted to tackle the problem mentioned above using information on the number of people that will be in the buildings and obtained a preliminary assessment of a number of premises of different kinds and different sizes. He corrected these figures by assuming that the number of fires occurring in a given occupancy would be approximately proportional to the square root of the number of people that would be at risk. He then obtained a relationship between mean values of target probability and size of building, expressed in terms of number of people at risk.

The relationship mentioned above suggested that, for small buildings with less than 15 people, the target probability of having a fire which could kill more than 5 people was about 5×10^{-7} per annum. This probability (per annum) was about 1×10^{-6} for a building with 15–100 people, 2×10^{-6} for a building with 100–500 people and 4×10^{-6} for a building with more than 500 people. Similar calculations produced target probabilities for a fire which could kill more than 15, 100 or 500 people for buildings of the four population sizes. Zero probabilities were assigned for cases such as more than 15 fatalities in a building with less than 15 people. In general, the probabilities (per annum) ranged from about 10^{-6} for premises where 5 or more people may be killed in a fire to 10^{-7} to 10^{-8} for premises where 100 or more may be killed.

The levels of societal risk are generally expressed on an F-n curve similar to that in Figure 4.1.

When judging probabilistic risk assessments against the above criteria, the predicted level of risk should generally be significantly below the criterion. If the predicted level of risk only just satisfies the criterion then care should be taken to ensure that the assumptions made in the study clearly err on the side of safety.

Some regulators have attempted to use the 'value of a life' to assess whether to invest in safety. Different methods have been developed for assessing the monetary value of a life (Ramachandran 1998). The value used can be taken from damages awarded in courts (based on future earnings) and a notional sum for pain, grief and suffering. This results in values in the order of £500,000 to £1million per life. This approach does not take account of the preferences of the people at risk nor the value of non-income earners.

Alternatively, the behaviour of individuals in the market place and/or social survey questionnaires can be used as a value of life through an empirical economic evaluation. This approach usually leads to a value that is approximately an order of magnitude higher than that driven by court compensation payments.

Absolute criteria do not necessarily take account of the actual estimate or recorded level of the risk and so they are an expression of perception. However, for that reason they are important as they allow decision makers to take account of people's perceptions of risk in a highly effective but largely qualitative way.

Absolute criteria in fire engineering set specific levels of risk so that risks can be assessed as negligible, tolerable or intolerable. The main advantage of absolute criteria is that they provide a benchmark for judging acceptability. The main disadvantage is that there is little or no link between the current standards in the codes and the absolute criteria.

4.3 Financial objectives

4.3.1 Introduction

Chapter 1 discussed corporate governance and in conjunction with this, an organisation or facility can decide, given its investments, competitive position, insurance cover, contingency plans etc., that it can tolerate certain levels of loss or interruption with certain return periods. These are usually expressed in terms of a financial loss per year or level of financial loss and a frequency (or return period).

Fires cause fatal and non-fatal injuries to occupants of buildings and inflict direct material damage to the buildings and their contents. Some fires may also cause indirect/consequential losses, such as loss of production, profits, employment and exports although, at the national level, these losses do not contribute significantly to total fire loss. This is due to the fact that the loss of a specific unit of productive capacity may be spread among the remaining capacity in the nation. On average per year, fires occurring in the United Kingdom cause direct material damage amounting to about £1200 million and indirect loss amounting to about £120 million. The direct and indirect losses in the UK represent about 0.21 per cent of the Gross Domestic Product.

Target levels have to be set for acceptable limits to life and property damage caused by fires. Target levels for life risk were discussed in the previous section. This section is concerned with target levels for property damage and financial losses to owners of particular industrial and commercial properties. Most of these losses can be claimed from fire insurance companies if the properties are adequately insured for direct and consequential losses. However, there is a need to set an acceptable limit to property damage, since the damage caused by a large fire can seriously disrupt or even bankrupt the industrial or commercial activity of a property owner.

4.3.2 Target level for property damage

A property owner can set an acceptable limit to property damage in terms of physical extent of fire spread. For example, they may set an acceptable upper limit for the probability of a fire spreading beyond the room of origin. This limit may be, say, 0.05 such that, at most, only 5 per cent of the fires occurring in the property may spread beyond the room of origin. This limit would depend on the type of room, e.g. office, storage room, production area in an industrial building, assembly or customer area in a department store. Life safety can also be considered in the determination of this limit.

To achieve the target probability of fire spreading beyond the room of origin, a property owner may adopt adequate passive and active fire safety measures. Some of these measures may be in addition to those required under fire regulations. For example, the fire resistance periods of the structural boundaries of compartments may be 90 minutes, although only 60 minutes resistance is required under the fire regulations. Also, the sizes of compartments may be limited to the maximum size permissible for an unsprinklered compartment but sprinklers installed in these compartments act as additional protection. Fire regulations permit larger sizes for sprinklered compartments.

The property owner can also set an acceptable upper limit to property damage in terms of area damage. The chance (probability) of area damage in an actual fire exceeding this upper limit should be a small value. Based on the probability distribution of area damage, the acceptable upper limit for area damage can be determined with due regard to consequences in terms of damage to life and property. This criterion has been adopted in determining a design fire size for smoke ventilation systems. For covered sprinklered shopping complexes, for example, $10m^2$ has been suggested as an acceptable design fire – see Morgan and Chandler (1981). This corresponds to a heat output of 5MW or 0.5MW per m^2. The chance of damage in sprinklered public areas of retail premises exceeding $10m^2$ was estimated to be 0–4 per cent. In these areas, in the absence of sprinklers,

14–17 per cent of fires were estimated to be larger than $10m^2$. In non-public areas of retail premises, the design fire size of $10m^2$ was exceeded in about 12 per cent of all fires in sprinklered premises and in about 19 per cent in unsprinklered premises. Carrying out a similar analysis based on the probability distribution of area damage, Bengtson and Ramachandran (1995) determined design fire sizes for underground facilities. Determination of design fire size is discussed in more detail in Chapter 6.

Determination of an acceptable limit for area damage would also depend on the density per m^2 of financial value at risk in the building structure and its contents. The financial loss in a fire can be assumed to be the product of the value density per m^2 and area damage in m^2.

Depending on assets and other economic factors, a property owner may be able to allocate in their annual budget, a certain amount of money to meet financial losses in small fires which will not disrupt their business activity. The property owner may be able to accept self-insurance for such small losses less than a 'deductible' level. But they should purchase insurance cover for losses exceeding this level. Losses exceeding a higher threshold level might disrupt their business activity permanently or temporarily for a long time. To determine the lower and higher threshold levels mentioned above for direct material damage to the building and contents, the property owner should take account of consequential losses and obtain insurance cover accordingly. Losses between the two levels might cause only a minor disruption of the business activity, with small consequential losses.

To determine the amounts which should be allocated annually for self-insurance and insurance, the property owner should carry out a risk assessment and estimate the annual frequency (number) of fires likely to occur in their property for each of the three classes demarcated by the lower and higher threshold levels mentioned above. The risk assessment would also enable the property owner to determine fire prevention and protection measures which will reduce the risk. This assessment would provide inputs to a cost-benefit analysis of fire safety measures and self-insurance and insurance requirements. Based on such an analysis, the property owner can determine an economically optimum package of fire protection and insurance strategies, taking into account tax allowances, savings in insurance premiums and other benefits for fire protection measures included in the package. This economic problem was discussed in detail by Ramachandran (1998).

4.3.3 Target levels for consequential loss

A property owner should take into account the consequential losses such as loss of production or profits etc. when determining the level of additional safety measures above those required under fire regulations. Apart from some general factors such as repairing fire-hit premises, using more power and materials and hiring additional people, consequential losses may be affected by special factors concerned with a particular industrial or commercial activity. Specialised equipment such as that controlled electronically or by computer and tailor-made driers or centrifuges used in the manufacture of pharmaceuticals, if damaged by fire, cannot be replaced easily and quickly. Loss of laboratory facilities may seriously interrupt testing and quality control programmes. In some industries, even minor fire damage at a critical point can cause a long period of interruption. Factors affecting consequential losses in different industries were discussed in detail by Ramachandran (1998).

For the reasons mentioned above, it is necessary to identify parts or areas of an industrial or commercial property which have potentially high risks of consequential losses. For each of such parts or areas of a property, target values should be specified for damage and associated probability, depending on the magnitude of consequential loss. To achieve these targets, adequate fire prevention and protection measures should be determined, taking into account the costs and benefits.

Target levels for consequential losses can be determined by applying utility/disutility theory techniques (Ramachandran 1998) which take into account factors such as the assets of a property owner and their attitude to fire risk, i.e. extent of risk-aversion in addition to fire damage. The disutility function of the property owner can be constructed by considering different levels of direct property damage. As discussed in the previous section, a range for these levels can be identified by determining the minimum damage, a loss less than which will cause no disruption to the business activity and the maximum, a loss exceeding which might cause serious disruption.

4.3.4 Target levels for structural failure

During the 1970s, civil engineers started adopting a limit state approach based on reliability methods, instead of a safety factor approach, and target safety probabilities for designing structures. For example, a range of 10^{-6} to 10^{-14} per annum was suggested for annual failure probabilities for reinforced concrete and steel buildings. Such results for buildings with 10 to 1000 people at risk were based on consequences in terms of multiple fatalities. These structural safety concepts were extended to fire safety design during the 1980s to propose hypothetical target probabilities for failure of primary structural members during a fire. Based on a lifetime of 50 years for a building, a target failure probability of the order of 10^{-8} per annum or less was estimated for a situation in a building where a collapse might bring about the deaths of hundreds of people. See Table 4.3.

Collapse of a structure during a fire occurs very rarely. A structural element (wall, floor or ceiling) of a compartment (or room) is more likely to experience thermal failure and allow a fire to spread to an adjacent compartment or area. If such a failure occurs with probability P_c, one or more deaths might occur with a probability P_d. Consistent with the target probability set for P_d, a target value may be determined for P_c the probability of compartment failure.

As discussed earlier, with a change in the notation, P_c is the product of P_A and P_B where P_A is the probability of flashover or serious fire and P_B the 'conditional

Table 4.3 Societal risk criteria for structural collapse due to fire

	General acceptable risk for 10 or more deaths (deaths per building year)	General acceptable risk for 100 or more deaths (deaths per building year)
Deaths per building year	5×10^{-7}	5×10^{-8}
Deaths per occupant year	1×10^{-8}	1×10^{-9}

probability' of compartment failure given flashover, i.e. $P_C = P_A \times P_B$. Depending on P_A and the target set for P_C, a target value may be assigned for P_B and the fire resistance required to meet this target determined accordingly.

In the CIB report (1986) 'Design Guide: Structural Fire Safety', illustrative examples are given for P_B with $P_C = 10^{-6}$ per annum for different values of P_A and for different compartment sizes. P_A is the product of annual probability of fire occurrence and the probability of fire developing into a large size.

Using the financial damage techniques and data in this book, it is possible to estimate the risk of damage that may result from a fire. This information may then be used to estimate potential monetary losses and enable cost-benefit analysis to be undertaken to establish the relative value of installing additional or alternative fire protection measures.

These financial criteria in terms of levels of loss or interruption should be set in conjunction with the organisation concerned and/or their financiers or insurer.

4.3.5 Cost-benefit analysis

For financial risks and acceptance criteria such as ALARP, cost-benefit analysis can be crucial in assessing the acceptability of risks. Various methods can be used in cost-benefit analysis of fire protection measures (Ramachandran 1998). What follows is a case study of a bus garage to illustrate the use of cost-benefit analysis in fire risk analysis (Charters 1998).

The first step was to determine the total costs of the sprinkler installation. This not only included the initial installation costs, but also covered the annual running costs. The following list, whilst not meant to be exhaustive, covers the main costs included in this case:

- design fees;
- installation/construction;
- commissioning/training of staff;
- maintenance/running etc.

The capital cost for the sprinkler system was £25,000 with an annual maintenance cost of £100. The benefits of the new installation were then listed, including:

- reduced property loss;
- reduced consequential losses;
- reduced insurance premiums;
- improved life safety etc.

The benefit rates from the quantified fire risk assessment were added to the difference in insurance premium to give the total benefit rate of £2,500 per garage year.

This is the figure used in the investment appraisal. Table 4.4 shows the discounted cash flow over a 30-year period. The discount factor used is 10 per cent. This is the norm for commercial premises and is spread over a 30-year life span (the normal life of the sprinkler system). The financial figures in Table 4.4 do not represent those of any particular garage or operator, but may be typical of some circumstances.

The cost benefit analysis showed a small positive net present value at the end of 30 years. The positive figure indicated that, strictly speaking, the installation of bus garage sprinkler systems did not represent a good investment. However, the smallness of the value indicated that this was a marginal case. In the light of the risk assessment, the bus operator decided that they had sufficient redundancy and diversity of bus supply through ownership (in several garages), leasing and buying and insurance not to require bus garage sprinklers. However, the risk assessment had highlighted several other areas, such as fire safety management and the separation of the IT centre that were much more cost-effective and these were implemented.

This study to assess the benefits of installing sprinkler systems in bus garages indicated that there were business continuity and property protection benefits to the operator. However, the cost-benefit analysis and the operator's contingency plans meant that there was no cost-benefit or consequence case for installing sprinklers the bus garage. As a result of the risk assessment, the operator did implement other forms of safeguard and fire precaution.

4.4 Other objectives

Other fire safety objectives can include the protection of heritage and the protection of the environment. Fire risk acceptance criteria may be provided by relevant guidance either explicitly or implicitly. Often there are no absolute criteria and so these can only be agreed by consensus between all the relevant regulators and stakeholders.

Table 4.4 Discounted cash flow for bus garage sprinkler system

Year	Capital cost (£)	Annual cost (£/yr)	Total cost (£/yr)	Savings (£/yr)	Net costs/ savings (£/yr)	Discount factor (10%)	NPV of costs/ savings (£)	Cumulative NPV
0	25000		25000	0	25000	1	25000	25000
1		100	100	-2500	-2400	0.9091	-2182	22818
2		100	100	-2500	-2400	0.8265	-1983	20835
3		100	100	-2500	-2400	0.7513	-1803	19032
4		100	100	-2500	-2400	0.6830	-1639	17392
5		100	100	-2500	-2400	0.6209	-1490	15902
...								
26		100	100	-2500	-2400	0.0839	-201	3014
27		100	100	-2500	-2400	0.0763	-183	2831
28		100	100	-2500	-2400	0.0693	-166	2664
29		100	100	-2500	-2400	0.0630	-151	2513
30		100	100	-2500	-2400	0.0573	-138	2375
Total			28000	-75000	-47000		2375	

References

Bengtson, S and Ramachandran, G (1995), Design fires in underground facilities, *Book of Abstracts, First European Symposium on Fire Safety Science*, ETH, Zurich.

Charters, D A (1998), Fire safety at any price?, *Fire Prevention* 313, October, 12–15.

CIB W14 (1986), Design guide: structural fire safety, *Fire Safety Journal*, 10, 2, 77–137.

Fryer, L S and Griffiths, R F (1979), United Kingdom Atomic Energy Authority, Safety and Reliability Directorate, Report SRD R149, UKAEA, Harwell.

Health and Safety Executive (1988), *The Tolerability of Risk from Nuclear Power Stations*, HMSO, London

Health and Safety Executive (1989), *Quantified Risk Assessment: Its Input into Decision Making*, HMSO, London

Morgan, H P and Chandler, S E (1981), Fire sizes and sprinkler effectiveness in shopping complexes and retail premises, *Fire Surveyor*, 10, 5, 23–28.

Ramachandran, G (1998), *The Economics of Fire Protection*, E & F N Spon, London.

Rasbash, D J (1984), Criteria for acceptability for use with quantitative approaches to fire safety, *Fire Safety Journal*, 8, 141–158.

Rasmussen, N C (1975), *An Assessment of Accident Risks in US Commercial Power Plants*, WASH 1400, US Nuclear Regulatory Commission, Washington, DC.

5 Initiation

5.1 Frequency of ignition/probability of fire starting

5.1.1 Global estimation

During a short period, say a year, fires do not occur frequently in a particular building. Hence, it is necessary to consider a group or type of buildings with similar fire risks to evaluate the annual frequency of ignition or probability of fire starting. This frequency or probability, for a given type of building, will increase with the number of ignition sources and, hence, with the size of the building.

As discussed in Section 3.2.1, the annual frequency of ignition or probability of fire starting would increase approximately with the size of a building according to a 'power' relationship. The building size may be expressed in terms of total floor area, A, in square metres or total financial value, V, at risk in the building and contents. This relationship is described in Equation (3.40) where F(A) is the annual frequency of ignition or probability of fire starting in a building of total floor area, A. The parameter α in this equation denoting the power is generally less than unity (one) for the reasons discussed in Section 3.2.1 – see also Table 3.4 with $\alpha=b$ according to equation 3.35.

The values of the parameters K and α in Equation (3.40) are 'global' estimates applicable to a given type of building. The values of these two parameters for buildings in the United Kingdom given in Table 3.4 were estimated more than 20 years ago. Hence, these parameters need to be re-evaluated by carrying out again surveys of buildings at risk and analysing relevant statistical data relating to fires which occurred during recent years. This exercise should be carried out by a national organisation for different types of buildings and, if possible, for major groups or fire risk categories within each type.

The probability of an accidental (not arson) fire starting in a building depends on the presence or absence of causes or sources which can be classified into two broad groups – human and non-human. The first group consists mainly of children playing with fire, e.g. matches, careless disposal of matches and smokers' materials and misuse of electrical and other appliances. The second group includes defects in, or faulty connections to, appliances using electricity, gas and other fuels. The appliances may be further classified according to cooking, space heating, central heating and other uses. This group also includes causes such as mechanical heat or sparks in industrial buildings, natural occurrences and spontaneous combustion. Some materials in a building, e.g. latex foam and finely powdered rubber, could be ignited even by a low energy smouldering sources.

The nature and number of ignition sources and materials will vary from one part of a building to another. In an industrial building, for example, three major parts can be identified – production, storage and other areas. Given that a building is involved in a fire, the conditional probabilities reflecting the relative or comparative risks due to various causes in different parts of the building can be estimated from group statistics such as those in Table 5.1. In this example, if a fire occurs, the conditional probability of fire starting due to, say, smoking materials in the stores/stock room is 0.0129 (= 15/1162).

5.1.2 Particular buildings

The conditional probabilities of fire starting, based on figures such as those in Table 5.1, would pertain to an average or reference building in the type or risk category considered. For a particular building in any type or risk category, an estimate of the conditional probability (given fire) for the ith cause in the jth part of the building is given by

$$I_{ij} P_{ij} \qquad\qquad\qquad (5.1)$$

where P_{ij} is the probability for this cause and part revealed by figures such as those in Table 5.1. The parameter I_{ij} will be assigned the value zero if the ith cause is totally absent in the jth part of the building considered for risk evaluation. If the cause is present, I_{ij} should be given a positive value depending on the extent to which this cause can be responsible for starting a fire in the jth part; this value can be greater than unity. A value equal to unity can be assigned if the building is similar to the 'average building' in this respect.

Taking smokers' materials as an example, it should be possible to determine, for the building type considered, a global quantitative measure, S_{ij}, denoting the exposure of, say, a storage area at risk from fire due to such a cause. This measure may be the total number of cigarettes etc consumed by all smokers per day, per m² of floor area. A similar quantitative measure, s_{ij}, should be evaluated with respect to the consumption of smoking materials in the storage area of the particular building subjected to risk evaluation. Then the ratio s_{ij}/S_{ij} is an estimated value of the parameter I_{ij} for the particular building considered. This ratio should be adjusted to take account of factors such as smoking lobbies and publicity measures, e.g. notices, circular letters making people aware of the risk of fire due to smoking materials. The assignment of a value to the parameter I_{ij} has to be somewhat subjective, with its accuracy depending on the extent and accuracy of relevant information used in the calculations.

Table 5.1 Spinning and doubling industry – places of origin of fires and sources of ignition

Sources of ignition	Production and maintenance		Assembly	Storage areas		Other areas	Miscellaneous areas	Total
	Dust extractor (not cyclone)	Other areas		Store/stock room	Loading bay, packing dept			
A Industrial appliances								
Dust extractor (electrical)	14	3	–	–	–	–	–	17
Dust extractor (other fuels)	12	–	–	–	–	–	–	12
Other appliances (electrical)	6	111	–	–	–	–	–	117
Other appliance (other fuels)	–	22	–	1	–	–	2	25
B Welding and cutting equipment	–	10	–	6	–	–	7	23
C Motor (not part of other appliances)	–	7	–	–	–	–	–	7
D Wire and cable	1	12	–	–	–	–	2	15
E Mechanical heat or sparks (electrical)	27	194	–	–	–	–	–	221
Mechanical heat or sparks (other)	52	387	–	2	–	–	–	441
F Malicious or intentional ignition	–	9	–	3	–	–	3	15
Doubtful	–	13	–	7	–	–	–	20
G Smoking materials	2	29	1	15	1	–	7	55
H Children with fire e.g. matches	3	4	–	12	2	4	5	30
J Others	4	29	2	3	2	–	12	52
K Unknown	11	78	–	14	–	–	9	112
Total	132	908	3	63	5	4	47	1162

Each possible cause or source of ignition in each part of the building considered should be identified and its I_{ij} value estimated. The aggregate probability of fire starting for the building is then

$$F(A) \sum_i \sum_j I_{ij} P_{ij} \qquad (5.2)$$

where $F(A)$ is the global probability of fire starting in a building of total floor area A (in m^2) estimated by Equation (3.40). The value given by the double summation part, excluding $F(A)$ in Equation (5.2), can be greater or less than unity depending on the extent to which the various causes are present or absent in the building. It will be equal to unity only if the building considered is more or less identical to the average characteristics of the underlying population of buildings in regard to causes or ignition sources.

The method described above was proposed by Ramachandran (1979/80, 1988). It is similar to the allocation approach used in fire risk assessments of nuclear power plants – see Apostolakis (1982).

5.1.3 Special factors affecting frequency of fire occurrence

5.1.3.1 Types and total amounts of fuel used

By comparing the number of fires in buildings attributed to various types of fuel, with the total amount of fuel used, it is possible to estimate correlations and predict trends. The number of fires caused by a type of fuel over a period of years may be plotted against the number of units of this fuel used during that period. This procedure followed by Chandler (1968) gave an approximately straight line correlation for electrical fires occurring in the UK between the years 1956–1966. When this trend was extrapolated, it was estimated that 25,500 fires would occur in 1970 associated with an output of (210×10^9) kWh of electricity. The number of fires which actually did occur in 1970 was in good agreement with the estimated number.

For gas, the number of fires during 1957–1966 did not vary linearly with the amount of town gas sold. In fact, the trend showed there was a reduction in the fire frequency per 10^9 therms of gas sold. However, extrapolating on the trend did predict 7,000 fires in 1970 and the number which occurred was 7,100. Fires due to solid fuel showed a reduction in number because of a reduction in the amounts of solid fuel sold. With oil, the fire frequency per million tons of oil delivered dropped through the period 1955–1966. This was apparently due to the advent of central heating which is much safer than portable oil heating.

5.1.3.2 Socio-economic factors

Various studies carried out in the United States demonstrated that fire incidence (with its consequences in terms of deaths and injuries) is related, though this does not necessarily imply causality, to a combination of factors – see, for example,

Bertrand and McKenzie (1976), Munson and Oates (1977) and Gunther (1975). These factors reflected poor and substandard housing, overcrowding, social class, race, lack of family stability and proportions of the young or elderly in the population.

In the United Kingdom, a detailed analysis by Chandler (1979) in relation to fires in London, revealed strong correlations between fire incidence and housing and social factors. The housing factors included tenancy (owner occupied, private rented etc) and the lack of amenities. People who own their homes might be expected to be more careful than those who live in rented properties. The social indicator most strongly correlated with fire incidence was proportion of children in care, which was thought to reflect family instability. Fire frequency appeared to be independent of rateable value of a property per person and the age distribution of the population. However, the incidence of casualties was generally highest among the young and elderly. Updated results of London analysis were included in a later study by Chandler, Chapman and Hollington (1984) with reference to Birmingham and Newcastle-upon-Tyne.

A summary of the results obtained in the UK studies mentioned above, along with other human aspects of fires in buildings, is contained in a paper by Ramachandran (1985). As pointed out in this paper, stress could be a primary factor responsible for carelessness leading to increasing fire incidence. Stress could result from socio-economic factors such as poor housing conditions and low income, both of which are interrelated. Personal circumstances force people into sharing facilities or overcrowded households. High fire incidence may be a function of an unquantifiable lack of social identity and stability.

5.1.3.3 Weather conditions

Severe weather conditions during winter can cause an increase in number of fires and fire casualties. Chandler (1982), for example, analysed data for fires in dwellings in the UK during the severe winter of 1978–1979. He found that in all regions, temperature and vapour pressure were significantly correlated with total fires and fires due to space heating, electric blankets and wire and cable, including leads. The same was true of life risk fires due to space heating. Fires due to cooking appliances were not generally influenced by severe weather conditions.

Chandler's analysis suggested that in the temperature range 0°C down to –5°C there were an extra 30 fires per week in England and Wales for every degree drop in temperature. This result was in general agreement with the assessment based on fire frequencies for the 1962–1963 winter – see Gaunt and Aitken (1964). Fires involving casualties, rescues or escapes were dependent on temperature and even more so those fires due to space heating.

5.2 Probable rate of fire growth

5.2.1 Introduction

Successful prediction of the course of a fire provides an indication of the size of the fire at a given time, the rate of fire growth, the time available for escape or suppression, the type of suppressive action that will be effective and other attributes that define fire risk in a particular type of building with known materials and ignition sources. The ability to predict the course of the fire would enable one to predict the effect of changes in the initial conditions, design of the building and passive and active fire protection systems, materials and ignition sources. Thus, a fire safety engineer would be able to select the best combination of design features, materials and fire protection devices, providing desirable or acceptable levels of life safety and property protection compatible with economic, amenity and aesthetic requirements.

Hence, a central problem in the design of a building for fire safety and provision of fire protection measures is to predict the development of a fire in the building as a function of time and estimate the rate at which the fire grows in the room of origin and subsequently spreads to other parts of a building. This rate depends initially on the heat output from the material or object first ignited, apart from other factors such as fire load, room dimensions and ventilation. The subsequent rate of fire growth depends on the heat transfer (spread) from the object first ignited and heat output and heat transfer properties of other objects within and outside the room of fire origin.

To estimate the rate of fire growth as a function of time and space, several deterministic mathematical models, based on scientific theories relating to heat output and other physical quantities, have been developed and validated in the light of experimental data. These models are mainly of three types – zone, field and simulation models. Computer software packages have also been developed for these models.

Several factors cause uncertainties in the patterns of development of actual (not experimental) fires in a building and hence, in the rate of fire growth. These uncertainties can be evaluated to some extent by using the computer software package of a deterministic model and performing simulations for different fire scenarios. Based on such simulations, it might be possible to calculate the average and other statistical parameters providing inputs for estimating, in probabilistic terms, the rate of growth of an actual fire occurring in a building with several objects. Such an exercise would, however, be time-consuming and expensive.

A considerably cheaper and more realistic method for estimating fire growth rate in probabilistic terms is provided by a non-deterministic statistical model based on data relating to real fires compiled by the fire brigades. One such model is the exponential model discussed in Section 3.3.2. This model, as argued in Section 3.3.2, is more realistic than the T^2 curve based mostly on data provided by fire tests. The exponential model is applicable to the period of fire growth after the onset of 'established burning', when more and more objects in a room are involved, whereas the T^2 model is really applicable to the initial stage of fire growth when only one object is involved. The exponential model directly takes into account the uncertainties governing the development of a real fire in a building. The rate of fire growth estimated by this model is applicable to a wider (general) range of fire conditions than a restricted range to which a T^2 curve or a deterministic model is applicable.

This section is concerned with the application of the exponential model for estimating the average and other statistical parameters of the rate of fire growth within and beyond the room of origin. The scenarios discussed for any occupancy type are those generated by different areas of fire origin and intervention by fire brigade and sprinklers. A statistical method is described for estimating the probabilistic upper confidence limit for the fire growth rate for predicting the worst case scenario. Some examples are given for illustrating the application and use of the model in fire safety engineering problems.

5.2.2 Stages of fire growth

A fire in a room or compartment usually starts (first stage) from a very small ignition source and, as the floor area affected gradually increases, it becomes sizeable but still small. After some time, established burning occurs with sustained growth and steady burning (second stage) during which the rate of increase of fire size suddenly changes and the phase of rapid growth occurs. The fire will continue to grow in the second stage until it reaches a maximum intensity (flashover) when most of the objects (fuel) in the room are completely involved. During the fully developed post-flashover period (third stage), the fire can penetrate the structural barriers of the room of origin and ignite objects in the adjacent rooms. The fourth stage is the 'decaying period' when the rate of fire growth decreases and the fire burns out.

The pattern of fire growth described above generally applies to 'free burning' fires not involving any fire fighting. Some of these fires may undergo a decaying process during any of the stages mentioned above due to the consumption of all available fuel or other reasons. This process, leading to an eventual burnout of a fire, is referred to as self-termination or self-extinguishment in fire science literature. Fire incidence data contain information on some of these small fires which were reported to the fire brigade but were out on or before the arrival of the brigade. Some small fires not reported to the fire brigade are extinguished by sprinklers, if installed or by first-aid means such as portable fire extinguishers, buckets of water or sand and smothering. The growth of a developing fire can be retarded, controlled or terminated at some stage by the successful operation of sprinklers, if installed. Fire brigades extinguish all the fires which develop beyond the 'infant mortality' stage and are not put out by sprinklers, portable fire extinguishers or other first-aid means.

5.2.3 The exponential model

The exponential model is applicable to the second and third stages of fire development described above. This is the period when a fire grows steadily after the commencement of established burning. During this period, heat output from a fire increases exponentially with time – see Thomas (1974) and Friedman (1978). This hypothesis is supported by experimental results – see Labes (1966).

The area damaged in a fire is approximately proportional to heat output. Hence, Ramachandran (1980) proposed the exponential model in Equation (3.42) for making the best use of statistical information on real fires attended and compiled by the fire brigades in the United Kingdom. This equation is reproduced below in order to facilitate the discussion in this section:

$$A(T) = A(0) \exp(\theta\, T) \tag{5.3}$$

where $A(T)$ is the total floor area (m^2) damaged in T (minutes) counted from the commencement of established burning.

The parameter $A(0)$ is the floor area initially ignited at time $T = 0$ when established burning commences. The parameter θ measures the increase in the value of logarithm of area damage, to base e, for a unit increase in the value of T. Thus, θ quantifies the rate of fire growth per minute.

Consider any type of building and area of fire origin such as the production or storage area of an industrial building and the customer area of a department store. For any such fire risk classification, an average value of $A(0)$ can be estimated by carrying out a fire load survey. An average value of $A(0)$ can also be estimated for a particular building by considering the objects in the area considered and the floor area occupied by these objects. Fire tests may be able to provide an estimate for the average value of $A(0)$.

As discussed in Section 3.3.2, an average value of $A(0)$ can also be estimated by analysing area damage, $A(T)$, and corresponding duration of burning, T, for a sample of fires which occurred in the risk category considered. By fitting a straight line to logarithm of $A(T)$, base e, and T, the values of logarithm of $A(0)$, base e, and θ can be estimated graphically or by applying the least squares method. This analysis would provide the average or expected value for logarithm of $A(0)$ and hence for $A(0)$. It would also provide an estimate for the average value of θ.

The value of θ estimated by a linear regression analysis as described above is an average value for the entire duration of burning. An average value of θ can be estimated for each of the component periods constituting the total duration of burning. This was explained in Section 3.3.2 with reference to four periods involving detection or discovery of a fire, calling of fire brigade, fire brigade arrival at the scene of the fire and the time taken by the brigade to bring the fire under control. To estimate the value of θ separately for the four periods mentioned above, the model in Equation (3.42) or Equation (5.3) was expanded, by expanding the term θT in the exponential to

Table 5.2 Fire growth parameters

Industry	Production			Storage			Other		
	$A(O)\,(m^2)$	θ_A	θ_B	$A(O)\,(m^2)$	θ_A	θ_B	$A(O)\,(m^2)$	θ_A	θ_B
Food, drink, tobacco	0.504	0.020	0.013	0.694	0.017	0.049	0.327	0.042	0.026
Chemicals and allied	0.225	0.038	0.033	0.628	0.048	0.035	0.218	0.027	0.044
Metal Manufacture	0.341	0.033	0.026	1.160	0.017	0.045	0.425	0.032	0.041
Mechanical, instrument and electrical engineering	0.248	0.038	0.038	0.619	0.018	0.072	0.225	0.042	0.045
Textiles	0.304	0.047	0.029	1.793	0.037	0.037	0.215	0.032	0.053
Clothing, footwear, leather and fur	0.723	0.038	0.064	1.346	0.025	0.039	0.315	0.028	0.075
Timber, furniture etc	0.485	0.046	0.046	0.949	0.037	0.052	0.566	0.030	0.037
Paper, printing and publishing	0.213	0.044	0.052	0.985	0.027	0.044	0.235	0.023	0.060

$$\theta_1 T_1 + \theta_2 T_2 + \theta_3 T_3 + \theta_4 T_4 \tag{5.4}$$

With the logarithm of $A(T)$ as the dependant variable, a multiple regression analysis was performed to estimate the values of θ_1, θ_2, θ_3 and θ_4.

In another study, Ramachandran (1986) added the first three time periods mentioned above to denote the period t_A until the arrival of the fire brigade at the scene of the fire: $t_A = T_1 + T_2 + T_3$. The fourth period, T_4, relating to the arrival time to the time when the fire was fought and brought under control by the brigade was redefined as t_B. The fire growth parameters for the above two periods may be denoted by θ_A and θ_B. Estimates for these two parameters are reproduced in Table 5.2 for some industrial buildings and three areas of fire origin – production, storage and other areas. The table also contains estimates for $A(0)$.

Fire fighting by the brigade can be expected to reduce the rate of fire growth. This hypothesis is supported by figures in Table 5.2 for the production areas of some industries; in these cases θ_B is less than θ_A. But for the storage and other areas of most of the industries, θ_B is greater than θ_A. These results do not indicate that the fire brigades are ineffective in controlling the rate of fire growth. They indicate, perhaps, the fact that a fire would be growing fast when the fire brigade arrives at the scene of the fire. The fire size at that time can be expected to be bigger than the average size before the arrival of the brigade. The results in Table 5.2 do not take into account the size of the fire at the time of brigade arrival, data for which were not available. Fires would normally continue to grow faster and faster if they are not extinguished by the fire brigade.

The results in Table 5.2 also do not take into account the interaction between the time periods t_A and t_B. If t_A is reduced due to early discovery or detection of a fire and/or due to quick attendance by the fire brigade, the fire will be in its early stage of growth when fire fighting by brigade commences and hence can be controlled quickly. A reduction in t_A would, therefore, shorten the control time, t_B, as well thus reducing the total duration of burning, T, and area damage. If installed, sprinklers also would reduce t_A and t_B. Sprinklers have a high probability of extinguishing a fire without the need for fire brigade intervention. If a sprinkler system operates but does not extinguish a fire, it can control the spread of the fire and reduce its rate of growth.

A fire requiring fire brigade intervention would be a fully developed fire with growth rate increasing as a function of time before the arrival of the brigade and for some time after its arrival. Likewise, a fire in which the sprinkler system is activated would have survived the 'infant mortality' stage and commenced established burning. The growth rate would be increasing in such fires as well. Fires which are not extinguished by sprinklers would develop into bigger fires requiring fire brigade action.

For the reasons discussed in the above three paragraphs, the conflicting effects of several factors confound the application of the exponential model expanded on the basis of an equation such as Equation (5.4). Such confounding effects produce interactions between independent variables in a multiple regression analysis and lead to the statistical problem known as 'multicollinearity.' Taking into account this problem, the additive model in Equation (5.4) needs to be modified but this is not possible, since fire brigade statistics do not contain the data required for this purpose. A modified model can be developed and applied by combining fire brigade data with data provided by experiments and simulations based on deterministic models.

In another application of the exponential model of fire growth, Ramachandran (1992) attempted to construct a general picture of fire development by using fire statistics and evaluating fire growth rates for the following four main scenarios:

1   fires which are extinguished (or self-terminated) by first-aid methods without any intervention by sprinklers or fire brigades;
2   fires extinguished by sprinklers;
3   fires in buildings with sprinklers extinguished by fire brigades;
4   fires in buildings without sprinklers extinguished by fire brigades.

In addition, for each of the four categories mentioned above, the fire growth parameter θ was estimated separately for each of the following three extent of fire spread classifications:

1   confined to item first ignited;
2   spread beyond item but confined to room of origin;
3   spread beyond room of origin.

For the above investigation, only summary (not raw) data for fires during 1984–1986 were available for analysis. These tables contained, for each scenario, only means and standard deviations for logarithm of area damage and time variables. Hence, it was not possible to estimate the parameter $A(0)$ in Equation (5.3) which was assumed as one square metre. Under this assumption, the mean value of θ was estimated by the ratio:

$$\bar{\theta} = \bar{y}/\bar{T} \tag{5.5}$$

where $\bar{y}$ is the mean value of logarithm (base e) of area damage, $A(T)$ and $\bar{T}$ is the mean value of total duration of burning, T. Equation (5.5) follows from Equation (5.3) with $A(0) = 1$.

The investigation described above was carried out for four industrial groups of buildings, retail distributive trade, wholesale distributive trade and office buildings. As an example, results for the industrial group Paper, Printing and Publishing are given in Table 5.3 to show the variation in the expected (average) value of the fire growth parameter θ for different scenarios. The corresponding 'doubling times' (minutes) are shown within brackets. As defined in Equation (3.43) the doubling time is given by $(0.6931) / \theta$.

For each fire-fighting scenario and for each area of fire origin, the figures in Table 5.3 provide some indication of the increase in the rate of fire growth as fire spreads within the room of origin. The growth rate for a fire spreading beyond the room of origin and the overall growth rate for the building are affected by the fire resistance of the structural barriers of the room. The frequencies of such fires were small and even nil in small fires which did not require the intervention of the fire brigade or sprinklers and in growing fires extinguished by sprinklers.

In storage and other areas equipped with sprinklers, the fire brigades had been effective in reducing the probability of fire spreading beyond the room of

Table 5.3 Expected values of the fire growth rate θ for different scenarios – paper, printing and publishing industries, UK

Area of fire origin	Extent of fire spread	Fire-fighting scenario			
		No sprinklers No fire brigade	No sprinklers Fire brigade	Sprinklers No fire brigade	Sprinklers Fire brigade
Production	(a)	0.0385 (18.0)	0.0138 (50.2)	0.1523 (4.6)	0.0235 (29.5)
	(b)	0.0668 (10.4)	0.0392 (17.7)	0.1080 (6.4)	0.0576 (12.0)
	(c)	– (–)	0.0459 (15.1)	– (–)	0.0376 (18.4)
	(d)	0.0440 (15.8)	0.0307 (22.6)	0.1330 (5.2)	0.0431 (16.1)
Storage	(a)	0.0347 (20.0)	0.0235 (29.5)	– (–)	0.0369 (18.8)
	(b)	0.0397 (17.4)	0.0471 (14.7)	– (–)	0.0276 (25.1)
	(c)	– (–)	0.0378 (18.3)	– (–)	– (–)
	(d)	0.0366 (18.9)	0.0401 (17.3)	– (–)	0.0297 (23.4)
Other areas	(a)	0.0128 (54.2)	0.0091 (76.2)	0.0349 (19.9)	– (–)
	(b)	0.0536 (12.9)	0.0316 (21.9)	0.0509 (13.6)	0.0397 (17.5)
	(c)	– (–)	0.0381 (18.2)	– (–)	– (–)
	(d)	0.0240 (28.9)	0.0301 (23.0)	0.0437 (15.9)	0.0392 (17.7)

Notes:
– no or few fires (less than 5)
(a) Confined to item first ignited
(b) Spread beyond item but confined to room of origin
(c) Spread beyond room of origin
(d) Building (overall)
The figures within brackets are 'doubling times' in minutes.

origin. Structural fire resistance and Fire Service fire fighting, acting together, appear to have reduced the fire growth rate in the production areas equipped with sprinklers. In most of the scenarios in which the fires were confined to the room of fire origin, whether the buildings were equipped with sprinklers or not, fire brigade action had reduced the growth rate in fires requiring the intervention of the brigade.

The figures in Table 5.3 provide some support to the fact that, in buildings with sprinklers, the rate of fire growth will be increasing before and for some time after sprinklers are operated. Sprinklers have to deal with fires growing in size.

5.2.4 Maximum rate of fire growth

Every fire individually provides an estimate of the growth rate as given by the ratio

$$\theta = Y / T \tag{5.6}$$

where Y is given by

$$Y = y - \log_e A(0) \tag{5.7}$$

and y is the logarithm (base e) of area damage, $A(T)$, $A(0)$ the area initially ignited and T the total duration of burning of the fire. In the model discussed in the previous section, $A(0)$ is one square metre such that $Y = y$. Under this assumption, the ratio in Equation (5.6) reduces to

$$\theta = y / T \tag{5.8}$$

In a population of fires, the growth rate θ has an expected or average value given by the 'ratio estimate'

$$\bar{\theta} = \bar{Y} / \bar{T} = (\bar{y} - \log_e A(0)) / \bar{T} \tag{5.9}$$

where $\bar{Y}$ and $\bar{T}$ are the mean values of Y and T and $\bar{y}$ the mean value of $\log_e A(T)$. With $A(0) = 1$, Equation (5.9) reduces to Equation (5.5). According to statistical theory (Frishman, 1975), the standard deviation of θ, as defined in Equations (5.6) and (5.8) is the square root of

$$\sigma_\theta^2 = (\sigma_Y^2 \cdot \bar{T}^2 - \sigma_T^2 \cdot \bar{Y}^2) / \bar{T}^2 (\sigma_T^2 + \bar{T}^2) \tag{5.10}$$

where σ_Y^2 is the variance of Y and σ_T^2 the variance of T. The variance of Y is the same as the variance, σ_y^2, of y since $A(0)$ is a constant.

Equation (5.10) reduces to

$$\sigma_\theta^2 = \frac{\sigma_Y^2(1-\rho^2)}{\sigma_T^2 + \overline{T}^2} = \frac{\sigma^2}{\sigma_T^2 + \overline{T}^2} \qquad (5.11)$$

where ρ is the coefficient of correlation between Y and T and σ^2 is the 'residual' variance. The standard deviation of the residual error, σ, can be estimated by performing the regression analysis based on the equation

$$\log_e A(T) = \log_e A(0) + \theta\,T \qquad (5.12)$$

which follows from Equation (5.3). The residual error is the difference between the logarithm of the observed value of area damage in a fire and the corresponding value for that fire, with duration of burning, T, predicted or estimated by Equation (5.12) according to the estimated value of θ.

As discussed in the previous section with reference to the results in Table 5.3, due to the non-availability of raw data, it was not possible to carry out a regression analysis to estimate the standard deviation, σ, of the residual error. Hence, the formula for σ_θ^2 given in Equation (5.10) was used to estimate the maximum rate of fire growth providing the worst case scenario. The maximum rate is estimated by the upper confidence limit, u_θ given by

$$u_\theta = \overline{\theta} + t\,\sigma_\theta \qquad (5.13)$$

where $\overline{\theta}$, as defined in Equations (5.5) and (5.9), is the average value of θ for individual fires defined in Equation (5.6). If θ is assumed to have a normal distribution in a population of fires, the random variable t has a standard normal distribution. Under this assumption, the probability of growth rate in a real fire exceeding the maximum rate u_θ given by Equation (5.13) is 0.025 if $t = 1.96$, 0.01 if $t = 2.33$ and 0.005 if $t = 2.58$. For any desired probability level for the maximum growth rate, the corresponding value of t can be obtained from a table of the standard Normal distribution.

The figures given in the example in Table 5.3 are the expected values, $\overline{\theta}$, of the growth rates, θ, for individual fires. With $t = 1.96$ in Equation (5.13), the corresponding estimates for maximum values of individual growth rates are given in Table 5.4 together with corresponding estimates for doubling times. As mentioned above, for any scenario described in this table, the probability of growth rate in a real fire exceeding the maximum rate given in the table is 0.025.

5.2.5 Individual growth rate and average growth rate

As mentioned at the beginning of the previous section, every fire provides an estimate of the growth rate according to Equation (5.6). This individual rate, θ, varies randomly in a population of fires and fluctuates around the expected or average value, $\overline{\theta}$, estimated by Equation (5.9). The fluctuation of θ is according to the standard deviation σ_θ estimated by Equation (5.10).

Table 5.4 Maximum values of the fire growth rate θ for different scenarios – paper, printing and publishing industries, UK

Area of fire origin	Extent of fire spread	Fire–fighting scenario			
		No sprinklers No fire brigade	*No sprinklers Fire brigade*	*Sprinklers No fire brigade*	*Sprinklers Fire brigade*
Production	(a)	0.2045 (3.4)	0.0523 (13.3)	0.4213 (1.6)	0.0637 (10.9)
	(b)	0.1777 (3.9)	0.0646 (10.7)	0.1845 (3.8)	0.0868 (8.0)
	(c)	– (–)	0.0758 (9.2)	– (–)	0.0613 (11.3)
	(d)	0.1995 (3.5)	0.0792 (8.8)	0.3250 (2.1)	0.0848 (8.2)
Storage	(a)	0.2216 (3.1)	0.0597 (11.6)	– (–)	0.2116 (3.3)
	(b)	0.0888 (7.8)	0.0748 (9.3)	– (–)	0.1060 (6.5)
	(c)	– (–)	0.0698 (9.9)	– (–)	– (–)
	(d)	0.1659 (4.2)	0.0721 (9.6)	– (–)	0.0897 (7.7)
Other areas	(a)	0.05 (13.9)	0.0299 (23.2)	0.1049 (6.6)	– (–)
	(b)	0.1330 (5.2)	0.0681 (10.2)	0.0898 (7.7)	0.0418 (16.6)
	(c)	– (–)	0.0761 (9.1)	– (–)	– (–)
	(d)	0.0853 (8.1)	0.0542 (12.8)	0.0979 (7.1)	0.0671 (10.3)

Notes:
–   no or few fires (less than 5)
(a) Confined to item first ignited
(b) Spread beyond item but confined to room of origin
(c) Spread beyond room of origin
(d) Building (overall)
The figures within brackets are minimum values of doubling times in minutes corresponding to maximum values of the fire growth rates.

Each sample of fires provides an estimate of the average rate $\bar{\theta}$. In samples of fires, $\bar{\theta}$ can be expected to vary randomly according to its standard deviation $\sigma_{\bar{\theta}}$ given by

$$\sigma_{\bar{\theta}} = \sigma / \sqrt{n}. \; \sigma_T \tag{5.14}$$

where σ is the standard deviation of the 'residual' error mentioned with reference to Equation (5.11) and n the number of observations (fires) in the sample analysed. The parameter σ_T is the standard deviation of the total duration of burning, T.

If $\bar{\theta}$ is assumed to have a normal distribution, the maximum value of $\bar{\theta}$ in repeated samples is given by

$$u_{\bar{\theta}} = \bar{\theta} + t \, \sigma_{\bar{\theta}} \tag{5.15}$$

where the variable t has a standard Normal distribution. In repeated samples, the probability of average growth rate, $\bar{\theta}$, exceeding $u_{\bar{\theta}}$ given by Equation (5.15) is 0.025 if $t = 1.96$, 0.01 if $t = 2.33$ and 0.005 if $t = 2.58$. It may be observed that, if only one sample is available for analysis, the expected value of the average rate $\bar{\theta}$ is the same as the expected value, $\bar{\theta}$, of individual rate, θ.

The distinction between the maximum values of individual and average growth rates can be explained with the aid of the following simple example. Consider the following sample of n ($= 5$) observations (x): 4, 9, 12, 15, 20. Calculations show that the mean ($\bar{x}$) and standard deviation σ_x of x are 12 and 5.4. The standard deviation $\sigma_{\bar{x}}$ of the mean $\bar{x}$ is equal to

$$\sigma_x / \sqrt{n} = 5.4 / \sqrt{5} = 2.41$$

Hence, assuming a normal distribution, the maximum value of the individual observation x is

$$\begin{aligned} u_x &= \bar{x} + t \sigma_x \\ &= 12 + 1.96 \times 5.4 = 22.58 \end{aligned}$$
if $t = 1.96$.

The maximum value of the average value $\bar{x}$ is

$$\begin{aligned} u_{\bar{x}} &= \bar{x} + t \, \sigma_{\bar{x}} \\ &= 12 + 1.96 \times 2.41 = 16.72 \end{aligned}$$
if $t = 1.96$.

The distinction between the maximum values of individual and average growth rates was discussed by Bengtson and Ramachandran (1994) in an investigation concerned with fire growth rates in underground facilities such as railway properties, public car parks, road tunnels and subways and power stations. The

Table 5.5 Growth rate and doubling time

Occupancy type	Number of fires	Area initially ignited A(0) (sq metres)	Average growth rate in all fires (β)			Growth rate in an individual fire (θ)		
			Expected value	Standard deviation	Maximum rate	Expected value	Standard deviation	Maximum rate
Railway properties								
Fires in all places	776	1.0002	0.0376 (18.4)	0.0021	0.0417 (16.6)	0.0376 (18.4)	0.0352	0.1066 (6.5)
Fires in public places	214	0.8	0.0454 (15.3)	0.0029	0.0511 (13.6)	0.0454 (15.3)	0.029	0.1022 (6.8)
Fires in basement	66	0.82	0.0273 (25.4)	0.0039	0.0349 (19.9)	0.0273 (25.4)	0.0226	0.0716 (9.7)
Public car parks								
Fires in all places	692	1.02	0.0362 (19.1)	0.0025	0.0411 (16.9)	0.0362 (19.1)	0.0318	0.0985 (7.0)
Fires in basement	165	1.26	0.0366 (18.9)	0.0058	0.0480 (14.4)	0.0366 (18.9)	0.0327	0.1007 (6.9)
Road tunnels and subways	107	1.00	0.0220 (31.5)	0.0024	0.0267 (26.0)	0.0220 (31.5)	0.0176	0.0565 (12.3)
Power stations	115	0.93	0.0208 (33.3)	0.0029	0.0265 (26.2)	0.0208 (33.3)	0.021	0.0620 (11.2)

The figures within brackets are corresponding doubling times in minutes.

results obtained in this investigation are reproduced in Table 5.5, where the average growth rate is denoted by β.

As one would expect from Equations (5.10) and (5.14), the standard deviation σ_θ of individual growth rate is higher than the standard deviation $\sigma_{\bar{\theta}}$ of average growth rate. Individual growth rates have wider fluctuations around the expected value than average growth rates. The maximum value of individual growth rate represents more realistically the worst case scenario.

5.2.6 Applications

As discussed in Section 3.3.2 and previous sections of this chapter, the exponential model provides a tool for estimating the area likely to be damaged in a fire as a function of total duration of burning until the fire is extinguished by first-aid means, sprinklers or fire brigades. The damage can be reduced by early discovery of the fire by occupants of the building or by early detection by automatic detectors. Early discovery or detection should be followed by quick action to call the fire brigade and, if possible, fight the fire with the aid of first-aid means such as portable fire extinguishers, buckets of water or sand and smothering.

The model can also provide some statistical basis for estimating the reduction in area damage and the consequential reduction in life risk due to quick response and early attendance by the fire brigade to arrive at the fire scene and commence fire fighting. The model may be useful in evaluating the economic value of bringing the fire under control quickly by improving the fire-fighting tactics adopted by the fire brigades. For the reasons mentioned above, the exponential model can provide inputs to fire brigade problems concerned with the determination of appropriate fire cover for a geographical area in terms of number, location and size (manpower and equipment) of fire stations.

In Section 3.3.2 and previous sections of this chapter, the variable $A(T)$ was used to represent the area damaged by direct burning. This information can be used to estimate the rate of growth of a real fire in terms of heat output – see Ramachandran (1995). The rate, (dL/dT), at which fire load (L) in a compartment is destroyed in a fire can be expected to be equivalent to $\dot{m}$ (kg/sec), the rate at which fuel mass is consumed. The fire load contained in A square metres of floor area is approximately $A\,\bar{L}$ where $\bar{L}$ (kg/m²) is the fire load density.

It follows, therefore, that

$$\dot{m} = (dL/dT) = \bar{L} \cdot (dA/dT) \tag{5.16}$$

where (dA/dT) can be estimated by Equation (5.3) with T in seconds:

$$\frac{dA}{dT} = A(0).\theta.\exp(\theta T) \tag{5.17}$$

Hence,

$$\dot{m} = \dot{m}o \exp(\theta T) \tag{5.18}$$

where

$$\dot{m}o = \overline{L} \, A \, (0) \, \theta \tag{5.19}$$

is the loss rate of fuel mass at the initial time corresponding to the commencement of established burning. Also, as shown in Equation (3.58), the rate of heat output $\dot{Q}$, is directly proportional to $\dot{m}$.

In deriving the above equations it has been assumed that $\dot{m}$ and $\dot{Q}$ increase exponentially with duration of burning since the commencement of established burning. In the next section, an application of these equations is discussed with reference to the time of occurrence of flashover.

Based on the rate of growth of fire in terms of heat output as discussed above, the rate of growth of smoke can be estimated by ascertaining the correlations between the two rates. Quantity of smoke produced is correlated with heat output. Smoke can be expected to grow exponentially with time faster than heat with a growth parameter two or more times the parameter θ for heat development. Rate of growth of smoke can also be estimated directly, to some extent, using data on total area damage instead of area damage by direct burning. Fire incident reports compiled by the fire brigades in the UK also contain information on total area damage but this includes water damage in addition to smoke damage.

Butcher (1987) attempted to establish the relationship between fire area and heat output. He used the results of a series of large-scale fire tests staged at the Fire Research Station, UK in 1966, in which a selection of fire loadings and two levels of window opening were considered. The size of the fire compartment was 85.5m². Time and temperature information for these tests was available from which Butcher derived the time–temperature curve for a compartment with the largest fire load density of 60kg/m². The value of heat output estimated from this curve was combined with the progressive area increases obtained by using the results on exponential fire growth produced by Ramachandran (1986). The heat output thus obtained for each fire area, at the appropriate time, was integrated to provide a value for the total heat output for the growing and spreading fire for any time value in the fire's history.

Based on the above analysis, Butcher showed that, for the example considered by him, it would take 22 minutes for a spreading and growing fire to reach a heat output of 5MW. He questioned the validity of using, for heat output, a constant value of 5MW over an area of 10m² or 0.5MW per m² for designing smoke ventilation systems. Since there can be an appreciable delay in a fire reaching the full 5MW heat output, calculating smoke temperature from the 5MW value can give a false picture of smoke movement for the early period of fire when the occupants of a building are attempting to escape.

As pointed out by Butcher, the concept of a constant design fire size such as 5MW over 10m² area or 0.5MW per m² for designing smoke ventilation systems is somewhat misleading. The final fire size attained is less relevant if escape of occupants takes place during the early stages of a fire. The design fire size need not have a single value. It should be a function of time in order to consider the interaction between smoke movement and the movement of escaping occupants. Such a functional relationship between fire size and time would also provide a tool for determining the optimum time for the operation of detectors and sprinklers to facilitate safe evacuation. There are also uncertainties involved in the operating times of detectors, sprinklers and smoke ventilation systems. Hence, there is a need to determine a probability–time-based design fire size for these fire protection systems. This problem is investigated in the next chapter.

The fire growth rate derived from statistics of real fires, as discussed in this section, involves a number of materials or objects and structural elements of a building. The expected and maximum values of this growth rate can, therefore, provide tools for testing whether the growth rates for different scenarios generated by a deterministic model satisfactorily reflect real fire situations. These tools can be improved by combining the statistically determined fire growth rates with those estimated by experiments involving physical quantities such as heat output. For this purpose, it would be necessary to estimate first a composite growth rate for a room or a building based on experimental results which are generally for individual materials. This might be a complex exercise. Statistical and experimental growth rates can be merged by applying Bayesian statistical technique – see Ramachandran (1998).

Fire growth rates can be estimated for different materials ignited first in order to describe the early stage in the development of a fire before, say, the arrival of the fire brigade at the fire scene – see Ramachandran (1986). Fire spread from material to material in a room would depend on room dimensions, ventilation and environmental conditions apart from the arrangement (overcrowding etc.) of the materials. Taking into account these and other factors, the statistical growth rate for any scenario, such as those discussed in this section, should be modified or adjusted.

5.3 Probability of flashover

5.3.1 Introduction

Generally, the structural elements of a compartment would only be affected if a fire grows into a fully developed stage. This stage, defined as 'flashover', is reached when the atmosphere temperature at the ceiling exceeds 520°C and the heat output attains a certain high level. This level depends on factors such as total area of the compartment walls, floor and ceiling, area of window (opening), height of window and heat transfer (loss) coefficient of the wall. Formulae based on scientific theories are available for estimating the heat output required for the occurrence of flashover in a compartment – see, for example, Walton and Thomas (1988).

The time taken by a fire to reach the flashover stage depends on the rate at which the fire grows. This rate depends on whether the compartment is protected by sprinklers or not, apart from the factors mentioned above and the combustible nature of materials or objects in the compartment. The rate also depends on whether the compartment has or does not have an installed ventilation system. Bengtson and Laufke (1979) have suggested a model for estimating the time to flashover for different room volumes and for the two cases, with and without an installed ventilation system.

The heat output produced in a compartment at the time of flashover would have destroyed a certain floor area of the compartment. By evaluating this area, the time to flashover since the onset of established burning can be estimated by applying the exponential model in Equation (5.3) together with Equation (3.58) and Equations (5.16) to (5.19).

The models discussed above (including the exponential model) are essentially deterministic in nature. In a real fire, area damage, duration of burning and heat output or fire severity are all random variables due to uncertainties caused by several factors. Hence, an element of probability governs the occurrence of flashover and the time of its occurrence. To estimate this probability, Ramachandran has developed three models which are discussed briefly below.

5.3.2　*Exponential model of fire growth*

The first model (Ramachandran, 1995) provides a mechanism for coupling deterministic rate for increase in heat output in a compartment with exponential fire growth rate based on area damage in real fires. Applying this model to the deterministic rate of heat output required for causing flashover, the time of occurrence of flashover and the total floor area damaged at that time are predicted. The probability distribution of area damage is then used to provide an estimate or probability of flashover. Essential features of this model are as follows.

As discussed earlier and expressed in Equation (3.58), the rate of heat output, $\dot{Q}_f$ (in kw), at the time of flashover, is directly proportional to the rate of consumption of fuel mass, $\dot{m}_f$ (in kg/sec) at the time of flashover:

$$\dot{Q}_f = \dot{m}_f \, \Delta H \tag{5.20}$$

where ΔH is the effective heat of combustion of the fuel, usually assumed to have the value 18,800 kilojoules per kilogram (kJ/kg). Hence, applying the exponential model it is reasonable to assume that

$$\dot{Q}_f = \dot{Q}_o \exp (\theta_h \, T_f) \tag{5.21}$$

where:

$\dot{Q}_f$　=　rate of heat output at the time of flashover
$\dot{Q}_o$　=　rate of heat output at initial time, T_o
θ_h　=　increase per second in the value of logarithm (base e) of rate of heat output $\dot{Q}$
T_f　=　time (in seconds) taken for the occurrence of flashover since the commencement of established burning at the initial time, T_o.

As in Equation (5.20),

$$\dot{Q}_o = \dot{m}_o \, \Delta H \qquad (5.22)$$

where, following Equation (5.19),

$$\dot{m}_o = \bar{L} \, A(0) \cdot \theta_a \qquad (5.23)$$

The parameter $\dot{m}_o$ is the loss rate of fuel mass at the initial time, $\bar{L}$ (kg/m^2) is the fire load density and $A(0)$ (in m^2) is the area initially ignited at the commencement of established burning. The fire growth parameter θ_a is the increase per second in the value of logarithm (base e) of area damage. The value of θ_a can be estimated by $(\theta/60)$ where θ is the value of the fire growth parameter estimated by Equation (5.3) in units of minutes.

The value of θ_h in Equation (5.21) may be equated to θ_a in Equation (5.23) i.e. $\theta_h = \theta_a$, assuming that heat output grows exponentially in time at the same rate as for area damage. The worst case scenario may also be considered by using the maximum value of θ_a, estimation of which has been discussed in Section 5.2.4.

For any compartment, the rate of heat output Q_f, required for the occurrence of flashover with a temperature of 520°C can be estimated by the following formula suggested by Walton and Thomas (1988):

$$Q_f = 610 \, (h_K \cdot A_T \, A_o \cdot \sqrt{h_o} \,)^{\frac{1}{2}} \qquad (5.24)$$

where

h_K	=	wall heat loss coefficient [(kW/m)/K]
A_T	=	total area (m^2) of the compartment excluding area of window opening
A_o	=	area of window opening (m^2)
h_o	=	height of window opening (m).

If the fire load density of the compartment is $\bar{L}$, the rate of heat output at initial time, $\dot{Q}_o$, can be estimated with the aid of Equations (5.22) and (5.23) and an assumed or estimated mean value for $A(0)$, the area initially ignited at the commencement of established burning.

With values for Q_f, $\dot{Q}_o$ and θ_h estimated as discussed above, the time, T_f, taken for the occurrence of flashover can be estimated with the aid of Equation (5.21). The probability of occurrence of flashover can then be estimated by considering the probability attached to the time, T_f for the occurrence of flashover and the corresponding (cumulative) area, A_f, destroyed. With $T = T_f$ in seconds, A_f is given by Equation (5.3) with $\theta = \theta_a$ calculated for time units in seconds. The probability of damage exceeding A_f can then be estimated by evaluating the parameters of the probability distribution of area damage for the compartment and occupancy type considered. This probability is equivalent to the probability, P_f, of flashover. According to analyses of data for actual fires attended, particularly by fire brigades in the UK, area damage has a log normal or Pareto probability distribution.

Consider, as an example, an office room of width 3.6m, depth 6.1m and height 3m. It has a self-closing fire-resistant door which need not be considered as an opening. The room has a window (opening) of height 1.5m and width 1.2m. The wall-lining material is 0.016m gypsum plaster on metal lath. The wall heat loss coefficient, h_k, may be assumed to have the value 0.03 [(kW/m)/K] for the period after the occurrence of established burning. With $h_k = 0.03$, $A_T = 100.32m^2$, $A_o = 1.8m^2$ and $h_o = 1.5m$, the heat output rate, $\dot{Q}_f$, required for the occurrence of flashover is 1570kW according to Equation (5.24). Using Equation (5.20), the loss rate of fuel mass at the time of flashover, $\dot{m}_f$, is 0.0835 kg/sec.

For an office room, the fire load density, $\bar{L}$, is 23kg/m². For an unsprinklered office room, according to an analysis carried out by Ramachandran (1992), the fire growth parameter θ, defined in Equation (5.3) in terms of minutes, has an average (expected) value of 0.04 and maximum value of 0.06. Hence, in terms of seconds, the fire growth parameter θ_a, defined in Equation (5.23), has the average value of 0.0007 and maximum value of 0.001. The values mentioned above are applicable to the period after the occurrence of established burning. It is assumed that established burning would occur when fire load on one square metre

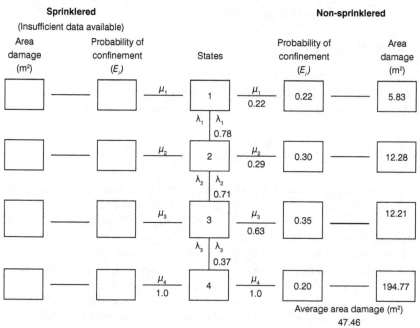

States

1   Confinement to items first ignited
2   Spread beyond items first ignited, but confinement to contents of room of fire origin
3   Spread beyond items first ignited, but confinement to room of fire origin with involvement of structure
4   Spread beyond room of fire origin

Figure 5.1 Probability tree for rooms and offices

is consumed or destroyed by fire such that $A(0) = 1$. A fire is unlikely to reach the flashover stage if established burning does not occur. About 22 per cent of fires in office rooms do not spread beyond the item first ignited – see Figure 5.1.

Consider the worst case scenario in an actual fire in an office room with the maximum value of 0.001 (per second) for θ_a. With $\bar{L} = 23\text{kg/m}^2$, $A(0) = 1$ and $\theta_a = 0.001$, $\dot{m}_o = 0.023\text{kg/s}$ according to Equation (5.23). Then, from Equation (5.22), $\dot{Q}_f = 432\text{kW}$. Hence, with $\theta_h = \theta_a$, from Equation (5.21),

$$Q_f = 432 \exp (0.001\ T_f) \qquad (5.25)$$

Therefore, from Equation (5.26), with $\dot{Q}_f = 1570$ kW, T_f may be estimated to have the value 1290 seconds or 21.5 minutes. Hence, if a fire in an office room reaches the established burning stage with a minimum damage of one square metre, it will take a further 21.5 minutes to reach the flashover stage.

When flashover occurs, from Equation (5.3), with $\theta = 0.06$, $A(0) = 1$ and $T = 21.5$, an area of 3.63m^2 would have been affected by direct burning (heat). According to Ramachandran (1993b), area damage (d) in a fire in an office room has a Pareto probability distribution:

$$Q(d) = c\ d^{-\lambda} \qquad (5.26)$$

where $Q(d)$ is the probability of area damage exceeding d. For an unsprinklered office room, $C = 0.84\text{m}^2$ and $\lambda = 0.67$ such that the probability of damage exceeding 3.63m^2 or probability of flashover is 0.35.

Area damage in a sprinklered office room has a Pareto probability distribution of the form in Equation (5.26) with $C = 0.28\text{m}^2$ and $\lambda = 0.70$. Hence, for a sprinklered office room, the probability of damage exceeding 3.63m^2 or probability of flashover is 0.11.

5.3.3 *Event probability tree model*

The event tree model was discussed in detail in Section 3.2.1. It was also discussed in Section 3.4.3 with reference to the stochastic state transition model (STM) – see Figure 3.18. Figure 5.1 is an event tree relating to an office room. The probabilities E_i ($i = 1, 2, 3, 4$), λ_i ($i = 1, 2, 3$) and μ_i ($i = 1, 2, 3, 4$) in this figure were defined in Section 3.4.3. The values of these probabilities are based on an analysis of fires attended by fire brigades – see Ramachandran (1992).

The parameters λ_i and μ_i are limiting values to which these probabilities for the ith stage would tend over a period of time. They are constants and are not functions of time since the start of the fire. Figures for area damage are averages for the four states (the figure 12.21m^2 is the combined average for the second and third states).

The product ($\lambda_1 . \lambda_2$) estimates the probability of a fire spreading beyond the item first ignited to other items in the room but remaining confined to the room. This probability (product) may be considered as the probability of flashover in the statistical sense. Under this assumption, it may be seen from Figure 5.1 that the probability of flashover for an office room without sprinklers is 0.55 (= 0.78 × 0.71). The probability λ_3 (= 0.37) is the conditional probability of thermal failure of the structural boundaries of the room given that flashover has occurred. The product $\lambda_1 . \lambda_2 . \lambda_3$ (= 0.20), as indicated in the figure for the parameter E_4 for the fourth state, is the probability of fire spreading beyond the room of origin but confined to the building of origin. Fires spreading beyond the building of origin were not considered in this model.

The probability of flashover provided by an event tree model, such as Figure 5.1, is an average or expected value for all office rooms. It applies to an 'average' or 'reference' office room with average dimensions, fire load, ventilation conditions and physical characteristics, discussed in Section 5.3.2. Some research, data collection and analysis are needed to adjust the global probability of flashover given by the product $(\lambda_1 . \lambda_2)$ for a particular room in an occupancy type, say, an office room with known dimensions and other characteristics mentioned above. This problem will not arise if the method proposed in Section 5.3.2 for a particular room is applied together with the probability distribution of area damage.

Based on event trees such as Figure 5.1, the probabilities of flashover have been estimated for a few industrial buildings and areas of fire origin (Ramachandran, 1992) and areas of non-industrial buildings (Ramachandran, 1999). The results obtained in these two reports are reproduced in Table 5.6. The parameter K in the last column is the ratio between the probabilities of flashover in unsprinklered and sprinklered rooms. This ratio can be used to determine the reduction in fire resistance that can be allowed for a sprinklered room – see Section 12.2.1.5.

5.3.4 Probability distribution of area damage

For any type of building and room of fire origin, the global probability of flashover can also be estimated directly by evaluating the probability distribution of area damage. As mentioned in the previous section, the global probability would only be applicable to a room with average characteristics relating to heat loss coefficient, room dimensions and ventilation factor. It is not applicable to any particular room.

In this global approach, the necessary first step is to determine, for any building type and room of fire origin, a design area damage, A_f, affected by direct burning (heat) which is likely to cause flashover. Flashover has been defined in fire science literature as the beginning of the fully developed stage of a fire when the atmosphere temperature at the ceiling exceeds 500°C. As discussed in Section 5.3.2, a certain rate of heat output, $\dot{Q}_f$, is required to cause flashover.

Suppose, for example, for a range of room sizes and other characteristics of an office room, the design value for $\dot{Q}_f$ has been estimated to be 2200kW. The corresponding design value for the rate of consumption of fuel mass, $\dot{m}_f$, is 0.117 kg/s since, from Equation (5.20), $\dot{m}_f = \dot{Q}_f/\Delta H$ where ΔH = 18,800 kJ/kg is the effective heat of combustion of the fuel. The rate $\dot{m}_f$ is applicable to a short interval of time around the time T_f of flashover. For an exponential model

Table 5.6 Probability of flashover

Occupancy type, area of fire origin	Sprinklered building			Unsprinklered building			Parameter K*
	λ_1	λ_2	Prob. of flashover	λ_1	λ_2	Prob. of flashover	
Textile industry							
Production	0.28	0.36	0.10	0.57	0.44	0.25	2.50
Storage	–	–	0.28	0.81	0.78	0.63	2.25
Other areas	0.34	0.35	0.12	0.58	0.57	0.33	2.75
Chemical etc industry							
Production	0.20	0.20	0.04	0.61	0.41	0.25	6.25
Storage	0.37	0.41	0.15	0.81	0.67	0.54	3.60
Other areas	0.29	0.31	0.09	0.61	0.61	0.37	4.11
Paper etc industry							
Production	0.23	0.26	0.06	0.49	0.39	0.19	3.17
Storage	0.29	0.34	0.10	0.78	0.72	0.56	5.60
Other areas	0.28	0.36	0.10	0.73	0.71	0.52	5.20
Timber etc industry							
Production	0.36	0.39	0.14	0.80	0.69	0.55	3.93
Storage	0.38	0.42	0.16	0.84	0.83	0.70	4.38
Other areas	0.30	0.40	0.12	0.76	0.76	0.58	4.83
Retail trade							
Assembly	0.28	0.32	0.09	0.67	0.60	0.40	4.44
Storage	0.24	0.25	0.06	0.82	0.76	0.62	10.33
Other areas	0.27	0.41	0.11	0.71	0.68	0.48	4.36
Wholesale trade							
All areas	0.26	0.42	0.11	0.85	0.75	0.64	5.82
Office premises							
Rooms used as offices	–	–	–	0.78	0.71	0.55	–
Other areas	0.21	0.43	0.09	0.65	0.65	0.42	4.67
All areas	0.42	0.60	0.25	0.63	0.84	0.53	2.12
Hotels							
All areas	0.15	0.47	0.07	0.70	0.71	0.50	7.14
Pubs, clubs, restaurants							
All areas	0.41	0.63	0.26	0.74	0.84	0.62	2.38
Hospitals							
Other areas	0.13	0.92	0.12	0.46	0.65	0.30	2.50
Flats							
Other areas	0.10	0.40	0.04	0.47	0.79	0.37	9.25

Probability of flashover $= \lambda_1 \cdot \lambda_2$
K* = Ratio between the probabilities of flashover in unsprinklered and sprinklered rooms

$$\dot{m}_f = m_f \cdot \theta_a \tag{5.27}$$

where m_f is the total mass equal to the total fire load (L_f) affected by heat at the time of flashover. Hence, $L_f = 117\text{kg}$ ($= 0.117/0.001$) since, as discussed in Section 5.3.2, the maximum value of θ_a for an office room is 0.001. Since $L_f = A_f$. $\overline{L}$ where $\overline{L} = 23\text{kg/m}^2$ is the fire load density for an office room, the total floor area, A_f, affected at the time of flashover is 5.1m^2.

According to Equation (5.3), with $\theta = 0.06$ in terms of minutes, area damage of 5.1m^2 and flashover would occur in 27.1 minutes since the commencement of established burning. This result also follows from Equation (5.25) with $\dot{Q}_f = 2200$ according to which $T_f = 1628$ seconds.

Equation (5.27) follows from the fact that the mass m contained in A m^2 is equal to A $\overline{L}$ and hence, at time T, the mass destroyed is given by

$$m(T) = \overline{L}\, A(T) = \overline{L}\, A(0) \exp(\theta T) \tag{5.28}$$

Therefore

$$\dot{m}(T) = \frac{dm}{dt} = \overline{L}\, A(0) \cdot \theta \cdot \exp(\theta T)$$

$$= m(T)\,\theta \tag{5.29}$$

Having determined the design area damage, A_f, for flashover, the next problem is to estimate the probability of damage exceeding A_f. Using Equation (5.26) for Pareto probability distribution, the probability of damage in an unsprinklerd office room exceeding 5.1m^2 may be calculated to be 0.28. This probability (0.28) may be considered to be the global probability of flashover for an office room without sprinklers. The global probability of 0.55 for flashover to occur in an office room without sprinklers, given in Table 5.6, has been estimated purely from fire statistics and has not been adjusted for room dimensions, ventilation, fire load and other physical characteristics.

In the example considered above, the rate of heat output of 2200kw at the time of flashover has been estimated to be over a floor area of 5.1m^2. This is equivalent to about 432 kW per m^2. This result has also been obtained in Section 5.3.2 where, for the office room considered, the rate of heat output of 1570 kw at the time of flashover was estimated to be over 3.63m^2. This statistical property is a feature of the exponential model according to which the heat output rate $\dot{Q}_f$ per unit area is a constant and is equal to $\dot{Q}_f/A(0)$. For the example (office room) considered, as expressed in Equation (5.25), $\dot{Q}_o = 432$ kw and $A(0) = 1$. It follows that the loss rate of fuel mass $\dot{m}$ per unit area is also a constant and equal to $\dot{m}_o/A(0) = 0.023\text{kg/s}$ for the example considered.

5.3.5 Hotel bedroom

Lawson and Quintiere (1986) performed an analysis of fire growth in the bedroom of a hotel. The authors considered a particular scenario in which a fire started from a cigarette lighter in the centre of a bed with polyurethane foam mattress. The room door was left one-quarter open as the occupant fled the room. For mass loss rate of fuel, the authors used the following exponential function for times up to 250 sec:

$$\dot{m} = \alpha \exp(\alpha t), \alpha = 0.03 \tag{5.30}$$

where $\dot{m}$ was expressed in g/sec and t in seconds. In kg/sec, the Equation (5.30) may be rewritten as

$$\dot{m} = 0.00003 \exp(0.03t) \tag{5.31}$$

Mass loss rates for times from 250 secs to 300 secs were calculated using a steady state burning formula for wood cribs.

It may be inferred from the above analysis that a steady state, or established burning, would occur in 250 secs when the fuel mass loss rate, $\dot{m}$, reaches a value of 0.05424 kg/sec according to Equation (5.31). This value is $\dot{m}$ in the exponential model discussed in this chapter. Using the value of 15.7 kJ/g or 15,700 kJ/kg used by the authors for the parameter ΔH for polyurethane foam mattress, from Equation (5.22), the value of $\dot{Q}_0$ at the time of commencement of established burning may be calculated to be 851.6kW. From Equation (5.29), the mass $\dot{m}_0$ involved at the time of established burning is 1.808 kg $(= \dot{m}_0/\theta)$. This mass would occupy an area of 0.11m^2 since the average fire load density, $\bar{L}$, for a hotel bedroom is 310MJ/m^2 or 17 kg/m^2 approximately, according to European data – see CIB W14 Design Guide for Structural Fire Safety (1986). Data are not available for the fire load density of hotel bedrooms in the USA.

According to the above analysis, the area affected, $A(0)$, at the time of commencement of established burning is 0.11m^2. This value may be applicable for the particular scenario considered. A higher value may be applicable for the average value of $A(0)$ over a range of scenarios for fires starting with the ignition of different objects in a hotel bedroom. A value of one square metre has been assumed for $A(0)$ in the exponential model discussed in this chapter. The corresponding value for $\dot{m}_0$ for an office room has been estimated as 0.023 kg/s. By carrying out a fire load survey it would be possible to estimate a more accurate value for $A(0)$ for any occupancy type and room or area of fire origin. This would give more accurate values for $\dot{m}_0 (= A(0) \cdot \bar{L})$ and $\dot{m}_0$.

For hotels in the UK, data on area damage and duration of burning are available for estimating the fire growth parameter θ in Equation (5.3) and hence for θ_a in Equation (5.23) or for θ_h in Equation (5.21). But this investigation has not yet been carried out. Area damage in UK hotels has a Pareto probability distribution. For unsprinklered areas of hotels, the parameter λ in Equation (5.26) has the value of 0.66 for assembly areas, 0.77 for bedrooms and 0.64 for storage and other areas – see Ramachandran (1993b). Respectively for these three areas, the parameter c in Equation (5.26) has the values of 0.71m^2, 0.54m^2 and 0.38m^2. For storage and other areas of hotels with sprinklers, the values of the parameters λ and c are 0.63 and 0.11 respectively.

5.3.6 Compartment size

If the size (floor area) of a compartment is increased, the total fire load in the compartment would also increase, which will increase the potential for a fire to reach a high level of severity. Hence, one can expect the probability of flashover to increase with an increase in compartment size. In Section 5.3.3, using the event tree in Figure 5.1 the probability of flashover in an unsprinklered office room was estimated to be 0.55. This global probability is applicable to an average room size of 50m² estimated from a sample of fires in office buildings. In Section 5.3.2, the probability of flashover in an unsprinklered office room of floor area 22m² was estimated to be 0.35. In estimating this probability, room dimensions including height, ventilation and other factors, were considered but these factors were not taken into account in the event tree method.

It can, however, be argued that the probability of flashover would decrease with increasing size of a room or compartment. Ramachandran (1990) provided statistical support for this hypothesis. The larger the floor area of a compartment, the longer it takes, generally, for a fire to involve a number of objects and produce sufficient heat to cause flashover. The extra time thus available would increase the chance of extinguishment of the fire by first-aid means or by the fire brigade. A larger room, generally, has a greater non-uniformity in the arrangement of objects and hence in that of fire load and lesser degree of overcrowding of objects. Probability of fire spread would decrease with increasing distances between objects.

In the particular unsprinklered office room of floor area 22m² considered in Section 5.3.2, the value of 0.35 estimated for the probability of flashover is associated with heat output produced by an area damage of 3.63m². In Section 5.3.4, the probability of flashover for an unsprinklered office room was estimated to be 0.28. This probability is for a design value of 2200kW for the heat output rate required for causing flashover. This design value is associated with an area damage of 5.1m². The flashover heat output rate of 2200kW and area damage of 5.1m² can be expected to apply to a compartment larger than 22m² with a flashover heat output of 1570kW and area damage of 3.63m².

As discussed above, the total floor area damaged, d_f, when flashover occurs can be expected to increase with increasing compartment size. The increase can, perhaps, be quantified approximately by the following 'power' function:

$$d_f = CA^\beta \tag{5.32}$$

where A is the compartment size. With $C = 1$ and $d_f = 3.63m^2$ for $A = 22m^2$, the value of β in Equation (5.32) can be estimated to be 0.42. With this value of β in Equation (5.32), a damage of 5.1m² at flashover would occur in a compartment of size 48m².

According to the statistical property of a probability distribution, the probability of damage in a fire exceeding any value specified for the damage, d, would decrease with increasing values for d. This would be apparent from

Equation (5.26) for Pareto distribution. It may, therefore, be inferred that the probability of flashover would decrease with increasing area damage for increasing compartment size. In a larger compartment, area damage at flashover would be higher but the probability of this damage level being exceeded or probability of flashover would be lower. This paradox can be explained by the fact that, due to uncertainties caused by several factors, area damage in a real fire is a random variable with a probability distribution. It is unrealistic to assume that area damage in a real fire can be predicted in exact terms satisfying deterministic formulae. The concept of probability distribution of area damage has been applied in the determination of design fire size for designing smoke ventilation systems – see Section 6.3.

5.3.7 Time before established burning

It may be desirable to consider the duration of burning since the start of the fire before the occurrence of established burning. This initial stage of fire growth, although small in size, can be very variable in length of time; it can last for hours or it can be over in minutes – see Butcher (1987) and Ramachandran (1988). Ignoring such uncertainties, one may consider the deterministic model such as in Equation (5.30) for the very early stage of fire growth. According to this equation, for the scenario considered, it would take 250 seconds or 4.2 minutes before established burning occurs.

Alternatively, one may consider the following T^2 curve:

$$\dot{Q} = \alpha T^2$$

$$(5.33)$$

For an office room with medium fire growth, $\alpha = 0.012$. Hence, on average it would take about 190 seconds or 3.2 minutes for a fire in the room considered in Section 5.3.2 with an initial heat output rate, $\dot{Q}_o$, of 432 kW.

5.4 Probable damage in a fire

5.4.1 Introduction

As discussed in Section 5.1, the probability of fire starting or frequency of fires occurring during a period is the first component of fire risk. The second component, which is the subject matter of this section, is the probable damage if and when a fire occurs. The damage can be measured in terms of any of the following four attributes:

1 Extent of spread
2 Floor area destroyed
3 Financial loss
4 Life loss – number of fatal and non-fatal casualties.

The probable damage in a fire in a building would depend on the levels of passive fire protection measures such as compartmentation, fire resistance of structural boundaries (walls, floor, ceiling) and means of escape facilities (number and widths of staircases, travel distance etc.) and on the presence or absence of active measures such as automatic fire detection and sprinkler systems. The probable damage is also affected by the successful operation and reliability of the fire protection measures when a fire occurs and their effectiveness in reducing the damage.

5.4.2 Extent of fire spread and floor area damaged

The probable damage in a fire can be estimated by considering different categories of fire spread and the probabilities and average or expected floor area damage associated with these categories. The probabilities and area damage can be estimated with the aid of statistical data provided by real fires. Such statistics produced by the fire brigades in the UK enable the extent of fire spread to be classified as follows:

1   Confined to item first ignited;
2   Spread beyond item but confined to room of origin:
    * confined to contents only;
    * structure involved.
3   Spread beyond room but confined to floor of fire origin;
4   Spread beyond floor but confined to the building of fire origin;
5   Spread beyond the building of fire origin.

A fire starting in a room can spread upward to the next floor without involving the entire floor of origin. It is not possible to estimate the number of such cases. Hence, in the example shown in Table 5.7, the third and fourth categories have been combined to denote the event of fire spreading beyond the room of origin but confined to the building of origin. Fires spreading beyond the building of origin have not been included in this table.

A fire can spread beyond the room of origin without involving the structural boundaries of the room but such cases are rare occurrences. A fire generally reaches the post-flashover stage, involving all the contents in a room, and then spreads beyond the room by attacking the structural boundaries and causing their thermal failure. Fire spread due to the collapse or destruction of the structural boundaries only occurs very rarely.

For each category of spread, the area damage shown in Table 5.7 is the average value for the category. The percentage figure for each category denotes the probability attached to the category and to the corresponding average area damage.

In the case of a sprinklered building, the percentage figures include one-third of fires in these buildings which were estimated to be extinguished by the system, but not reported to the fire brigades (Rogers, 1977). In other words, fire brigades

Table 5.7 Fire extent of spread – textile industry, UK

Extent of spread	Sprinklered			Un-sprinklered		
	Percentage of fires	Average area damage (m²)	Financial loss (£)	Percentage of fires	Average area damage (m²)	Financial loss (£)
Confined to item first ignited	72	4.43	3278	49	4.43	3278
Spread beyond item but confined to room of fire origin:						
i) contents only	19	11.82	8747	23	15.04	11130
ii) structure involved	7	75.07	55,552	21	197.41	146,083
Spread beyond room	2	1000.00	740,000	7	2000.00	1,480,000
Average damage		30.69	22710		187.08	138,440

Notes:
Fires considered in the case of sprinklers are those in which the system operated.
In the case of sprinklers, the percentage for the first category, confined to item first ignited, includes one-third of fires extinguished by the system but not reported to the fire brigade.
The financial loss per m² is £740 at 1999 prices updating for inflation, Home Office estimate of £225 per m² at 1978 prices for textile industry (Maclean, 1979)

only attend two-thirds of fires in sprinklered premises. Most of the unreported small fires in sprinklered buildings were confined to the items first ignited. No insurance claims were made for compensating the financial losses in these small fires. It is apparent that, if they operate, sprinklers would increase the probability of a fire being confined to the item first ignited and thus reduce the probability of the fire spreading beyond the room of origin. Consequently, as shown in Table 5.7, sprinklers reduce considerably the overall damage expected in a fire. In many fires, the heat generated may not be sufficient to activate the sprinkler heads. There are also other causes for non-operation of sprinklers in fires – see Section 8.3.2.

Fire statistics collected by the US Fire Administration provide figures for probabilities and dollar losses for different categories of fire spread. In a study concerned with residential fire loss, Gomberg, Buchbinder and Offensend (1982) estimated dollar losses for different spread categories which were the same for both sprinklered and non-sprinklered buildings. They differentiated the probabilities of extinction to reflect the effectiveness of sprinklers. Their study also included the effectiveness of smoke detectors and life loss (fatalities and injuries).

Gomberg, Buchbinder and Offensend (1982) used probability trees to assess the final extent of flame spread and the consequences in terms of dollar loss and life loss. Three possible levels of spread were considered – confined to the object of origin (O), spread beyond this object but confined to part of the room of origin (< R) and spread beyond room (≥ R). Figure 5.2 is an example reproduced from this study. The 'suppression size' in this figure denotes the fire size at the start of a suppression activity. As with UK fire statistics, the US database does not provide probabilities for suppression size since only the final size after a fire was extinguished is recorded in fire reports. Hence, expert judgement was used to assess the suppression size.

The overall floor area expected to be damaged in a fire can also be estimated directly by fitting the probability distribution of area damage instead of evaluating the probabilities and average damage for each of the extent of spread categories. This distribution is log normal or Pareto as discussed in Section 3.3.4.

5.4.3 Financial loss

As shown in Table 5.7, the expected value for financial loss in a fire can be determined approximately by estimating the expected value for area damage (m²) and using an estimate for financial loss per m². Data on fire insurance claims (losses) compiled by insurance companies, particularly for large losses, if collated with fire brigade reports on such fires, can provide estimates for financial losses per m² for different occupancy groups and risk categories.

Insurance sources can also provide data for estimating the probability distribution of financial loss which, as discussed in Section 3.3.4, is log normal or Pareto. By identifying and fitting an appropriate distribution to observed values of losses, the expected value, standard deviation and other parameters of fire loss can be evaluated for any occupancy group or risk category. For this estimation, standard statistical techniques can be applied if loss data are available for all fires

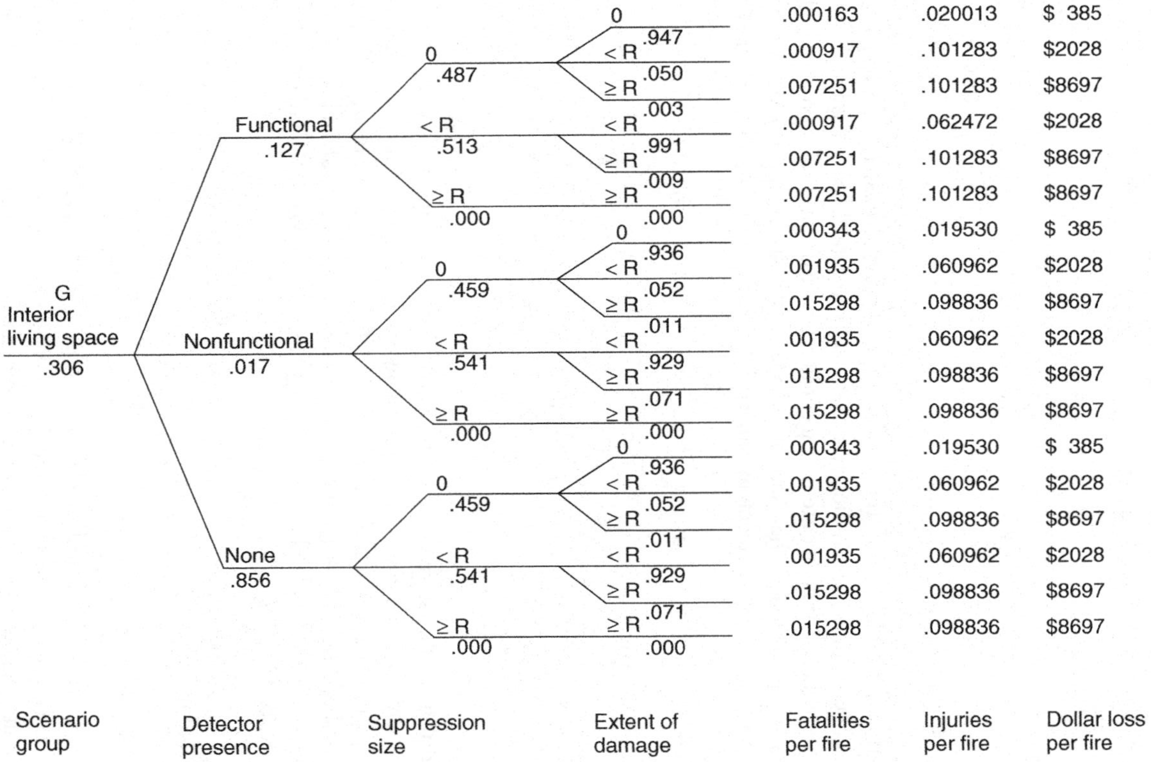

Figure 5.2 Sprinkler alternative probability tree

Scenario group: G Interior living space .306

Detector presence:
- Functional .127
- Nonfunctional .017
- None .856

Suppression size	Extent of damage	Fatalities per fire	Injuries per fire	Dollar loss per fire
0 .487	0 .947	.000163	.020013	$ 385
	<R .050	.000917	.101283	$2028
	≥R .003	.007251	.101283	$8697
<R .513	<R .991	.000917	.062472	$2028
	≥R .009	.007251	.101283	$8697
≥R .000	≥R .000	.007251	.101283	$8697
0 .459	0 .936	.000343	.019530	$ 385
	<R .052	.001935	.060962	$2028
	≥R .011	.015298	.098836	$8697
<R .541	<R .929	.001935	.060962	$2028
	≥R .071	.015298	.098836	$8697
≥R .000	≥R .000	.015298	.098836	$8697
0 .459	0 .936	.000343	.019530	$ 385
	<R .052	.001935	.060962	$2028
	≥R .011	.015298	.098836	$8697
<R .541	<R .929	.001935	.060962	$2028
	≥R .071	.015298	.098836	$8697
≥R .000	≥R .000	.015298	.098836	$8697

or extreme value techniques (Section 3.3.5) if data are only available for large losses.

The financial loss per m^2 may be assumed to be equal to financial value v per m^2 at risk in a building and its contents. Under this assumption, if the floor area destroyed in a fire is D m^2, the financial loss is Dv. In the estimation of the value density, v per m^2, the implicit assumption is that the total financial value V at risk in the building and contents is spread uniformly over the floor area, A, of the building such that $v = V/A$.

Using the value density, v, the power function in Equation (3.41) in terms of floor area A can be transformed to

$$D(V) = c' \, V^\beta \tag{5.34}$$

where

$$c' = c \, v^{-\beta} \tag{5.35}$$

Equation (5.34) expresses the fact that the financial loss, $D(V)$, in a fire increases approximately according to a power of the total financial value, V, at risk in the building and its contents. According to statistical studies on actuarial problems in fire insurance, the value of the power, β, is less than unity for most of the building types – see Ramachandran (1970, 1979/80, 1988) and Benktander (1973).

The power function in Equation (3.40) can also be transformed to

$$F(V) = k' \, V^\alpha \tag{5.36}$$

where

$$k' = k \, v^{-\alpha} \tag{5.37}$$

V is the total value at risk and v the value density per m^2. Equation (5.36) expresses the fact that the annual probability of frequency of fire occurrence increases approximately according to a power of the total financial value, V, at risk in the building and its contents. According to the statistical and actuarial studies mentioned above and in Section 3.3.1 the value of α is less than unity for most of the building types.

5.4.4 Life loss

5.4.4.1 Location of fire casualties

According to Table 5.8, based on UK Fire Statistics for the period 1978 to 1991, most of the casualties in single occupancy dwellings were found in the room of fire origin. A high percentage of the casualties were also found elsewhere on the floor

Table 5.8 Location of casualties – single and multiple occupancy dwellings

	Number of persons	
Whereabouts of casualty and occupancy type	Fatal	Non-fatal
Single occupancy dwellings		
Room of origin of fire	3539 (58.2)	26259 (44.6)
Elsewhere on floor of origin	1216 (20.0)	15500 (26.3)
Floors above floor of origin of fire	1267 (20.8)	14835 (25.2)
Floors below floor of origin of fire	54 (1.0)	2330 (3.9)
Total	6076 (100.00)	58924 (100.0)
Multiple occupancy dwellings		
Room of origin of fire	2347 (66.8)	15353 (35.0)
Elsewhere on floor of origin	823 (23.4)	18245 (41.6)
Floors above floor of origin of fire	330 (9.4)	9066 (20.6)
Floors below floor of origin of fire	16 (0.4)	1233 (2.8)
Total	3516 (100.0)	43897 (100.0)

The numbers within brackets denote the percentage number of casualties
Source: Fire Statistics United Kingdom, 1978–1991

of origin or floors above the floor of origin. A comparatively smaller number of casualties were found in floors below the floor of fire origin. Location of casualties in multiple-occupancy dwellings had a similar pattern except that an almost equal number of non-fatal casualties were found in the room and floor of fire origin.

It is understandable that occupants in the floors above the floor of fire origin have greater risk than those in floors below. Fire, smoke and toxic gases generally spread upwards and are more likely to be encountered by people in upper floors if they remain in their places of occupation or attempt to escape to safe places in or outside the building involved in fire. People in lower floors have a greater chance of avoiding combustible products and escaping safely.

It is apparent that, while fire is a major threat to occupants in its immediate vicinity, it is generally smoke and toxic gases which pose a greater threat than flame (heat) to occupants who are remote from the fire. Smoke and fumes travel faster than fire to occupied areas and escape routes. Even a small fire can generate considerable amounts of smoke and other combustible products and threaten a greater number of occupants outside the room of origin. Building fires spread mostly by convection (advance of flame, smoke and hot gases) rather than by the destruction of the structural boundaries of a room.

5.4.4.2 Nature of injuries

Consistent with the location of casualties, a high percentage of fatalities in the room of fire origin were caused by burns, apart from gas or smoke, which was the major cause, accounting for more than 50 per cent of the fatalities in dwellings – see Table 5.9. Statistical studies and surveys carried out in the UK in the 1970s revealed that not only were a large proportion of fatal and non-fatal fire casualties being reported in the category 'overcome by smoke and toxic gases' rather than heat and burns but also that there was a four-fold increase in the former category between 1955 and 1971.

It is an accepted fact that toxic products of combustion are the major causes of incapacitation and death in fires (Berl and Halpin, 1976; Harland and Woolley, 1979). In many fires, death or injury is not due to the immediate toxic effects of exposure to these products but results from the victim being prevented from escaping due to irritation and visual obscuration caused by dense smoke or to incapacitation caused by narcotic gases. Consequently, the victim remains in the fire and sustains fatal or non-fatal injury due to a high dose of toxic products, inhaled during the prolonged exposure, or due to burns. High percentages of fatalities and non-fatal casualties are trapped by smoke or fire, because they are unaware of the fire (asleep etc.) or for other reasons. Survivors from fires may also experience pulmonary complications and burn injuries which can lead to delayed death.

Table 5.9 Fatal casualties in dwellings by whereabouts of casualties and cause of death

Whereabouts of casualty and occupancy type	Cause of death		
	Overcome by gas or smoke*	Burns or scalds	Other or unknown causes
Single occupancy			
Room of origin of fire	1653	953	328
Floor of origin of fire	731	133	116
Elsewhere	868	130	118
Total	3252	1216	562
Multiple occupancy			
Room of origin of fire	1217	510	200
Floor of origin of fire	504	66	67
Elsewhere	217	37	40
Total	1938	613	307

Source: Fire Statistics United Kingdom 1978–1988
*Including cases where burns and overcome by gas or smoke were joint causes of death.
A breakdown of figures for causes of death as in the table has not been published for the years 1989 to 1991.

Increasing fire risk due to smoke and other combustion products led to the commencement of intensive research on combustion toxicology during the 1970s. These studies have ranged from fundamental laboratory-based thermal decomposition experiments to large-scale fires with comprehensive gas analysis, bioassay and detailed pathology of fire victims. Models developed include the 'mass loss' model and the 'fractional effective dose' model. These models require as inputs the rates of generation of life-threatening combustion products and estimate the times when tenability limits are exceeded, resulting in incapacitation or death. Purser (2002) carried out a very detailed review of various studies and models on toxicity assessment of combustion products.

5.4.4.3 Causes of fatal and non-fatal casualties in fires

The majority of fatalities in accidental fires in dwellings are due to causes such as careless disposal of smokers' materials, ignition of matches – mostly by children playing with them, incidents with space heating – mainly misuse or placing articles too close to them and misuse of cooking appliances. Electricity is the major fuel in regard to deaths caused by the misuse of space heaters or cooking appliances. A contributory factor for fire deaths during severe winter is the use of portable heating appliances to supplement central heating.

The leading causes of non-fatal casualties in accidental dwelling fires are careless disposal of smokers' materials and misuse of cooking appliances, mostly those using electricity and some using gas. The next most common specific causes are electric and gas space heating appliances, electrical wiring, electric blankets and bedwarmers. Television sets, washing machines and dishwashers are other minor causes of non-fatal casualties.

5.4.4.4 Materials first ignited

During the past few years, fires in which textiles, upholstery and furnishings were the materials first ignited, accounted for a large number of deaths. Within this group, the major items were bedding, upholstery or covers and clothing. According to a study carried out in 1976, the chance of a fatality in fire involving furniture and furnishings was twice that in other fires in houses. The majority of these fires involved upholstery and bedding and were caused by smokers' materials, electric blankets and space heating. The main hazard appeared to be to people in armchairs and beds, using potential sources of ignition (smoking, space heating etc.) failing to respond to a fire in their vicinity through being asleep or otherwise incapacitated.

During the past three decades, smoke-related casualties in homes have increased partly due to an increase in the use of synthetic materials in furnishings and upholstered furniture. This is also partly due to changes in living styles which have led to more furnishings and upholstery material being used in homes.

Change from natural to synthetic materials has brought certain benefits in fire performance. Natural materials tend to be prone to smouldering from small ignition sources, particularly when in contact with a lighted cigarette, whereas synthetic materials tend to be more resistant to this type of ignition. However, the synthetic fabrics are mainly thermoplastics and when subjected to a flame can burn rapidly with the fabric 'melting' to expose the flammable infill fibres and foams. Natural fabrics (wool, cotton etc.) tend to form carbonaceous chars during flame exposure which can act as an effective barrier to the penetration of fire.

5.4.4.5 *Casualty rate per fire*

Casualty rate per fire is a simple but very useful yardstick for measuring life risk due to fires in any type of building. This rate is the ratio between the number of fatal or non-fatal casualties and the number of fires. Data for estimating these rates for any type of building are available from Fire Statistics United Kingdom, published annually by the Home Office.

According to UK fire statistics, both the fatal and non-fatal casualty rates per fire for dwellings do not vary significantly from year to year. In fact, there is some indication that the fatality rates are gradually declining over the years. This is perhaps due to effective performance of fire fighting and protection strategies including fire safety codes, regulations and standards. This is also due to the fact that the number of fires, particularly in multiple-occupancy dwellings, has gradually increased over the years but the number of deaths has not increased. This result indicates the need for increasing fire prevention activities aimed at reducing the frequency of fire occurrence.

The non-fatal casualty rate per fire in dwellings has gradually increased over the years.

The casualty rate per fire indicates the probability of one or more people becoming a casualty in the event of a fire occurring. It should be multiplied by the annual frequency (number) of fires occurring to express it on an annual basis.

As discussed in Section 3.3.4, the number of deaths likely to occur in a fire is a random variable, taking integer values according to a discrete (discontinuous) probability distribution. Well-known examples of such a distribution are Poisson and negative binomial. The extended form of Poisson described in Equation (3.49) provides an estimate of the probability of x deaths occurring in a fire if occupants of a building are exposed for t minutes to lethal conditions caused by a combustion product.

The parameter δ in Equation (3.49) quantifies the increase in the probability of death for every extra minute of exposure to lethal conditions caused by a combustion product. If, on average, occupants of a building are exposed to such conditions for $\bar{t}$ minutes, the probability of x deaths occurring in a fire is given by Equation (3.49) with $t = \bar{t}$. For a particular building, the values of δ and $\bar{t}$ should be estimated for each combustion product such as heat or smoke. It may be possible to estimate these values for a particular building by performing simulations with models discussed by Purser (2002) or other deterministic models. The results for different combustion products may then be combined and used in Equation (3.49) to provide estimates of the probabilities of death for different values of x.

The parameter λ in Equation (3.46) quantifies the increase in the fatality rate per fire, P, for every minute of delay in discovering or detecting the existence of a fire in a building. The value of λ was estimated to be about 0.0007 for single and multiple-occupancy dwellings. The value of the parameter K denotes the fact that the fatality rate or probability of one or more deaths occurring would be about 0.0015 even if a fire in these buildings is detected immediately after ignition, i.e. if $D = 0$. Hence, with an average discovery time of 15 minutes, the overall fatality rate in these buildings was about 0.012 [$= 0.0015 + (15 \times 0.0007)$]. The fatality rate would have reduced to 0.0022 ($= K + \lambda$) if all the single and multiple occupancy dwellings were fully protected by automatic detection systems.

According to Ramachandran (1993a), the value of λ for non-fatal casualties was 0.0042 for single-occupancy dwellings and 0.0047 for multiple-occupancy dwellings. Respectively, for these two types of dwellings, the value of K was 0.0938 and 0.1092. Hence, with average fire discovery times of 9 and 10.6 minutes, the overall non-fatal casualty rate was 0.1314 for single-occupancy dwellings and 0.1592 for multiple-occupancy dwellings. If all these dwellings were protected by automatic detection systems, the non-fatal casualty rate per fire would have reduced to 0.098 for single-occupancy dwellings and 0.1139 for multiple-occupancy dwellings.

5.4.4.6 Other measurements of life risk

The $f(N)$ relationship for investigating life risks due to various types of hazards was discussed in Section 4.2.2. This function expresses the annual frequency of N or more fatalities. There is a need to update the $f(N)$ curves using recent data on the frequency distribution of number of fire deaths. Such data are available for several countries.

Fatal Accident Frequency Rate (FAFR) is another measurement of life risk which is the number of fatalities that occur during a hundred million man-hours of exposure to an occupation or activity. FAFR has been calculated for various industrial occupations such as nuclear and chemical industries and non-industrial activities such as travelling by bus, train, car or air, canoeing and rock climbing. In regard to fire safety in the chemical industry, for example, FAFR for the industry as a whole has been estimated to be 4. On this basis, it has been arbitrarily proposed that no single activity which any person is carrying out should contribute more than 10 per cent to the FAFR, i.e. 0.4 (Kletz, 1976).

North (1973) calculated rough values of FAFR for many occupancies in the UK. These only covered the years 1967 to 1969 and, in cases where deaths were infrequent, the estimates had wide confidence limits. For fire deaths in a dwelling, the FAFR was 0.19 and in hotels 3.6. The latter figure was probably distorted by some serious multiple-fatality fires that happened during 1967 to 1969.

An American report (Balanoff, 1976) allowed estimates to be obtained for FAFR for firemen on or following activities at the fireground. The report indicated

86 deaths per 100,000 firemen per annum which provided an estimate of 42 for FAFR. About half of the deaths were due to heart failure and 10 per cent due to each of building collapse, burns and smoke inhalation.

5.4.5 Total performance

In some cases it would be useful to express the probable damage (P) in the event of a fire occurring as the product of two components Q and C, where Q is the probability of occurrence of an undesirable event and C is the probable consequences or damage if the undesirable event occurs:

$$P = Q \times C$$

The total (joint) performance (effectiveness) and not the individual performance of a building design and installed fire protection measures should ensure that the value of P will not exceed a level acceptable to society. The acceptable level would depend on consequences in terms of life loss and injury, and property damage.

Undesirable events would include:

1   fire spreading beyond room of origin;
2   'failure' of a fire-resistant compartment;
3   visual obscuration due to smoke;
4   incapacitation due to burns and toxic gases.

References

Apostolakis, G (1982), Data analysis in risk assessments, *Nuclear Engineering and Design*, 71, 375–381.

Balanoff, T (1976), *Fire Fighter Mortality Report*, International Association of Fire Fighters, Washington, DC.

Bengtson, S and Laufke, H (1979/80), Methods of estimation of fire frequencies, personal safety and fire damage, *Fire Safety Journal*, 2, 167–180.

Bengtson, S and Ramachandran, G (1994), Fire growth rates in underground facilities. *Proceedings of the Fourth International Symposium on Fire Safety Science*, International Association of Fire Safety Science, Ottawa.

Benktander, G (1973), Claims frequency and risk premium rate as a function of the size of the riskl *ASTIN Bulletin*, 7, 119–136.

Berl, W G and Halpin, B M (1976), Fire related fatalities: an analysis of their demography, physical origins and medical causes, *Fire Standards and Safety*, ASTM STP 614, American Society for Testing and Materials, Philadelphia, PA.

Bertrand, A L and McKenzie, L S (1976). *The Human Factor in High Risk Urban Areas: A Pilot Study in New Orleans, Louisiana*, US. Department of Commerce, Washington DC.

Butcher, G (1987), The nature of fire size, fire spread and fire growth, *Fire Engineers' Journal*, 47, 144, 11–14.

Chandler, S E (1968), *Estimated Fire Frequencies In Buildings Based On Expected Fuel Usage*, Fire Research Note 716, Fire Research Station, Borehamwood.

Chandler, S E (1979), *The Incidence of Residential Fires in London: The Effect of Housing and Other Social Factors*, Information Paper IP 20/79. Building Research Establishment, Fire Research Station, Borehamwood.

Chandler, S E (1982), The effects of severe weather conditions on the incidence of fires in dwellings, *Fire Safety Journal*, 5, 1, 21–27.

Chandler, S E, Chapman, A and Hollington, S J (1984), Fire incidence, housing and social conditions – the urban situation in Britain, *Fire Prevention*, 172, 15–20.

CIB W14 (1986), Design guide: structural fire safety, *Fire Safety Journal*, 10, 2, 77–137.

Friedman, R (1978), Quantification of threat from a rapidly growing fire in terms of relative material properties, *Fire and Materials*, 2, 1, 27–33.

Frishman, F (1975), On the arithmetic means and variances of products and ratios of random variables, *Statistical Distributions in Scientific Work, Volume 1, Models and Structures*, D Reidel, Dordrecht.

Gaunt, J E and Aitken, I S (1964), *Causes of Fires in Dwellings in London, Birmingham and Manchester and their Relationship with the Climatic Conditions during the First Quarter of 1963*, Fire Research Note 538, Fire Research Station, Borehamwood.

Gomberg, A, Buchbinder, B and Offensend, F J (1982), *Evaluating Alternative Strategies for Reducing Residential Fire Loss – The Fire Loss Model*, Report NBSIR 82-2551, National Bureau of Standards Centre for Fire Research, Gaithersburg, MD.

Gunther, P (1975), Fire-cause patterns for different socioeconomic neighborhoods in Toledo, Ohio, *Fire Journal*, 3, 52–58.

Harland, W A and Woolley, W D (1979), *Fire Fatality Study, University of Glasgow*, Information Paper IP 18/79, Building Research Establishment, Fire Research Station, Borehamwood.

Kletz, T A (1976), The application of hazard analysis to risks to the public at large, World Congress of Chemical Engineering, Session A5, Amsterdam, July 1976.

Labes, W G (1966), The Ellis Parkway and Gary Dwelling burns, *Fire Technology*, 2, 4, 287–297.

Lawson, J R and Quintiere, J (1986), Example illustrating slide rule estimates of fire growth, *Fire Technology*, 22, 1, 45–53.

Munson, M J and Oates, W E (1977), *Community Characteristics and Incidence of Fire: An Empirical Analysis*, Princeton University Press, Princeton, NJ.

North, M A (1973), *The Estimated Fire Risk of Various Occupancies*, Fire Research Note 989, Fire Research Station, Borehamwood.

Purser, D A (2002), Toxicity assessment of combustion products, Chapter 2–6, *SFPE Handbook of Fire Protection Engineering*, National Fire Protection Association, Quincy, MA.

Ramachandran, G (1970), *Fire Loss Indexes*, Fire Research Note 839, Fire Research Station, Borehamwood.

Ramachandran, G (1979/80), Statistical methods in risk evaluation, *Fire Safety Journal*, 2, 125–145.

Ramachandran, G (1980), *Economic Value of Automatic Fire Detectors*, Information Paper IP27/80, Building Research Establishment, Fire Research Station, Borehamwood.

Ramachandran, G (1985), The human aspects of fires in buildings – a review of research in the United Kingdom, *Fire Safety Science and Engineering*, ASTM STP 882 American Society for Testing and Materials, Philadelphia, PA.

Ramachandran, G (1986), Exponential model of fire growth, *Fire Safety Science: Proceedings of the First International Symposium*. Hemisphere Publishing Corporation, New York.

Ramachandran, G (1988), Probabilistic approach to fire risk evaluation, *Fire Technology*, 24, 3, 204–226.

Ramachandran, G (1990), Probability based fire safety code, *Journal of Fire Protection Engineering*, 2, 3, 75–91.

Ramachandran, G (1992), *Statistically Determined Fire Growth Rates for a Range of Scenarios; Part 1: An Analysis of Summary Data. Part 2: Effectiveness of Fire Protection Measures: Probabilistic Evaluation*, unpublished report to the Fire Research Station, Borehamwood.

Ramachandran, G (1993a), Early detection of fire and life risk, *Fire Engineers Journal*, 53, 171, 33–37.

Ramachandran, G (1993b), Fire resistance periods for structural elements – the sprinkler factor, Part 3, *Proceedings of the CIB W14 International Symposium on Fire Safety Engineering*, University of Ulster, Jordanstown.

Ramachandran, G (1995), Heat output and fire area, *Proceedings of the International Conference on Fire Research and Engineering*, SFPE, Orlando.

Ramachandran, G (1998), *The Economics of Fire Protection*, E & F N Spon, London.

Ramachandran, G (1999), *Reliability and Effectiveness of Sprinkler Systems for Life Safety*, unpublished report to the Fire Research Station, Borehamwood.

Rogers, F E (1977), *Fire Losses and the Effect of Sprinkler Protection of Buildings in a Variety of Industries and Trades*, Current Paper CP 9/77, Building Research Establishment, Fire Research Station, Borehamwood.

Thomas, P H (1974), *Fires in Model Rooms: CIB Research Programmes*, Current Paper CP32/74, Building Research Establishment, Fire Research Station, Borehamwood.

Walton, W D and Thomas P H (1988), Estimating temperatures in compartment fires, Section 2, Chapter 2, *SFPE Handbook of Fire Protection Engineering*, 1st edn. National Fire Protection Association, Quincy, MA.

6 Design fire size

6.1 Introduction

Fire protection measures cannot be expected to cope with all possible sizes of fires. The best that can be achieved is to design these measures around a large or maximum size that is likely to be encountered. Determination of such a size, defined as the 'design fire size', is fundamental to the design of fire detection, sprinkler and smoke ventilation systems, particularly for buildings where the fire safety engineering approach is more common, e.g. retail premises, shopping complexes, atrium buildings. The design fire size is expressed in terms of the amount of heat output likely to be produced in a large fire. This amount depends on fire load and other factors affecting the growth of fire in the space considered.

Deterministic formulae and models supported by experimental results are available for calculating the heat output in a large fire in a compartment of given dimensions, fire load and ventilation factor. However, due to uncertainties caused by several factors, and the possibility of several scenarios, probabilities are attached to quantities of heat output produced in an actual (not experimental) fire over the period of fire growth. The Technical Report ISO/TR 13387-2: 1999(E) describes a systematic approach to the identification of significant fire scenarios that need to be considered in fire safety design. Clark and Smith (2001) established a database of specific fire characteristics for a series of realistic fire scenarios based on results of experimental studies.

The chance of heat output in the actual fire exceeding a large value is small. An acceptable value for this small chance or probability can be determined for any type of property by considering consequences in terms of damage to life and property.

Heat output is correlated with area damage (Ramachandran, 1995a). Hence, the probability of heat output exceeding a specified large value can be ascertained by estimating the probability for the corresponding large value for area damage. Statistical data, compiled by the fire brigades in the UK, provide information on area damage in real fires and its probability distribution. This information has been used for estimating design fire sizes for shopping complexes, railway properties, public car parks, road tunnels and power stations. These design sizes and their estimation are discussed in Section 6.2.

The design sizes mentioned above are based on the assumption of a steady-state fire and are constants independent of time. They do not take into account the interactions between smoke and escaping occupants. Hence, such design sizes for designing fire protection systems have been considered to be unrealistic by some experts, particularly for protecting life. There is a need to develop probability–time based design sizes which are more realistic. A framework for this purpose is outlined in Section 6.3.

6.2 Current concept of design fire size

As mentioned in the previous section, the design fire size currently used for designing a fire protection system for any type of occupancy is a constant quantity and not a function of time (duration) since the commencement of ignition. It is based on a large value for area damage in an actual fire, the chance of exceeding which is small. Such an extreme (large) value can be considered as average area damage during the post-flashover period of a fire in a compartment. The heat output corresponding to this average area damage is estimated and used in designing a fire protection system. The following is a summary of studies based on the concept of constant design fire size.

Morgan and Chandler (1981) analysed statistical data for fire damaged areas in sprinklered retail premises. The results obtained by these authors are shown in Table 6.1. Since the sample size (number of fires) analysed was small, it was considered that a zero probability that a fire in a public area would exceed $10m^2$ was not credible. Accordingly an 'informed guess' was made that less than 4 per cent would exceed $10m^2$ in public areas. The heat output corresponding to $10m^2$ of area damage was assumed to be 5 MW which was equivalent to 0.5 MW per m^2. Based on this study, heat output of 5 MW over a fire area of $10m^2$ was suggested as an acceptable design fire for designing smoke ventilation systems for covered shopping complexes. Morgan and Chandler also concluded that, in the absence of sprinklers, 14 to 17 per cent of fires in public areas of retail premises was larger than $10m^2$ of area damage. In non-public areas, the design size of $10m^2$ was exceeded in about 12 per cent of fires in sprinklered premises and 19 per cent in others.

Table 6.1 Horizontal fire-damaged area in retail premises, 1978

Area (m²)	Number of incidents		
	Public areas	Non–public areas	Total
0 or minimal	27	28	55
1–10	25	41	66
11–50	–	8	8
51–100	–	–	–
Over 100	–	1	1
Total	52	78	130

The results of Morgan and Chandler (1981), discussed above, were confirmed to some extent by Law (1986) who analysed data on 'area of direct burning' for public and storage areas of sprinklered retail premises. The data related to fires which occurred during the period 1984–87. She also analysed 1981–87 data on number of sprinkler heads operating in public and storage areas (area damage records were not available for 1981–83). The results obtained by Law are reproduced in Table 6.2 which indicate that area damage in sprinklered public areas exceed 10m² in about 7 per cent of fires. More than four sprinkler heads operate in less than 5 per cent of incidents. Accordingly, it appears that the probability of heat output exceeding 5 MW is less than 5 per cent. More than ten heads operate in less than 2 per cent of fires.

The design principles for smoke ventilation in enclosed shopping centres outlined in the report by Morgan and Gardner (1990) are based upon a 12m perimeter (3m × 3m) 5 MW fire. Factors affecting the design fire size are discussed in this report. Should a different design fire be considered, for whatever reasons, the equations, figures etc. given in this report may no longer apply and advice should be sought from experts. Other fire sizes have occasionally been specified by designers for both sprinklered and unsprinklered shops. The problem of unsprinklered shops has been discussed in detail by Gardner (1988), who has shown the importance of considering 'flashover' in such units and the consequent need to consider potentially very large fires.

The procedure of selecting a fixed size of fire for designing smoke ventilation systems has been adopted for occupancies other than retail premises which are also commonly associated with atrium buildings such as offices and hotel bedrooms (Hansell and Morgan, 1994). An atrium can be defined as any space penetrating more than one storey of a building, where the space is fully or partially covered. Most atria within shopping centres may be considered as part of the shopping mall and treated accordingly. For offices, Morgan and Hansell (1985) suggested a design size of 16m², 14m perimeter if sprinklered and 47m², 24m perimeter if unsprinklered. For hotel bedrooms, Hansell and Morgan (1985) recommended the floor area of the largest bedroom as the design size. The corresponding design heat

Table 6.2 Area of direct burning and number of sprinkler heads opening for different fractiles, sprinklered retail premises

Fractile (%)	Public areas		Storage areas	
	Area of direct burning (m²)	Number of sprinkler heads opening	Area of direct burning (m²)	Number of sprinkler heads opening
20	3	2	6	3
10	7	3	12	4
5	16	4	50	4

Note: The fractile is the percentage number of fires in which the area of direct burning or number of sprinkler heads opening was exceeded.

outputs are approximately 1MW for sprinklered office, 6MW for unsprinklered office and 1MW for hotel bedroom. The hotel bedroom fire represents a fully developed unsprinklered fire. Where sprinklers are present, a value of 500KW (for a 6m perimeter fire) may be more appropriate.

In the above studies, the design size for an occupancy type has been identified from the frequency distribution (table or graph) of area damage. A better method would be to fit an appropriate probability distribution to such data. Statistical studies have shown that area damage in fires in most of the occupancies has either a log normal or Pareto probability distribution – see Section 3.3.4.

Ramachandran (1992a) found the Pareto distribution to be more appropriate than log normal in an investigation concerned with design fires in underground facilities. Design sizes were estimated for four types of occupancies: railway properties, public car parks, road tunnels and power stations. The results were based on data for the three years 1985–87 for the first two types of buildings and for the years 1979 and 1984–87 for the other two types. These statistics were provided by the fire statistics section of the Home Office, United Kingdom. The analysis is described below (a summary of the results was presented by Ramachandran and Bengtson (1995c) at the First European Symposium on Fire Safety Science held in Zurich in August 1995).

The Pareto distribution has the following form:

$$\phi\,(d) = k\,d^{-\lambda}\,;\,k = m^{\lambda} \tag{6.1}$$

where $\phi\,(d)$ is the probability of area damage exceeding d and m the minimum damage. The parameters k and λ are constants for any occupancy type. They also depend on factors such as whether a building is sprinklered or not. The parameters can be estimated by applying the least square method to fit the following straight line to data:

$$\log\phi(d) = \log k - \lambda\log d \tag{6.2}$$

If data on area damage, d, are available in the form of a frequency distribution table, $\phi(d)$ is the proportion of fires (out of the total number) exceeding d. Approximate values of k and λ can be obtained graphically by plotting the pairs of values $(\log\phi(d), \log d)$ for different values of d.

Table 6.3 contains the values of λ, k and m estimated for the four occupancies considered. The results relate only to fires in buildings without sprinklers. They also relate to area damaged by direct burning (fire area). Fires starting in lifts and stairs were excluded.

The design size can be estimated by assigning an appropriate value for $\phi(d)$. For example, if a level of 0.1 is acceptable for $\phi(d)$, the design size, D, from Equation (6.1) is given by

$$0.1 = k\,D^{-\lambda} \tag{6.3}$$

Table 6.3 Pareto parameters

Categories	Number of fires	Parameters			Average damage (m²)
		λ	k	m (m²)	
Railway properties					
Fires in all places	826	0.8084	1.4038	1.5213	21.71
Fires in public places	220	0.6818	0.6182	0.4939	22.58
Fires in basement	73	0.9371	0.2692	0.2465	5.04
Public car parks					
Fires in all places	709	1.2738	1.1575	1.1216	4.67
Fires in basement	165	1.2762	1.4950	1.3704	5.87
Road tunnels and subways	122	1.0094	0.5219	0.5251	3.3
Power stations	133	0.82	0.4118	0.3389	6.04

or by

$$D = (10k)^{1/\lambda}$$

If 0.05 is an acceptable value for $\phi(d)$ then

$$D = (20k)^{1/\lambda} \tag{6.4}$$

Area damage would exceed the value of D given by Equation (6.3) in 10 per cent of fires and the value given by Equation (6.4) in 5 per cent of fires. The design sizes based on these two equations are given in Table 6.4. They are applicable to premises without sprinklers.

Design fire sizes, expressed in terms of area damage, can be converted to heat output rate by applying the method discussed by Ramachandran (1995a). From Equations (5.3), (5.18) and (5.19), the loss rate of fuel mass, $\dot{m}$ (kg/sec) is given by

$$\dot{m} = \bar{L} \cdot \theta \cdot A(T) \tag{6.5}$$

where $\bar{L}$ is the average fire load density in kg/m², θ the fire growth parameter per second in the exponential model in Equation (5.3) and $A(T)$ the fire area (m²) destroyed in T seconds. The heat output rate, $\dot{Q}$ (kW), corresponding to $\dot{m}$, as shown in Equation (3.58) is given by

$$\dot{Q} = \dot{m}\Delta H \tag{6.6}$$

where ΔH is the effective heat of combustion of the fuel usually assumed to have the value 18,800 kilojoules per kilogram.

Table 6.4 Design fire size $(D) m^2$

Occupancy category	$\varphi(D) = 0.1$ (m^2)	$\varphi(D) = 0.05$ (m^2)
Railway properties		
Fires in all places	26	62
Fires in public places	14	40
Fires in basement	3	6
Public car parks		
Fires in all places	7	12
Fires in basement	8	14
Road tunnels and subways	5	10
Power stations	6	13

Consider now power stations for which the average fire load density, $\bar{L}$, is 600MJ/m² (or 33 kg/m²) according to Table A1.3.13 (page 116) of 'Design Guide: Structural Fire Safety' (CIB W14 1986). For this occupancy type, the expected and maximum values of θ (in minutes) in an individual fire are 0.0208 and 0.0620 (Bengtson and Ramachandran, 1994) – see Table 5.5. Expressed in seconds, the expected and maximum values of θ are 0.000347 and 0.001033.

According to Table 6.4, the design fire size for power stations is 6m² for $\phi(d)$ = 0.1. In this case, with $A(T) = 6m^2$, calculations would show the expected and maximum values of $\dot{m}$ are 0.0687 kg/sec and 0.2045 kg/sec. The corresponding expected and maximum values of $\dot{Q}$ are 1292 kW and 3845 kW. With the design size of 13m² for $\phi(d) = 0.05$, the expected and maximum values of $\dot{m}$ are 0.1489 kg/sec and 0.4432 kg/sec. The corresponding expected and maximum values of $\dot{Q}$ are 2799 kW and 8332 kW. Estimates for fire load densities are not available for the other three occupancy types mentioned in Tables 6.3 and 6.4.

Average fire load densities in shopping centres and department stores vary from 380 MJ/m² for textile items to 585 MJ/m² for items such as foods, furniture, carpets etc. – see Table A.1.3.5 (page 108) of 'Design Guide: Structural Fire Safety' (CIB W14 1986). Hence, one could assume an average fire load density of 480 MJ/m² or 26kg/m² for shopping centres. According to Ramachandran (1992b), for fires in assembly (customer) areas of retail premises extinguished by sprinklers, the average and maximum values of the fire growth parameter θ in minutes (Equation (5.3)) are 0.0608 and 0.1043. The corresponding values of θ in seconds (Equation (6.5)) are 0.001013 and 0.001738. The maximum value would represent the rate of growth in a fast-growing fire.

For a design fire size of 10m², with $A(T) = 10$ and $\bar{L} = 26$, inserting the values of θ (in seconds) mentioned in Equation (6.5), the average and maximum values of $\dot{m}$ are 0.2634 kg/sec and 0.4519 kg/sec. Hence, from Equation (6.6), the average and maximum values of heat output rate $\dot{Q}$ are 4952 kW and 8496 kW. These results provide support to the design heat output of 5MW currently used for ventilation systems in shopping centres. This value should be regarded as an average, since the maximum heat output can be 8.5MW.

According to Equation (6.5), $\dot{m}$ per m² estimated by $\dot{m} / A(T)$ is a constant, given by $\overline{L}.\theta$ which, in the above example for shopping centres, is 0.02634 kg/sec for a fire with average fire growth and 0.04519 kg/sec for a fast-growing fire. The corresponding constant values of $\dot{Q}$ are 495 kW/m² and 850 kW/m². Realistically, $\dot{m}$ and $\dot{Q}$ per m² are not constants since the fire growth parameter θ is not a constant during the period of fire development. But θ has been assumed to be a constant in the simple exponential model in Equations (5.3) and (6.5).

In an investigation concerned with the 'sprinkler factor' for reducing fire resistance requirement for a sprinklered compartment, Ramachandran (1993) fitted the Pareto distribution (Equation (6.1)) for area damage in fires in three types of occupancies – office buildings, retail premises and hotels. The results obtained by him are reproduced in Tables 6.5(a) and 6.5(b). The parameters λ_0 and k_0 apply to rooms without sprinklers and λ_s and k_s to rooms with sprinklers. M_0 and M_s are the minimum area damage, m, in Equation (6.1). Using the results in Tables 6.5(a) and 6.5(b), the design sizes, D, have been estimated and given in Table 6.6 for two values, 0.1 and 0.05, for $\phi(d)$ in Equation (6.1). The estimates are based on Equations (6.3) and (6.4).

Sprinklers, if they operate satisfactorily, would cool the heat and smoke produced, retard the rate of growth of heat and smoke and, hence, reduce the probability of area damage exceeding a specified level. Considering sprinklered office rooms, as an example, using Equation (6.1) with values $k = 0.2778$ and $\lambda = 0.6987$, probability of damage exceeding 24m² is 0.03 while the corresponding probability, according to Table 6.6, is 0.1 if the room is without sprinklers. Similarly, the probability of damage exceeding 68m² is 0.015 if the office room is sprinklered and 0.05 if the room is unsprinklered. For the reasons mentioned above, as one would expect, the design sizes for sprinklered rooms are considerably smaller than the design sizes for unsprinklered rooms.

Judging from the figures in Table 6.6, the design size of 47m² suggested by Hansell and Morgan (1985) for unsprinklered offices is reasonable, but the design size of 16m² suggested by these authors for sprinklered offices appears to be an overestimate. Ferguson (1985) expressed doubt about the choice of a 16m² design fire for office atrium buildings. It may be argued that it would be safer to provide a larger safety margin and use a higher estimate for the design fire size for any occupancy type and room type. A more realistic estimate for the design size for any type of occupancy and room should be evaluated by considering the consequences, particularly to life risk, and determining an appropriate level for the probability quantified by Equation (6.1).

In a letter to the editor of *Fire Safety Journal*, Law (1986) drew attention to the data used by Morgan and Hansell (1985). According to Law, the data exhibited the following 'exponential' relationship for large fires in offices of area 10m² or more:

$P = 60A^{-0.63}$ for sprinklered offices

Table 6.5(a) Pareto distribution of area damage – rooms without sprinklers

Parameters	Office buildings		Retail premises			Hotels		
	Office rooms	Other rooms	Assembly areas	Storage areas	Other areas	Assembly areas	Bedrooms	Storage and other areas
Number of fires	1860	4369	8207	5144	7194	518	1205	3821
λ_0	0.6686	0.7146	0.6947	0.7304	0.8936	0.6603	0.7734	0.6392
M_0 (m^2)	0.7749	0.4647	0.5968	1.1583	0.7942	0.5907	0.4543	0.2176
k_0	0.8432	0.5783	0.6987	1.1133	0.8139	0.7063	0.5432	0.3772

Table 6.5(b) Pareto distribution of area damage – rooms with sprinklers

Parameters	Office buildings		Retail premises			Hotels
	Office rooms	Other rooms	Assembly areas	Storage areas	Other areas	Storage and other areas
Number of fires	18	127	224	354	183	35
λ_s	0.6987	0.8711	0.8644	0.8858	0.6991	0.6310
M_s (m^2)	0.1599	0.2646	0.4156	0.2852	0.2142	0.0322
k_s	0.2778	0.3141	0.4681	0.3291	0.3405	0.1144

Table 6.6 Design fire size D (m²)

Occupancy type	Room type	Without sprinklers		With sprinklers	
		$\varphi(D) = 0.1$ (m²)	$\varphi(D) = 0.05$ (m²)	$\varphi(D) = 0.1$ (m²)	$\varphi(D) = 0.05$ (m²)
Office buildings	Office rooms	24	68	4	12
	Other rooms	12	31	4	8
Retail premises	Assembly areas	16	45	6	13
	Storage areas	27	70	4	8
	Other areas	10	23	6	16
Hotels	Assembly areas	19	55	–	–
	Bedrooms	9	22	–	–
	Storage and other areas	8	24	1.24	3.71

$$P = 180A^{-0.78} \text{ for non sprinklered offices} \qquad (6.7)$$

P is the proportion (%) of fires exceeding a given area A (m²). With A denoted by d and P by $\phi(d)$, Equation (6.7) is essentially the same as the Pareto distribution shown in Equation (6.1). Equation (6.1) is applicable to all fires, small and large, while Equation (6.7) is only applicable to large fires.

According to the figures in Tables 6.5(a) and 6.5(b), the values of the parameter λ for office rooms are 0.67 if unsprinklered and 0.70 if sprinklered. The corresponding figures of Law are 0.78 and 0.63, respectively, which appear to be unrealistic. From theoretical considerations one would expect the value of λ for a sprinklered room to be higher than that for an unsprinklered room. This hypothesis is supported by the figures in Tables 6.5(a) and 6.5(b) except those for 'other rooms' of retail premises. For storage and other areas of hotels, the estimate of λ for a sprinklered case is almost the same as the estimate for the unsprinklered case.

6.3 Probability–time based design fire size

6.3.1 Introduction

The loss rate of fuel mass $\dot{m}$, and rate of heat output, $\dot{Q}$, are not constants but are functions of the time elapsing since the start of ignition. These two rates would be generally small quantities during the initial stage of a fire involving the object or item first ignited. As the fire spreads to other combustible items in the room, there will be a progressive increase in the fire area and heat output. That is to say, the heat output of unit area of the fire will start small and it will increase with time until a final value such as $0.5MW/m^2$ is reached. After this, the heat output will decay as the fuel on that particular unit area is consumed.

For the reasons mentioned above, Butcher (1987) pointed out that a fire does not reach its fully developed size but grows and only reaches, say, the $0.5MW/m^2$ value after an interval of time has elapsed. The 5MW value of heat output over $10m^2$ of fire area is reached only after an appreciable delay. Hence, the concept of using a constant design size of 5MW heat output for a fire situation is somewhat unrealistic. Assigning the full 5MW per $10m^2$ or $0.5MW/m^2$ to the early stages of a fire, where occupants are attempting to escape, can cause a serious error. During this period, the heat of the fire will be low; hence the smoke temperature will be low and the buoyancy movement sluggish. Calculating the smoke temperature from the 5MW value for heat output could give a false picture of smoke movement for the early period of a fire. But, as argued by Holt (1987), safety margins may be badly eroded by adopting smaller design sizes.

Several factors can cause uncertainties (randomness) in the determination of a design size for any type of occupancy or room. The presence or absence of sprinklers is also a source of uncertainty. Factors influencing the design fire were discussed by Morgan and Gardner (1990). The uncertainties can be quantified by probabilities and statistical parameters such as mean and standard deviation.

For the reasons mentioned above, there is a need to develop a probability–time based model to determine more realistic and accurate design fire sizes. A framework for this purpose is outlined in the next subsection. The framework takes into consideration the interaction between movement of heat/smoke and movement of escaping occupants. In the following three subsections, the framework is applied to determine design fire sizes for detectors, sprinklers and ventilation systems.

6.3.2 Framework

The first step is to develop a fire growth curve which will realistically express the relationship between heat output and time. For this purpose, it is not unreasonable to assume that heat output does not increase significantly before the occurrence of established burning. Heat output would grow 'steadily' after the commencement of established burning. Such a steady growth is depicted by an exponential model, according to which heat output increases exponentially with time – see Section 5.2.3.

Heat output is positively correlated with fire area (Ramachandran, 1995a). Heat output is directly proportional to fire area or area damage by direct burning. Hence, as discussed in Section 5.2.6 (Equations (5.18) and (5.19)), the exponential model in Equation (5.3) can be used to estimate the loss rate of fuel mass as a function of time and then Equation (3.58) is used to obtain a similar relationship for the rate of heat output, $\dot{Q}$:

$$\dot{Q} = \bar{L} \cdot \theta \Delta H \cdot A(0) \exp(\theta T) \tag{6.8}$$

The parameters in Equation (6.8) have already been defined with reference to Equations (5.3), (6.5) and (6.6).

The next step is to estimate the design value for time, T, required for the operation of a fire protection system such as detectors, sprinklers or smoke ventilation system. This design time should be sufficient for the occupants of a building to escape to a safe place, without being killed or injured, before the escape routes are blocked by heat, smoke or toxic gases. For such a successful evacuation of any type of building, the time of operation, according to the current design heat output, may or may not be sufficiently safe.

Consider, for example, the design value of 5MW (5000kw) for $\dot{Q}$ currently used for smoke ventilation systems in shopping centres. As discussed in Section 6.2, for sprinklered customer areas of retail premises, the maximum value of the fire growth rate θ estimated by Ramachandran (1992b) is 0.001738 per second. In this estimation, one square metre was used for $A(0)$, the area ignited at the commencement of established burning. $\dot{Q} = 5000$ kW, $\bar{L} = 26$ kg/m^2, $\Delta H = 18,800$ kJ/kg and $\theta = 0.001738$, calculation based on Equation (6.8) would show that the minimum value of T applicable to a fast growing fire is 1020 seconds or 17 minutes. It may also be calculated that, with $\theta = 0.001013$ per second, $T = 2283$ seconds or 38 minutes for a fire growing at an average rate. For two scenarios, Butcher (1987) estimated the time taken by a fire to reach a heat output of 5MW as 16 and 22 minutes.

The values of T estimated above were based on statistically determined fire growth rates in real fires involving several objects, data for which were provided by fire brigades. They are not based on a deterministic model of fire growth applied to individual objects with parameter values estimated from experimental data. A question now arises as to whether time periods ranging from 16 to 38 minutes for the operation of a smoke ventilation system in a real fire (not experimental fire) provide an adequate safety margin for the successful evacuation of occupants. Should the system operate earlier at a lower heat output?

To answer the above question, it is necessary to consider the interaction between the movement of a combustion product and the movement of occupants – see Ramachandran (1993, 1995b) and Section 12.2. For safe evacuation, it is a fundamental condition that the total time, H, taken by an occupant or group of occupants to reach a safe place in the building, or outside the building, should be less than the time, F, taken by a combustion product, e.g. smoke, to travel from the place of fire origin and produce an untenable condition on the escape route selected. Some occupants may sustain fatal or non-fatal injuries if egress failure occurs with $H > F$.

As discussed in Section 3.7.3.2, the total evacuation time, H, is the sum of three component periods. The first period, D, is the time taken to discover the existence of a fire by an occupant (Section 8.1.1) or by an automatic detection system (Section 8.1.2). The second period, B, is the 'recognition time' or 'gathering phase' discussed in Section 9.1. The third period, E, is the time taken by an occupant to travel from their place of work and reach a safe place, such as the entrance to a protected staircase or exit to the outside of the building – see Section 9.4. Thus

$H = D + B + E$. The time, F, can relate to a particular combustion product such as heat, smoke or a toxic gas.

As discussed in Section 3.7.3.3, it would be necessary to estimate and use in a probabilistic evacuation model, the means (averages) and standard deviations of H and F over several fire and evacuation scenarios likely to occur in an actual fire. Then 'partial safety factors' can be estimated to derive the design value for H according to Equation (3.89). A similar calculation procedure would provide the design value for F (Equation (3.90)). The values of the partial safety factors for H and F should be such that the design criterion $H < F$ for successful evacuation is satisfied by the design value of H (Equation (3.91)).

The above model, based on partial safety factors, is a semi-probabilistic approach. A better probabilistic model for estimating the design value of H is provided by the 'Beta method', discussed in detail in Section 3.7.4. This method involves the probability distributions of H and F and their joint distribution in addition to their means and standard deviations. In this method, the design value of H is evaluated according to Equation (3.93) to meet a (small) target probability specified for egress failure. This target probability should take into consideration consequences in terms of fatal and non-fatal injuries likely to be sustained by the occupants attempting to escape from a fire.

Subject to the limit specified for the design value of the total evacuation time, H, appropriate design values can be estimated for the operation times of automatic fire detection systems (Section 6.3.3), sprinkler systems (Section 6.3.4) and ventilation systems (Section 6.3.5). The design value for sprinkler operation time should also take into consideration the fact that sprinklers have the potential to increase the value of F even if they do not extinguish a fire. Sprinklers would reduce the rate of growth of heat and smoke and thus delay the occurrence of an untenable condition on an escape route. The value of F can be infinity or very high if sprinklers extinguish a fire.

The total fire area, $A(T)$, likely to be damaged when a fire protection system operates according to the design time, T_d, can be estimated with the aid of Equation (5.3). The corresponding value of $\dot{m}$ will be given by Equation (6.5) and $\dot{Q}$ by Equation (6.6). All the parameters $A(T)$, $\dot{m}$ and $\dot{Q}$ depend on the fire growth rate θ which, in turn, depends on the occupancy type and whether the growth is slow, medium or fast. The probability attached to $A(T)$ in a real fire can be estimated if data are available for evaluating the probability distribution of area damage (log normal, Pareto etc.) in the presence of the fire protection system considered.

6.3.3 Detectors

Automatic detection systems are designed to detect heat and/or smoke from a fire in its early stages of growth, give an audible signal and call the fire brigade if directly connected to the brigade. Such a signal would enable first-aid fire fighting to commence early so that the fire could be controlled quickly and prevented from causing extensive damage. Unlike sprinklers, which both detect fires and actively participate in fire fighting, detectors are passive and play no role in fire control.

Although it is possible to calculate from test results the response time of a heat/smoke detector under known conditions of ceiling height, detector spacing and fire/smoke intensity (total heat/smoke release rate), the time of operation of a detector head in an actual fire depends on many factors. The time when a fire product, heat, smoke or radiation, reaches a detector head depends on the rate of spread of the product which is controlled by the room/building configuration and environmental conditions. The factors mentioned above cause uncertainties in the performance of a detector, which may or may not operate in an actual fire; if it operates, it may do so at a random time. Detectors would fail to operate if the heat or smoke generated is insufficient to activate the system.

According to Bengtson and Laufke (1979/80), operating times for heat detectors range from 2 minutes in 'extra high hazard' (XHH) occupancies such as plastic goods factory, to about 20 minutes for 'light hazard' (XLH) which includes flats and other residential premises. The operating times of smoke detectors range from 0.5 min (XHH) to 2.25 min (XLH) for wood materials and to 0.75 min (XLH) for polystyrene. For wood materials, glowing fires give out most smoke but for polystyrene, apparently flaming conditions produce sufficient smoke for a quicker response time. According to some tests relating to dwelling fires quoted by Custer and Bright (1974), detection times for smouldering upholstery fires are long, for both rate-of-rise and fixed temperature detectors. In another test, involving a rapidly developing fire in a trash barrel, the rate-of-rise detector operated at 2 minutes while the fixed temperature unit responded at 5 minutes and the photoelectric detector in 8 minutes.

Nash *et al.* (1971) carried out some tests involving high stacked storage, using various types of detectors. In a series of similar tests, heat detectors operated between 1 min 16 sec and 3 min 58 sec. On ignition, ionisation chamber detectors operated between 1 min 5 sec and 4 min 30 sec while optical detectors took over 3 minutes to operate. Infra-red detectors operated in about 3 minutes and laser beam detectors took about 5 minutes to operate, if well above the fire.

The studies mentioned above provide some indication of the operating times of detection systems currently used, according to designs recommended in fire safety codes and standards. For example, the normal operating temperature of a heat detector head would be 65°C. The first step is to investigate whether the operating time of an existing detector system, currently designed for an occupancy type, would be consistent with the design value for the total evacuation time, H, discussed in Section 6.3.2. This is to ensure the safe evacuation of the occupants. If the operating time of an existing detector system added to the other two time periods, B and E, would increase the value of H beyond its design value, this time should be reduced to a new design time for the operation of the detector system. This design time would provide, for example, an estimate of the operating temperature of a heat detector head, which should be used in the design of the detector head.

Practically, with the existing technology, a detector system for any part of a building, e.g. kitchen, office room, etc. can be designed according to any level of physical parameters such as temperature and heat output from a fire. But care should be taken to ensure that the detector does not increase the frequency or probability of occurrence of false or unwanted alarms. For example, if the sensitivity of a heat detector head is increased by lowering the operating temperature, the head may pick up signals given by spurious fires from sources such as cigarette smoking and cooking which are normal activities. On the other hand, if the sensitivity is decreased by increasing the operating temperature, this will increase the risk of genuine fires being undetected. Sufficient research has not yet been carried out to determine an operating temperature which will provide an optimum balance between the detection of genuine fires and non-detection or blocking of false alarms.

6.3.4 Sprinklers

Sprinklers are generally required to operate at an average temperature of 68°C but there are special requirements for certain occupancies and important aspects such as the flow of hot gases in fires, which can determine the siting of sprinkler heads to achieve acceptable operating times. As with detection systems, several factors cause uncertainties in the activation and operating times of sprinklers in actual fires, although scientific (deterministic) methods have been developed for estimating the response time. Based on factors such as rate of temperature rise, height of upper fire surface above the floor and height of the premises, Bengtson and Laufke (1979/80) estimated sprinkler operation times varying from 2.5 minutes for 'extra high hazard' occupancies to 16.8 minutes for 'light hazard'. The operation time of sprinklers in experimental fires have been estimated in several studies carried out particularly by the Fire Research Station, UK, Factory Mutual Research Corporation, USA and National Institute of Standards and Technology, USA. In some fires, the heat produced may not be sufficient to activate a sprinkler system.

As discussed in Section 6.3.3 with reference to detectors, it should be investigated first whether the operating time of sprinklers currently used is such that it will lead to a total evacuation time less than the design value of H. If this condition is not satisfied, the operating time of a sprinkler system should be reduced from its current level. However, the design value of H can be increased for a building protected by sprinklers since, as mentioned before, sprinklers have the potential to increase the time, F, taken by a combustion product to produce an untenable condition on an escape route. At the same time, it is necessary to ensure that with $H < F$, the safety margin provided by the difference $(H - F)$ is such that it would meet the target probability specified for egress failure. For some buildings, e.g. residential buildings, it may be necessary to install fast-response sprinkler systems which operate quicker than sprinkler systems currently used in industrial buildings. The design value for sprinklers should also take into consideration whether or not a detection system has been installed in the building.

6.3.5 Ventilation systems

Deterministic models and experimental data may be able to provide an estimate of the time of operation of a ventilation system designed for smoke from a fire with a heat output of, say, 5MW. But in a real fire, as pointed out in Section 6.3.2, the operation time of a 5MW ventilation system installed in a shopping centre can vary from 16 to 38 minutes, depending on several factors. It may be necessary to design the system according to a lower heat output operating much earlier.

It is arguable whether a vent should operate before the operation of a sprinkler, if installed, or after the operation of the sprinkler. There are indications from current research that the effects of venting on the opening of the first sprinklers and their capacity to control the fire are likely to be small. There are also indications that the earlier vents are opened, the more likely they would be effective in preventing smoke-logging in a sprinklered building. In the initial stage of fire growth, a vent should, perhaps, operate before a sprinkler if life safety is the dominant objective, e.g. in hotels, shopping centres, office buildings. In industrial buildings, the first sprinkler may operate before the opening of any vent. The operation times of vents and sprinklers can be appropriately adjusted subject to the condition that the total evacuation time, H, does not exceed its design value.

References

British Standard BS 7974 (2002), *The Application of Fire Safety Engineering Principles to the Design of Buildings*, British Standards Institute, London.

Bengtson, S and Laufke, H (1979/80), Methods of estimation of fire frequencies, personal safety and fire damage, *Fire Safety Journal*, 2, 167–180.

Bengtson, S and Ramachandran, G (1994), Fire growth rates in underground facilities, *Proceedings of the Fourth International Symposium on Fire Safety Science*, International Association of Fire Safety Science, Ottawa.

Butcher, G (1987), The nature of fire size, fire spread and fire growth, *Fire Engineers Journal*, 47, 144, 11–14.

CIB W14 (1986), Design guide: structural fire safety, *Fire Safety Journal*, 10, 2, 77–137.

Clark P and Smith D A (2001), Characterisation of fires for design purposes: a database for fire safety engineers, *Proceedings of the Interflam 2001 Conference*, Interscience Communications, London.

Custer, R L P and Bright, R G (1974), *Fire Detection: The State-of-the-Art*. NBS Technical Note 839. National Bureau of Standards, Washington DC.

Ferguson, A (1985), Fire and the atrium, *Architect's Journal*, 181, 7, 63–70.

Gardner, J P (1988), Unsprinklered shopping centres: design fire sizes for smoke ventilation. *Fire Surveyor*, 17, 6, 41–47.

Hansell, G O and Morgan, H P (1985), Fire sizes in hotel bedrooms – implications for smoke control design, *Fire Safety Journal*, 8, 3, 177–186.

Hansell, G O and Morgan, H P (1994), *Design Approaches for Smoke Control in Atrium Buildings*, Report BR258, Building Research Establishment, Fire Research Station, Borehamwood.

Holt, J E (1987), Fire growth and design size (Letter to the Editor), *Fire Engineers Journal*, September 1987, 26.

International Standards Organisation (1999), *Fire Safety Engineering – Part 2: Design Fire Scenarios and Design Fires*, Technical Report ISO/TR 13387-2 Fire Edition, 1999-10-15, ISO, Geneva .

Law, M (1986), Letter to the Editor, *Fire Safety Journal*, 10, 67–68.

Law, M (1995), The origins of the 5MW design fire, *Fire Safety Engineering*, April, 17–20.

Morgan, H P and Chandler, S E (1981), Fire sizes and sprinkler effectiveness in shopping complexes and retails premises, *Fire Surveyor*, 10, 5, 23–28.

Morgan, H P and Hansell, G O (1985), Fire sizes and sprinkler effectiveness in offices – implications for smoke control design, *Fire Safety Journal*, 8, 3, 187–198.

Morgan, H P and Gardner, J P (1990), *Design Principles for Smoke Ventilation in Enclosed Shopping Centres*, Report BR186, Building Research Establishment, Fire Research Station, Borehamwood.

Nash, P, Bridge, N W and Young, R A (1971), *Some Experimental Studies of the Control of Developed Fires in High-racked Storages by a Sprinkler System*, Fire Research Note 866, Fire Research, Borehamwood.

Ramachandran, G (1992a), *Fires in Certain Types of Buildings – Growth Rates and Design Sizes* Report submitted to Brandskyddslaget, Enskede, Sweden.

Ramachandran, G (1992b). *Statistically Determined Fire Growth Rates for a Range of Scenarios: Part 1: An Analysis of Summary Data. Part 2: Effectiveness of Fire Protection Measures. Probabilistic Evaluation* Report to the Fire Research Station, Borehamwood.

Ramachandran, G (1993), Fire resistance periods for structural elements – the sprinkler factor, *Proceedings of the CIB W14 International Symposium on Fire Safety Engineering*, University of Ulster, Jordanstown

Ramachandran, G (1995a), Heat output and fire area, *Proceedings of the International Conference on Fire Research and Engineering*, SPFE, Orlando, FL.

Ramachandran, G (1995b). Probability-based building design for fire safety. Part 1: *Fire Technology*, 31, 3, 265–275; Part 2: *Fire Technology*, 31, 4, 355–368.

Ramachandran, G and Bengtson, S (1995c), Design fires in underground facilities, *Book of Abstracts, First European Symposium on Fire Safety Science*, ETH, Zurich, August.

7 Fire spread beyond room of origin

7.1 Probability of fire spread beyond room of origin

A fire may spread beyond the room of origin if and when a structural member of the room – wall, floor or ceiling – reaches a limit state (condition) and experiences 'thermal failure' or 'collapse' in the extreme case. These two modes of structural failure can normally occur if and when 'flashover' occurs and the fire produces intense heat during the post-flashover stage. Due to uncertainties caused by several factors, a probability is attached to the occurrence of flashover in a real fire starting in a room. Evaluation of this probability has been discussed in detail in Section 5.3. A probability is also attached to the thermal failure of a structural member of a room if flashover occurs. Evaluation of this probability is discussed in Section 7.2.

In the event of a fire, collapse of the structural barriers of the room of origin rarely occurs according to fire statistics. Fires spread beyond the room of origin to an adjacent room or space mostly by convection (advance of flames and hot gases) by penetrating through a structural element, which may 'fail' thermally. A fire may also spread beyond the room of origin through a door or window left open or through some other opening. Such cases can also be considered as thermal failure of a structural element since the fire resistance of the element will reduce to zero if a door or window, forming part of the element, is left open.

As discussed above, the probability, Q, of a fire spreading beyond the room of origin due to the thermal failure of a structural element is the product of the following two components (see Ramachandran 1995):

A – probability of flashover
B – conditional probability of structural failure given flashover

$$Q = A \times B \tag{7.1}$$

As discussed in Section 5.3, statistics on real fires attended by fire brigades provide estimates for A, probability of flashover – see Table 5.6. This probability, as described in Section 5.3.3, is the product of the parameters λ_1 and λ_2 of a stochastic state transition model (STM) discussed in Section 3.4.3 – see also

Figures 3.19 and 5.1. These statistics also provide estimates for Q – see Table 7.1. Q is the product of the three parameters λ_1, λ_2 and λ_3 of an STM.

According to Equation (7.1), an estimate of B, the conditional probability of structural failure given flashover, is given by

$$B = Q/A \qquad\qquad (7.2)$$

Table 7.1 Probability of fire spreading beyond room of origin

Occupancy type		
Area of fire origin	*Sprinklered building*	*Unsprinklered building*
Textile industry		
Production	0.04	0.12
Storage	0.04	0.25
Other areas	0.04	0.15
Chemical etc industry		
Production	0.03	0.08
Storage	0.08	0.20
Other areas	0.03	0.17
Paper etc industry		
Production	0.03	0.08
Storage	0.05	0.21
Other areas	0.05	0.22
Timber etc industry		
Production	0.06	0.21
Storage	0.04	0.28
Other areas	0.04	0.25
Retail trade		
Assembly	0.02	0.13
Storage	0.02	0.22
Other areas	0.01	0.18
Wholesale trade		
All areas	0.03	0.23
Office premises		
Rooms used as offices	–	0.20
Other areas	0.01	0.11
All areas	0.03	0.15
Hotels		
All areas	Less than 1%	0.14
Pubs, clubs, restaurants		
All areas	0.07	0.17
Hospitals		
All areas	0.06	0.04
Flats		
All areas	0.01	0.07

The value of the probability B given by Equation (7.2) is based on fire statistics. This value, for the following two reasons, does not provide a satisfactory indication of the probability of thermal failure and of the fire resistance of the structural barriers of the room. A 'room' as recorded in the fire brigade reports on fires is not necessarily a fire compartment. The figures for number of fires that spread beyond the room of origin, estimated from fire brigade reports, include a small number that spread by destruction of barrier elements, although most of the fires would have spread by convection (advance of flames and hot gases) by penetration through the barrier elements.

Equation (7.1) provides a method for determining an acceptable level for B, the probability of failure of a structural element or compartment by specifying an acceptable level for Q, the probability of fire spreading beyond room (Ramachandran, 1995). An acceptable value for Q can be determined, with due regard to likely consequences in terms of damage to life and property, if a fire spreads beyond the room of origin. The above method would then provide an estimate of the fire resistance required for the structural elements to meet the acceptable level determined for the probability B. Subject to the value specified for Q, the fire resistance for a sprinklered compartment can be reduced, depending on the extent to which sprinklers are likely to reduce A, the probability of flashover. This would increase the value of B but the value of the product Q would be maintained at the same specific level.

Materials in a room involved in a fire can be expected to have an influence on the rate of fire growth and the probability of fire spreading beyond the room. The occurrence of established burning and, to some extent, the occurrence of flashover would depend very much on the material first ignited. This material is specified in the reports on fires attended by the fire brigades in the UK. Table 7.2, based on a past study (unpublished), is an example based on these data for the chemical and allied industries. Most of these fires occurred in unsprinklered buildings but some fires could have occurred in sprinklered buildings. The study did not consider separately sprinklered and unsprinklered buildings. For the industry considered, the overall probability of fire spreading beyond the room of origin is 0.07. Among known materials first ignited, the leading materials with probabilities exceeding the overall probability are, in order, paper (0.16), unspecified waste (0.15), packaging (0.13) and structure (0.10).

Other factors affecting the probability of fire spread beyond the room of origin are time of occurrence of the fire, i.e. day or night; age of the building, i.e. modern or old; and whether a building is single storey or multi-storey. To estimate the influence of these factors, Baldwin and Fardell (1970) applied the logit model (Equation (3.48)) discussed in Section 3.3.3. According to this study, fire brigade attendance time had no influence on fire spread but this factor appears to have some effect, as discussed in Section 10.1.

Table 7.2 Material ignited first – probability of fire spreading beyond room of origin (chemical and allied industries)

Material ignited first	Total number of fires	Number of fires spreading beyond room of origin	Probability of fire spread beyond room of origin
Gases	86	4	0.05
Liquids	436	29	0.07
Textiles	76	3	0.04
Furniture	5	–	–
Structure	30	3	0.10
Fittings	30	–	–
Food	13	–	–
Paper	106	17	0.16
Packaging	61	8	0.13
Lagging	107	1	0.01
Dust, powder, flour etc	96	4	0.04
Electrical insulation	279	15	0.05
Unspecified waste	13	2	0.15
Other	664	33	0.05
Unknown	82	29	0.35
Total	2084	148	0.07

7.2 Performance and reliability of compartmentation

7.2.1 Probabilistic approach

Fire resistant compartmentation has long been the core of fire safety measures specified in building regulations, codes and standards. A building is regarded as being composed of compartments, perfectly isolated from one another, and the spread of fire as taking place by successive destruction (or possibly thermal failure) of the compartment boundaries – walls, floor and ceiling.

If the boundaries of a compartment are of sufficient fire resistance, it is argued, the compartment will not 'fail' for a specified length of time, defined as fire resistant period, R. This performance will reduce the probability of fire spreading beyond the compartment within the period R and thus provide sufficient time for the response and control of the fire by the fire brigade.

Compartment failure can occur if and when a fire grows to a 'flashover' stage. During the post-flashover stage, heat or severity, S, measured in time units (minutes) can attain a high level and cause a progressive deterioration of the structural boundaries (walls, floor and ceiling) which might reach 'limiting states' and violate performance criteria relating to any of the following three aspects. The first aspect, stability, is one of the characteristics of the load-bearing capacity of a structural element. Strength and ductility are the other two characteristics. The second aspect, integrity, is concerned with the strength of a structural element to prevent the penetration (spread) of fire from the fire compartment to an adjacent compartment or space. The third aspect is the thermal insulation provided for a structural element to prevent the spread of a fire/heat from the fire compartment to an adjacent compartment or space.

To delay the time taken by a structural element to reach limiting states, the element is provided with a period of fire resistance, R (minutes), greater than a high level of severity, S. Calculation of such a design value for R can be based on partial safety factors (Section 3.7.2) which is essentially a non-probabilistic method.

A semi-probabilistic method, currently followed by fire safety engineers for estimating the maximum severity, is based on an analytical (deterministic) model such as the 'equivalent time of fire exposure', T_e, discussed in Section 3.7.3.2 – see Equation (3.78). T_e is estimated by a function of density, per m^2, q, of fire load (weight of combustible materials), thermal inertia, c, of the compartment boundaries and ventilation factor, w, estimated through Equation (3.79). Maximum severity is estimated by inserting for q in Equation (3.78), the value of fire load density corresponding to a high fractile value such as 80 per cent or 90 per cent of the fire load density distribution.

Consider, as an example, an office room of brick construction without sprinklers. The room has a floor area (A_f) of 21.96m^2 (= 6.1 × 3.6) and height 3m. The total area of all the bounding surfaces including the ventilation area is

$$A_T = 2(18.3 + 10.8 + 21.96) = 102.12m^2$$

With a height (h) of 1.5m and width 1.2m, the ventilation area (A_v) is 1.8m^2. The ventilation factor w in Equation (3.79) is

$$w = \frac{21.96}{(102.12 \times 1.8 \times 1.225)^{1/2}} = 1.4636$$

According to the CIB Design Guide (CIB W14 1986), the value of c for the office room considered can be taken as 0.07. Also, according to this publication, for office rooms, the fire load density corresponding to 80 per cent fractile of fire load density distribution is 570 MJ/m^2. Using the above values of the parameters in Equation (3.78), maximum severity may be estimated to be 58.4 minutes. Hence, the structural members of the office room considered should be provided with a fire resistance exceeding 59 minutes. The probability of compartment failure in this case is assumed to be 0.2, i.e. 20 per cent. It may be safer to consider a higher fractile of 90 per cent with a fire load density of 740MJ/m^2, such that the fire resistance exceeds the maximum severity of 75.8 or 76 minutes. The probability of compartment failure in this case may be assumed to be 0.1.

The method discussed above, used by some fire safety engineers, is simple but somewhat subjective since it involves an arbitrary selection of the fractile value and assumption about the probability of compartment failure. The probabilistic safety margin provided by this method is really an unknown quantity.

A better approach, used by some fire safety engineers, is a probabilistic method based on the probability distribution of fire severity, S, considered as a random variable in a real fire occurring in a compartment. Fire resistance, R, is regarded as a constant (not variable) in this method. This 'univariate' approach has been described in Section 3.7.4.2, using exponential and normal distributions.

Sufficient data may not be available to estimate the probability distribution of S. In such cases, the method based on Equation (3.78) can be used for the estimation of the mean, μ_s, and standard deviation, σ_s – see Ramachandran (1998a). For the example (office room) considered earlier, the mean value and standard deviation of fire load density, q, are respectively 420MJ/m^2 and 370MJ/m^2. With these values, calculations based on Equation (3.78) show that the values of μ_s and σ_s are respectively 43 minutes and 38 minutes. Normal or exponential distribution may be assumed.

A much better method is the 'bivariate' method regarding both fire severity, S, and fire resistance, R, as random variables in a real fire occurring in a compartment. This method, more commonly known as the Beta method, has been discussed in detail in Section 3.7.3.4. Under this method, the fire resistance required for a structural element or compartment is set equal to or greater than the value given by Equation (3.108) to meet a target failure probability specified through the safety index β.

The probabilistic methods discussed above recognise the fact that, in a real fire, S and R are random variables affected by uncertainties caused by several factors, some of which cannot be controlled. The performance of a structural element or compartment in a real fire would be different from its performance in a fire resistance test or a large-scale compartment test carried out under known and controlled conditions. Due to end and rotational restraints, often present in a building, the fire resistance attained in a real fire can be significantly different from the resistance achieved in a fire resistance test.

Fire resistance is also affected by weakness caused by penetrations, doors, windows or other openings in the structural barriers of a compartment. Pipes, cables, etc for central heating, television, telephone and other services generally pass through holes in walls, ceilings or floors. The openings around such holes should be well sealed, otherwise fire and smoke can spread through the openings, thus reducing the fire resistance of the structural element.

The fire resistance of a door, even if it is rated according to a test, is generally less than that of the wall on which the door is located. Consequently, the fire resistance of the wall will be less than the fire resistance for which it has been designed and tested. The fire resistance of the wall will be practically zero if the door is left open at the time of a fire; heat, smoke and toxic gases spread quickly through open doors. Doors, particularly in office and industrial buildings and department stores, are likely to be kept open for facilitating passage of people and goods and during warm weather conditions. Several mechanical devices are available for closing a door automatically and positively when a fire occurs. Such a device can be coupled with an automatic smoke detector system with a control unit (Langdon-Thomas and Ramachandran, 1970).

Apart from uncertainties governing the development of a real fire, there are also uncertainties associated with the values used for the parameters of a deterministic model. For example, uncertainties quantified by mean, standard deviation and probability distribution are associated with the parameters c and w in Equation (3.78) in addition to the uncertainties associated with the fire load density, q. The values of these parameters depend on the compartment size. It is doubtful whether Equation (3.78) would be applicable to a large compartment.

Uncertainties are also associated with the parameters of a model used for estimating the fire resistance, R, of a structural member. For example, according to an analysis of fire tests of thin wall steel members (Homer, 1979),

$$R = \left[\frac{f \cdot h \cdot m}{g} \right]^{0.8}$$

where h = thickness of insulation, m = total mass of insulation and steel structural member, g = average perimeter of protective material and f = a factor representing the insulation heat transmittance value for the material. Uncertainties are associated with the values used for f, h, m and g and the power (exponent) 0.8. Parameters not included in the model can cause some uncertainties.

Ramachandran (1998a) reviewed in detail the basic features of the probabilitistic methods discussed in this section and Sections 3.7.3 and 3.7.4 for determining the fire resistance required for a structural element or compartment. This simplified guide also contains a brief discussion about the full probabilistic approach, determination of the 'design point' based on first order (second moment) reliability theory and the application of extreme value theory. Further applications of extreme value theory and the reliability approach (next section) were discussed in a recent paper (Ramachandran 2003).

7.2.2 Reliability approach

For any target level specified for the probability of failure, the probabilistic methods discussed in the previous section can be applied to determine the fire resistance required for any of the six structural elements (four walls, floor and ceiling) or for the compartment as a whole. Two questions arise. Firstly, with a change in the notation, if the structural elements have fire resistance periods (minutes) of T_1, T_2, ..., T_6, what would be the fire resistance period of the compartment as a whole? In order to achieve a fire resistance period of T minutes for the compartment, what should be the fire resistance period of each of the six structural elements?

The above two questions can be answered by developing and applying reliability techniques to compartment failure. In reliability technology, a compartment is a 'system' composed of structural elements as 'components'. Reliability is defined as the probability that a component or system will perform its designed-for function without failure in a specified environment for a designed period. In structural fire protection, a structural element or compartment constructed in an environment such as an office building, department store, hotel or hospital should be reliable in not failing within a specified fire resistance period when a fire occurs.

Reliability of structural fire protection and of other fire protection devices, such as automatic detection and sprinklers, is an under-researched topic on which only very few investigations have been carried out so far – see Ramachandran (1998b). This problem involves complex statistical (probabilistic) analyses requiring large amounts of data which are not available at present. Attempts are, however, made in this section to provide a framework for carrying out a reliability analysis of structural (compartment) fire protection.

Consider first the thermal failure of a structural element or compartment. If the ith element ($i = 1, 2, ..., 6$) has a reliability (probability) $R_i(T_i)$ of not failing thermally within T_i minutes, $[I - R_i(T_i)]$ is the probability of failing after T_i minutes. In other words, there is a probability $[I - R_i(T_i)]$ that, after T_i minutes, heat will penetrate through the structural element to an adjacent compartment or space. Consider the simpler case with T_i equal to a constant T such that all the six structural elements have the same fire resistance, T according to fire resistance tests. But the elements may have different reliabilities, $R_i(T)$, due to differences in weakness caused by openings (doors, etc) and other factors.

Thermal failure of the compartment would occur if any of the six structural elements fail thermally and allow the fire to spread to an adjacent compartment or space. Under this assumption, the elements (components) are in a 'series' arrangement in reliability technology in regard to the compartment considered as a 'system'. In this case, according to a fundamental theorem in reliability theory, the reliability, $R_c(T)$, of the compartment in not failing within T minutes is the product of the reliabilities of the structural elements in not failing within T minutes.

$$R_c(T) = R_1(T) \cdot R_2(T) \cdot .... \cdot R_6(T) \tag{7.3}$$

It will be apparent from Equation (7.3) that the reliability of the compartment is likely to be less than the reliability of any of the structural elements, since the values of $R_i(T)$ ($i = 1, 2, ..., 6$) are all less than one. The value of $R_c(T)$ will be significantly reduced if the values of $R_i(T)$ are considerably less than one. If, for example, the reliabilities of the elements, for a given value of T, are 0.95, 0.96, 0.96, 0.97, 0.98 and 0.99, the reliability of the compartment will be 0.82. Such a low probability of 0.82 for compartment success and a high probability of 0.18 for compartment failure may not be acceptable due to consequences in terms of damage to life and property.

Suppose, for example, an acceptable minimum value for compartment reliability is 0.999 with a failure probability less than 0.001. Such a high level of reliability may be necessary due to the fact that the building considered is big with a large number of occupants. A compartment reliability of 0.999 can be achieved by providing fire resistance to the structural elements sufficiently high that the value of the product of their reliabilities exceeds 0.999. This condition can be met, for example, if the reliabilities of each of the four walls is 0.9998 and the reliabilities of the floor and ceiling are 0.9999:

$$(0.9998)^4 \cdot (0.9999)^2 = 0.999$$

To determine the fire resistance required for a structural element to meet a specified target level for reliability, consider a wall in the above example which should have a reliability of 0.9998. For this purpose, it may be considered desirable to apply the Beta method discussed in Section 3.7.4. Following this method, Equation (3.108) may be applied, assuming that both the fire resistance and fire severity have normal probability distribution.

For a success probability of 0.9998, straight interpolation of the figures for 0.9995 and 0.9999 in Table 3.18, would give an approximate value of 3.61 for β but a more accurate value is 3.54 according to a table of standard Normal distribution. If the compartment considered is an office room, as discussed in the example in Section 7.2.1, the values of the mean, μ_s, and standard deviation, σ_s, of fire severity are 43 and 38 minutes respectively. The coefficient of variation V_s is 0.88 (= 38/43). Such a high value for V_s indicates that the survey of fire loads in offices quoted in the CIB Design Guide (CIB W14 1986) had produced inaccurate estimates. Results with a coefficient of variation exceeding 0.2 are generally considered by statisticians to be inaccurate.

For purposes of illustration, assume that μ_s is 45 minutes and σ_s is 9 minutes with a coefficient of variation of 0.2. The mean fire resistance, μ_r, is the parameter to be estimated. For the coefficient of variation V_R, of fire resistance, 0.15 may be considered to be an acceptable value if data are not available for estimating this parameter. Accordingly, the standard deviation, σ_r, of fire resistance is $0.15\mu_r$. Then, according to Equation (3.108), an estimate for μ_r is given by

$$\mu_r = 45 + 3.54\left[(9)^2 + (0.15\mu_r)^2\right]^{\frac{1}{2}} \tag{7.4}$$

$$= 45 + 3.54\left[81 + 0.0225\mu_r^2\right]^{\frac{1}{2}}$$

Converting Equation (7.4) into a quadratic, the following equation would provide an estimate of μ_r:

$$0.718\mu_r^2 - 90\mu_r + 1009.9404 = 0 \tag{7.5}$$

Since the fire resistance should exceed 45 minutes, an estimate of its value is 113 minutes according to a solution for Equation (7.5). This value (113 minutes) for minimum fire resistance for a wall can be expected to meet a success probability, i.e. reliability, exceeding 0.9998 or failure probability less than 0.0002.

The value of the safety factor, θ, in Equation (3.113) is, hence, 2.51 (= 113/45) as also given by Equation (3.115) with $V_R = 0.15$, $V_S = 0.2$ and $\beta = 3.54$. If data are not available for estimating σ_r and σ_s, Equations (3.113) and (3.115) may be applied for estimating μ_r by estimating μ_s according to an equation such as Equation (3.78) and assuming realistic values for the coefficient of variation V_R and V_S for fire resistance and severity.

The value of β is 3.719 (Table 3.18) for a reliability or success probability of 0.9999 for the floor and ceiling. With values of 0.15 and 0.2 for the coefficients of variation V_R and V_S, of fire resistance and severity, Equation (3.115) would provide an estimate of 2.66 for θ in Equation (3.113). Hence, the floor and ceiling should be provided a fire resistance exceeding 120 minutes (= 2.66 × 45) to meet a reliability exceeding 0.9999 or failure probability less than 0.0001.

According to the probabilistic analysis discussed above, fire resistance exceeding 120 minutes for all the six structural elements can be expected to provide a reliability exceeding 0.999 for the compartment considered as an example. Similar calculations can be carried out to determine the fire resistance required for all the compartments in the office building considered or in any other type of building. A much higher level of fire resistance, with a very high reliability, would be required for the compartments of a building which is very big or tall with, say, over one or two thousand occupants. Such huge buildings would also require adequate means of escape, fire detection and suppression systems and fire warning and communication systems apart from structural fire resistance (Ramachandran 2008).

The formula in Equation (7.3) for components in a 'series' arrangement is only valid if the failure of one component does not lead to the failure of any other component in the same system, i.e. the components are independent. This may not be true in the case of thermal failure of a compartment, involved in a fire, with walls, floor and ceiling as components. Progressive deterioration of a wall under severe heat might, in some buildings, affect the performance of the floor or ceiling. A fire-resisting wall may be affected by the deflection of a beam in a fire – see Figure 7.1. Joints and other constructional features are likely to cause such a dependency.

There is a need to carry out further research to modify the formula in Equation (7.3), particularly for structural fire protection, to take account of the interactions between structural elements such as columns and beams with regard to their performance in a real fire. It is necessary to identify these interactions and quantify them in order to identify interactions which may exercise critical effects on the probability of compartment success or failure in a fire – see Ramachandran (1998c).

Only the 'thermal failure' of a compartment in a fire has been considered in the probabilistic and reliability analysis discussed so far in this chapter. There is also a need to investigate the 'collapse' (total destruction) of the structural barriers of a compartment in a fire, although this may be a rare event with a very low probability of occurrence. The risk or probability of occurrence of collapse should be reduced to a very low level, particularly for the compartments of a big or tall building with a

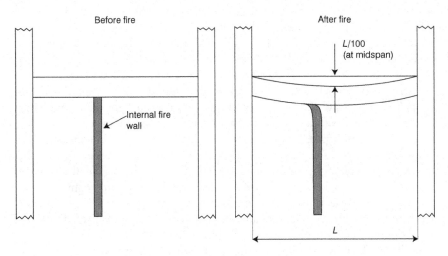

Figure 7.1 Effect of beam deflection on a fire-resisting wall

Source: British Standard BS 5950: Part 8, 1990, section four, page 15

large number of occupants. The application of extreme value theory for estimating the probability of collapse of a tall building against a catastrophic fire and the fire resistance required to reduce this probability to acceptable level was discussed by Ramachandran (2008).

Generally, the fire load contained in the normal contents (furniture etc) of a compartment is unlikely to produce a highly intense heat necessary to cause the collapse of the structural boundaries of the compartment during a short period after the occurrence of flashover. But the boundaries can collapse over a longer period of time after flashover due to progressive deterioration caused by intense heat. Collapse of the structural boundaries can occur immediately after the start of a fire, if the fire causes explosion by igniting a large volume of gas if present in the compartment.

A compartment would collapse if any of the structural elements collapse, particularly due to the interactions between the elements. This assumption appears to be reasonable and would lead to a 'series' arrangement of components (structural elements) in reliability technology with regard to the collapse of a compartment considered as a system. Probabilistic and reliability techniques need to be further developed to predict the occurrence of collapse of a compartment or a building due to fire or explosion.

7.3 Performance and reliability of building structure

As shown by the figures in Tables 7.1 and 7.2, depending on the occupancy type, area of fire origin and material ignited first, about 15 per cent of fires spread beyond the room of origin. This percentage would be reduced by a factor of more than two if the building was sprinklered. The spread, of course, would also depend on other materials involved in the room of fire origin and other materials in the floor of fire origin and in other floors in the building. A fire spreading beyond the room of origin can thus spread throughout the building. Some fires can spread beyond the building of origin and ignite other buildings in the immediate vicinity but the probability of occurrence of such a fire is small according to fire statistics.

A stochastic model such as network model, discussed in Section 3.4.4, can be applied to estimate the probability of fire spread throughout a building. Such an extreme event with a low probability would progressively occur over a long period in minutes, depending on the materials involved and the building size. But it can occur during a shorter period if, in addition to the ignition sources normally present in the building, a highly flammable fuel from a source outside the building is thrust into the building accidentally or deliberately. A recent example of this outside source is the large quantities of jet (aviation) fuel from the two aircraft which were hijacked by terrorists and crashed on the twin towers of the World Trade Center (WTC) in New York on 11 September 2001.

In the WTC disaster, as the burning jet fuel spread across several floors of the two buildings, it ignited much of the buildings' contents, causing simultaneous fires on several floors. Over a period of many minutes, this heat induced additional stresses into the structural frames, damaged by the impact of the aircraft, while simultaneously softening and weakening these frames. This additional loading and the resulting damage were sufficient to induce the collapse of both structures. The progressive collapse of the WTC towers was a result of the combined effects of the impacts of the aircraft and ensuing fires. Collapse of a building can also progressively occur over a short period of time (minutes) if an explosion occurs in the compartment of fire origin due to the ignition of a large volume of gas present in the compartment.

Deterministic and non-deterministic (probabilistic or stochastic) models developed so far do not appear to be capable of assessing the performance in a fire of the structure of an entire building, although the models can assess, as discussed earlier, the performance of the structure of individual compartments in the building. The performance of a building can, perhaps, be evaluated by further developing and applying the reliability technique discussed in Section 7.2.2. For this purpose, a building may be considered as a system composed of several compartments and floors as components. In this context, reliability may be defined as the probability of the entire building not failing thermally for a specified period of minutes. The structural members of the building should be designed to meet this specified reliability standard. This standard should take into consideration the likely damage to life and property if the structure fails. Estimating the performance or reliability of a building structure in not collapsing in a fire is a more complex problem.

The performance of a building structure, with regard to thermal failure or collapse of structural members, cannot be considered in isolation from other passive fire protection measures, e.g. means of escape facilities, and active measures, e.g. detection and sprinkler systems installed in the building. The total performance of the building should be evaluated, taking into account the interactions between all the structural and passive measures and active measures. The total performance should be such that the unreliability, i.e. failure probability, is less than a specified target probability. The target levels for structural failure and for other aspects such as life risk, property damage, consequential losses and societal risk have been discussed in detail in Chapter 4.

As discussed in Section 4.2.3, in determining the target levels and fire protection for a large building, consideration should be given to the probability of occurrence of a multiple-death fire which would depend on the number of people at risk in the building. Consideration should also be given to the probability of a fire spreading beyond the building of origin and involving other buildings in the neighbourhood, leading to a conflagration causing catastrophic damage to life and property and consequential losses. To reduce the probability of occurrence of a conflagration, the distances between buildings in an area should exceed a certain 'critical distance' and the street widths should be designed accordingly.

References

Baldwin, R and Fardell, L G (1970), *Statistical Analysis of Fire Spread in Buildings*, Fire Research Note 848, Fire Research Station, Borehamwood.

CIB W14 (1986). Design guide: structural fire safety. *Fire Safety Journal*, 10, 2, 77–154.

Homer, R D (1979), The protection of cold-form structural elements against fire, *Proceedings of International Conference on Thin-Wall Structures*, Wiley, New York.

Langdon-Thomas, G J and Ramachandran, G (1970), Improving the effectiveness of the fire check door, *Fire International*, 27, 73–80.

Ramachandran, G (1995), Probability-based building design for fire safety, Part 1, *Fire Technology*, 31, 3, 265–275; Part 2, *Fire Technology*, 31, 4, 355–368.

Ramachandran, G (1998a), *Probabilistic Evaluation of Structural Fire Protection – A Simplified Guide*, Fire Note 8, Building Research Establishment, Fire Research Station, Borehamwood.

Ramachandran, G (1998b), Reliability of fire protection systems, Advances in Reliability Technology Symposium, Manchester, UK, April.

Ramachandran, G (1998c), Reliability of structural fire protection, *Book of Abstracts*, Annual Conference on Fire Research, NISTIR 6242, National Institute of Standards and Technology, Gaithersburg, MD.

Ramachandran, G (2003), Probabilistic models for fire resistance evaluation, *Proceedings of the Conference on Designing Structures for Fire*, Society of Fire Protection Engineers and Structural Engineering Institute, Baltimore, MD.

Ramachandran, G (2008), Enhanced structural fire protection for a tall building against a catastrophic fire – probabilistic evaluation of performance and economic value, *Proceedings of the 7th International Conference on Performance Based Codes and Fire Safety Design Methods*, SPFE, Auckland.

8 Performance and reliability of detection, alarm and suppression

8.1 Detection

8.1.1 Human detection

As defined in Section 3.7.3.2, the total time taken by an occupant or group of occupants to evacuate a building involved in a fire is the sum of three periods demarcated sequentially by four critical events: ignition, discovery or detection of the fire; commencement of evacuation and reaching a safe place. The first period is the elapsed time from the start of ignition to perceive or discover the existence of the fire. This time period can be long if the building is not equipped with automatic fire detection systems. In such a building, the occupants have to rely on fire cues such as unusual smells or noises, e.g. breaking glass, which are generally ambiguous or misinterpreted (Canter, 1985).

As mentioned in Section 3.3.2, the fire brigades in the UK provide estimates of the fire discovery time according to four categories – discovered at ignition, discovered under 5 minutes after ignition, discovered between 5 and 30 minutes after ignition and discovered more than 30 minutes after ignition. While the discovery time is zero for the first category, average values of 2, 17, and 45 minutes can be used for human discovery times for the second, third and fourth categories if an automatic detection system is not installed. Based on the number of fires for these four categories for the 14-year period 1978–1991, the average human discovery time was 11 and 12 minutes for single and multiple-occupancy dwellings.

The statistics for 1978–1991 were used in estimating the relationship between the fatality rate per fire and discovery time discussed in Section 3.3.3 – see Equation (3.46). In this investigation, the first discovery time category, discovered at ignition, was not included, since the fatality rate for this category was higher than that for the second category. Excluding the first category, the average discovery time based on the other three categories was estimated to be 13 and 18 minutes for single and multiple-occupancy dwellings; these correspond to the overall fatality rate of 0.012 for both the types of dwellings.

In occupancies such as industrial premises, office buildings and department stores, a fire occurring during the day will be discovered soon after the start of ignition, since occupants and staff will be generally present. Fires occurring in these premises in storage areas and during night times may take longer times to be discovered if automatic detection systems are not installed. The discovery time may be reduced if the premises are patrolled by security personnel.

In hospitals, detection of a fire takes place relatively early in fire development as compared with other occupancies – see Ramachandran (1990). This is likely to be due to the more general spread of people throughout the building and the fact that there is always somebody awake, on duty. According to an analysis of fire statistics compiled by the National Health Service Estates, UK (1996), most fires (60 per cent) in hospitals are discovered by employees, 7 per cent by patients and about 5 per cent by visitors and passers-by. About 26 per cent of fires are detected by smoke detectors and 2 per cent by heat detectors. The vast majority (84 per cent) of fires in hospitals are detected within 5 minutes. In patient care areas, about 90 per cent of fires are detected within 5 minutes, about 32 per cent by smoke detectors, apparently due to a higher coverage by these automatic devices. Where ignition is 'deliberate', the patients themselves may raise the alarm. In non-patient care areas, only 73 per cent of fires are detected within 5 minutes, about 18 per cent by smoke detectors. This may be due to the lower number of staff and lower coverage by automatic detectors in some of these areas of hospitals.

While smoke and toxic gases pose a greater threat than fire to occupants who are remote from the place of fire origin, fire itself is a major threat to occupants in its immediate vicinity, who could be affected seriously by heat and flame. Even if a fire is discovered soon after ignition starts, it may be too late for people in the room of fire origin to attempt any fire fighting or escape. According to UK fire statistics, most of the casualties in single and multiple-occupancy dwellings were found in the room of fire origin. Figures for dwellings reveal that a high percentage (62 per cent) of fatal casualties in the room of origin were caused by burns apart from gas or smoke.

In cases where anyone is present at the ignition and detection stage they are nearly always incapacitated in some way by alcohol, mental or physical retardation, sleep or a combination of these conditions – see Ramachandran (1985). Those present, whose responses are not thus impaired, are usually arsonists. Studies of actual fires suggest that initial response to a fire may be triggered by any of the following four senses – sight, hearing, smell and touch. Sight is the most common, followed by sound, with smell and touch much less often. Visual detection of smoke and less frequently of flames predominates in domestic fires.

8.1.2 Automatic detection

8.1.2.1 Performance

Mathematical (deterministic) models have been developed to calculate the response time of an automatic heat or smoke detector under given conditions of ceiling height, detector spacing and fire/smoke intensity (total heat/smoke release rate). The parameters of these models are generally estimated with the aid of data provided by standard tests and research experiments with detectors, carried out under known and controlled conditions. However, satisfactory operation of a detector head in an actual fire occurring in a room and time of operation are random variables affected by uncertainties caused by several factors. These factors include the location of the seat of a fire in relation to the location of the detector head, the rate of growth of the combustion product, heat or smoke and environmental conditions such as humidity, temperature and ventilation prevailing in the room of fire origin.

Whether or not a smoke detector will respond depends on a number of factors which include smoke aerosol characteristics, aerosol transport, detector aerodynamics and sensor response. Smoke aerosol characteristics at the point of generation are a function of the fuel composition and the combustion state (smouldering or flaming) and include particle size and distribution, composition, colour and refractive index. Once smoke reaches the detector, the response depends on the aerodynamic characteristics of the detector and the type of sensor, ionisation or photoelectric.

According to a Swedish study (Bengtson and Laufke, 1979/80), operating times for heat detectors range from 2 minutes in 'extra high hazard' occupancies such as a plastic goods factory, to about 20 minutes for 'light hazard' such as flats and other residential premises. The operating times of smoke detectors range from 0.5 minutes for 'extra high hazard' to 2.25 minutes for 'light hazard' involving wood materials and to 0.75 minutes for polystyrene (light hazard).

In a series of tests carried out by the Fire Research Station, UK in 1970, heat detectors operated between 1 min 16 sec and 3 min 58 sec of ignition, ionisation chamber detectors operated between 1 min 5 sec and 4 min 30 sec, while optical detectors took over 3 minutes to operate. Infrared detectors operated in about 3 minutes and laser beam detectors took about 5 minutes to operate, if well above a fire.

According to Fire Statistics United Kingdom 1991, published by the Home Office, the proportion of fires discovered in less than 5 minutes in dwellings was 69 per cent for fires discovered by smoke detectors. In other words, the probability of a smoke detector operating within 5 minutes is 0.69. In other occupied buildings, the probability of a smoke detector operating within 5 minutes is 0.78.

Statistical data are lacking for evaluating the probability of a detector system operating in a real fire. A value of 0.8 for this probability has been assumed by Helzer *et al.* (1979) in a study concerned with the assessment of the economic value of different strategies for reducing upholstered furniture fire losses. Apart from other reasons, a detector head would fail to operate if the heat or smoke generated by a fire is insufficient to activate the system.

8.1.2.2 *Effectiveness*

Detectors do not actively take part in fire fighting. However, by detecting and informing the occupants of a building about the existence of a fire when the fire is in its early stages of growth, detectors would enable the early commencement of fire fighting by first-aid means such as buckets of water or sand and portable fire extinguishers and/or fire brigade. Consequently, the fire can be controlled and extinguished quickly before it spreads and causes extensive property damage.

According to Fire Statistics United Kingdom 1991, among fires in occupied buildings detected by smoke detectors, 67 per cent of fires are confined to items first ignited and 0.2 per cent spread beyond the building of fire origin. If smoke detectors are not installed in these buildings, only 36 per cent of fires would be confined to items first ignited and 2.5 per cent would spread beyond the building. For dwellings, the probability of a fire being confined to the item first ignited is 0.68 if a smoke detector has been installed and it operates and 0.41 if a smoke detector has not been installed. According to an analysis of statistics for the period 1960–1967 compiled by Cerberus, a Swiss manufacturer of ionisation detectors, the average fire loss in buildings in Switzerland protected by Cerberus detector systems was only one-third of the average loss in buildings without these systems.

Statistics in the UK have shown that fire brigades are able to control and extinguish fires quicker in buildings protected by automatic fire detection systems than fires in buildings without these systems. The fire brigade control time is reduced by about half a minute for every minute of early arrival of the brigade at the fire scene (Ramachandran, 1992). This reduction would vary from one type of building to another. The saving in control time due to detectors, together with the reduced time in detecting a fire, would considerably reduce the total duration of burning and area damage. The area damage will be further reduced, though not significantly, if the detector system is directly connected to the fire brigade – see Figure 3.11.

Figure 3.11 is based on the exponential model (Equation (3.42)) discussed in Section 3.3.2. This figure is an example (textile industry) showing the sizes of the fire in terms of area damage (m^2) at the times of fire brigade arrival and control for three cases – detector connected to the brigade, detector not connected to the brigade and detector not installed. The figure reveals a significant reduction in area (property) damage due to automatic detectors. The figure also shows that, in the absence of fire brigade intervention and attack, a fire in a textile industry building can burn for more than 54 minutes with a damage exceeding $140m^2$.

The important time in a fire situation is the first five minutes after the start of ignition when the occupants are attempting to escape. Early detection of a fire would enable the early commencement of evacuation which would increase the chance of occupants reaching a safe place before the escape routes become untenable due to heat, smoke or toxic gases. According to the study discussed in Section 3.3.3, and the parameter λ in Equation (3.46), for every minute saved in detection time (and hence in evacuation time), the fatality rate per fire in dwellings would be reduced by about 0.0007, i.e. 7 deaths per 10,000 fires.

Automatic detectors would reduce the fire discovery time in dwellings by about 14 minutes, assuming that they operate, on average, in one minute after the start of ignition. Consequently, if all the dwellings had been protected by automatic detectors, the fatality rate per fire would have reduced to about 0.002, i.e. 2 deaths per 1000 fires from 0.012, i.e. 12 deaths per 1000 fires, which was the average life risk level that prevailed during 1978–1991. This denotes a saving of 10 deaths per

1000 fires. With about 55,000 fires per year in dwellings during that period, 550 lives could have been saved every year if, particularly, smoke detectors had been installed which operated, on average, within one minute. According to a US study (Bukowski *et al.*, 1987), detectors would reduce the fatality rate per fire in single and two-family dwellings by a factor of two, from 0.0085 to 0.0043.

8.1.2.3 *Reliability*

Although automatic detectors are only required to operate in a proportion of fires not discovered by occupants, e.g. less than one-third of fires in hospitals, the system should be available for operation 'on demand' when a fire occurs. If the system is maintained in good working condition by frequent routine checks, it will be 'available' most of the time for operating on demand and raising an alarm when a fire generating sufficient heat or smoke breaks out. Unrevealed mechanical faults or defects will remain unrectified if the system is not checked frequently. Failure of electrical mains will lead to unavailability but this incidence only occurs rarely. The probability of a detector operating on demand whenever it is required to act is the first aspect of reliability of an automatic fire detector (AFD) system.

The second aspect of reliability of an AFD system is concerned with the system not triggering an alarm in a non-fire situation. Examples of such situations are dust, debris and insects in the sensing chamber and system fault. The system should not also trigger alarms in fires which have no potential to spread beyond the point of ignition. Examples of such fires are cooking smoke, bathroom water vapours and cigarette smoke. The two types of situations mentioned above are generally called false or nuisance alarms. Such alarms cause wastage of time and money, particularly to fire brigades whose response to genuine fires may be delayed due to unnecessary call-outs.

Major causes of failure of AFD systems to detect genuine fires are:

*   mechanical faults;
*   malfunction;
*   power surges;
*   power failure;
*   mechanical damage or abuse after installation; and
*   accumulation of dirt and dust.

For estimating the relative frequencies of failure due to different causes, 'global statistics' at the national level do not appear to be available for any country.

Since 1994, fire brigades in the UK provide in their fire reports (FDRI) information on whether a detection system was installed in the building involved in a fire, and if so, whether the system operated or not and on the reasons for non-operation. But the Home Office only processes data provided by a sample of fires to produce national statistics. Hence, it is difficult to use the fire statistics compiled by the Home Office to estimate reliably the probability of a detector operating in a genuine fire and the frequencies of non-operation due to different causes. The reports on fire incidents in UK hospitals compiled and analysed by National Health Service Estates do not contain information on fires in which AFD systems failed to operate and the causes of such failures.

As mentioned earlier, major causes of false or nuisance alarms are signals from cooking smoke, e.g. burnt toast, steam from boiling water in an electric kettle, bathroom water vapours, cigarette smoke, dust, debris and insects in the sensing chamber, system fault and lack of maintenance. Fry and Eveleigh (1975) analysed data collected in a survey on detector actuations carried out by UK fire brigades in 1968. They estimated a ratio of 11:1 between false and genuine calls to the fire brigades. They attended 11 false alarms to every genuine fire attended and extinguished by them. The ratio was 11:1 for heat detectors and 14:1 for smoke detectors. Mechanical and electrical faults, especially defective wiring of heads, accounted for 46 per cent of false calls. Ambient conditions, especially extraneous heat and smoke, accounted for 26 per cent of false calls. According to Davies (1984), 95 detector systems of a Swiss manufacturer gave 1329 false calls as opposed to 85 genuine alarms (a ratio of 16:1).

Reasons for nuisance alarms arising from smoke detectors in homes have been identified by the National Smoke Detector Project carried out by the US Consumer Product Safety Commission – see the Commission's final report (1993) on the first study, 'Smoke Detector Operability Survey – Report on Findings'. As pointed out in this report, power sources for a high percentage of smoke detectors in homes are intentionally disconnected because of nuisance alarms. The report also suggested several potential solutions to address this problem. Repeated false alarms for an organisation may cause a fire brigade to cancel connection facilities, thus exposing the organisation to increased fire risks.

In 1995, the Home Office, UK, introduced a reporting form FDR3, in which fire brigades were asked to furnish information on false alarms attended by them. In this form, false alarms are classified according to three main categories: malicious, good intent and due to apparatus. The third category has been further classified into the following four sub-categories – dust/thrips, system fault, unsuitable equipment or positioning and 'other'. The breakdown figures for these four sub-categories were furnished by 23 brigades in 1995 and 50 brigades in 1996. According to an analysis of these figures by Ramachandran (unpublished), 22 per cent of false alarms were due to dust, insects etc., 37 per cent due to system fault, 3 per cent due to unsuitable positioning and 38 per cent due to 'other'.

As mentioned earlier, statistical data are lacking for evaluating the 'on demand' probability of a detector operating in a genuine fire. In estimating this probability, it is not appropriate to include small fires in which the heat or smoke generated is not sufficient to activate the system. This success probability is as high as 0.95 according to some manufacturers, provided, of course, the system is maintained all the time in a satisfactory working condition. The failure probability is 0.05. Manufacturers usually use the exponential probability distribution to calculate the failure rate of their product per hour by carrying out laboratory tests and then convert the results to estimate the failure rate per year. Most of the manufacturers are reluctant to disclose the data and method adopted by them in the calculation of the failure rate.

If an item has an exponential lifetime probability distribution, 63 per cent of the specimens of the items would fail within a time t when tested in a laboratory; 37 per cent would survive beyond the time t. Test data would provide an estimate of t which is known in reliability theory as mean time to failure (MTTF) for a non-repairable item or mean time between failure (MTBF) for a repairable or replaceable item. MTTF or MTBF is the average time likely to lapse before the item fails after it is installed or replaced after it had failed. The reciprocal $(1/t)$, where t is MTTF or MTBF, is the failure rate which is usually expressed as failure rate per year.

The Weibull probability distribution can be applied to estimate the failure rate of an AFD system for detecting genuine fires. The failure rate for this distribution is not a constant but a function of time since the system has been installed or replaced. Formulae have been derived to estimate the parameters of this distribution and MTTF or MTBF. To estimate these parameters, data are required on the date of installation or replacement of each of the AFDS systems installed in an occupancy and the dates on which the system has failed. If such data are available, the Weibull distribution can be fitted to the data and the parameters estimated to assess whether the age of the system is a factor affecting the failure rate.

For some types of buildings, information may be available on the dates of occurrence of false alarms and types of fire detection systems producing these alarms. The Weibull distribution can be fitted to time periods between successive dates of false alarms to estimate the failure rate – see, for example, Peacock and Sutcliffe (1982). Failure, in this context, is the failure of an AFD system to block the communication of information about the false alarm to occupants of a building, control panels or to the fire brigade. A conventional heat or smoke detector does not have the capability to discriminate between genuine fires and false alarms. But such a capability has been incorporated in some more 'modern' computer controlled AFD systems such as analogue addressable systems, which have been developed during the past two decades. However, the reliability of these modern systems depends on the data collected by the sensors and the method (algorithm) adopted in the computer software (program) installed in the system.

In an addressable system, signals from each detector and each call point are individually identified at the control panel. Each circuit is a form of simple data communication rather than simply an electrical circuit. Within the software of an addressable system, the device identity can be converted into a pre-programmed location, which is then displayed on some form of text such as an LCD or vacuum fluorescent display. The control panel of a conventional heat or smoke detector system cannot identify the detector head from which a signal is communicated to a panel.

Among the conventional types, heat detectors are generally the most reliable in terms of component failure, since these devices respond directly to the presence of heat by a physical change in the detector operating elements. Heat detectors

~~do not react to smouldering fires~~ which do not generate heat. A heat detector does not have an open measurement chamber and is only affected by increases in temperature. The fire will only be detected when the temperature in the room of the fire origin has reached a threshold level. This simplicity leads to very few unwanted alarms. Heat detectors require practically no maintenance.

Since an optical type smoke detector has an open measurement chamber, it is susceptible to dirt and dust accumulation which will lead to lower sensitivity. An ion type smoke detector also has an open measurement chamber which must be cleaned as often as necessary. Otherwise the chamber will become dirty or dusty, leading to an increase in the sensitivity of the detector and unwanted alarms.

Addressable detection systems can be tested and checked by a printout of the measurement data in the central control unit. Detectors with measurement data which have changed beyond a given limit must be cleaned. The others will operate normally and can be used without any further inspection until the next routine check. For the reasons mentioned above, service and maintenance are simplified substantially in the case of addressable detectors, meaning far greater reliability than conventional systems.

The four major components of an AFD system are:

- detector heads;
- zone control panels;
- central control panel; and
- connection to fire brigade.

Each group of detector heads located in a specific place, e.g. ward in a hospital, are generally connected to a particular zone panel. The zone panels are connected to a central panel which is connected to the fire brigade.

Gupta (1984/85) estimated the hardware failure rates of components of an automatic fire detection (AFD) system at a psychiatric hospital by analysing the component structure and configuration of the system. The system comprised a distributed system of ionisation type smoke detectors, break-glass units and heat detectors. All of them were connected into various zone panels, which in turn were connected into a central control unit and three repeater panels. The control unit carried a fire area identity annunciation and was the means for the receipt of the alarm for activating the hospital's audible fire alarm and the transmission of a fire brigade call-out signal. The failure rates of electronic components were obtained from Military Standardization (1974).

For ionisation smoke detectors, Gupta estimated the total mode failure rate as 0.057 faults per year of which 0.04 faults per year were of the safe type and 0.017 faults per year were dangerous. Since these figures applied to a first class environment, factors such as air speed and humidity were taken into account and the failure rate was assessed to be 0.46 faults per year, which was eight times greater than for a first class environment. For break-glass unit, the total relevant failure rate was assessed to be 0.032 faults per year of which 0.018 faults per year were of the safe type and 0.014 faults per year were dangerous. The figures did not include the spurious alarm rate due to misuse of the system. For control units, Gupta estimated the total failure rate for unrevealed dangerous type as 0.06 faults per year – 0.042 faults per year for indicator control module and 0.018 faults per year for monitor unit.

Studies by Finucane and Pinkney (1988) showed overall failure rates for control units varied from 0.25 faults per year, up to one fault per year, with an unrevealed fail-to-danger rate of typically 0.1 faults per year. According to these authors the overall failure rates for all types of detectors were 0.1 faults per year with fail-to-alarm failure rates varying from 0.01 faults per year to 0.1 faults per year. For all types of detectors, Appleby and Ellwood (1989) estimated a failure rate of 0.02 per year. According to them, the failure rates of addressable systems can be expected to be considerably less than 0.02 per year.

Failure rates for electrical components and sub-components are available from data banks such as those maintained by UK Atomic Energy Authority (UKAEA) and AEA Technology and Reliability Analysis Center, New York. These data can provide some indication of the failure rates of AFD system components.

For all types of detectors, false alarm rate per detector per year has been estimated in some studies:

- 0.025 by Finucane and Pinkney (1988);
- 0.025 by Appleby and Ellwood (1989);
- 0.092 by Peacock and Sutcliffe (1982) for Lincolnshire hospitals;
- 0.044 by Bukowski and Istvan (1980) for USA hospitals;
- 0.06 quoted by Caffolla (1997) for St Paul's hospital, Vancouver;
- 0.047 by Japanese Association of Fire Science and Engineering (1982), for Japanese health care facilities; and
- 0.1 by Pearce (1986) for UK health care premises.

Gupta (1984/85) divided false alarms into four categories:

1   failure of equipment;
2   'non-fire' disturbances;
3   'external' effects;
4   'unknown' reason for alarm.

The second category included causes such as cigarette smoke, steam, dust and smoke/vapour from cooking. The third category included human error, water from leaks, power supply interruption or surge, electrical interference such as arcing or switching and birds, animals and insects.

Gupta analysed data collected from various sites on time periods between successive events for the categories mentioned above except the first. He fitted the Weibull probability distribution to these data in order to understand the statistical behaviour of the events. Parameters of the distribution were estimated using maximum likelihood and least square methods. The mean of the 'scale' parameter, interval between successive events, varied between 10 and 42 days for the second category ('non-fire'), 16 and 40 days for the third category ('external effects') and 14 and 60 days for the fourth category ('unknown'). The value of the 'shape' parameter in all the three categories was less than 1.0 (except in 'external effects' for one site) which indicated a decreasing failure rate for the occurrence of the events.

On the other hand, the mean duration between two successive false alarms at various sites varied between 5 and 15 days. This duration had little bearing on the type of site, according to results presented in the paper. Little variance in the 'shape' parameter observed for different sites within a given category indicated a common cause of false alarm. The explanation for this phenomenon was perhaps the attribution to the increasing effectiveness, efficiency and better maintenance policies of the fire officers of sites.

For any type or group of buildings, statistical data may be available for estimating the overall reliability of the whole of an AFD system in detecting a genuine fire or suppressing a false alarm. But the overall reliability of an AFD system for a particular building within a type or group depends on the reliabilities of the components and sub-components constituting the system and the types of connections between them. Estimation of the system reliability of an AFD system for a particular building is a complex problem on which practically no research studies have been carried out so far. In developing a model for this problem, the following general points may be considered.

In the case of detecting a genuine fire, the four major components of an AFD system, mentioned earlier, are in a 'series' arrangement in the context of reliability theory. Failure of any of the components to transmit information about the existence of a fire to the next component in the arrangement will lead to system failure For a series arrangement, the overall reliability of a system is the product of the reliabilities of the components. Failure of a detector head to operate when a fire occurs in the area protected by it may be considered to be system failure. However, the fire may be detected by any other head in the vicinity of the fire area or in the place considered. Hence, detector heads in a specified place may be considered to be in a 'parallel' arrangement. For a parallel arrangement the overall unreliability of a system is the product of the unreliabilities of the components. Unreliability is the probabilistic counterpart of reliability. But failure of one or more heads in a place will lead to a delay in commencing the evacuation and fire fighting and increase in life risk. The system is, therefore, a mixture of series and parallel arrangements of main components and sub-components.

In the case of false alarms, none of the detector heads in a place should respond to a signal from a non-fire situation for the system to succeed in not communicating the signal. Hence, the heads in a specified area may be considered to be in a series arrangement. Even if such a signal is transmitted to a zone panel, further communication of the signal can be blocked by this panel or by the central panel. The system would succeed if any of the four major components suppress information about a false alarm. Hence, the components can be regarded as being in a parallel arrangement.

Apart from the main components mentioned above, there are sub-components of an AFD system such as cables/wires, power supply and discriminator software (for an analogue addressable system).

Practically, any level of sensitivity to fire, in particular to signals of combustion, can be achieved with existing technology. However, if sensitivity is set at a high level, a detector may pick up signals given by spurious fires from sources such as cigarette smoking and cooking which are normal activities. On the other hand, a low level of sensitivity can increase the risk of genuine fires being undetected.

Some studies, e.g. Cholin (1975), have suggested that the gap between reliability in detection and rate of unwanted and false alarms may be closed to some extent by cross-zoning of detectors where the activation of the alarm is delayed until a second detector is activated. Another approach suggested by Custer and Bright (1974) is the use of multi-mode detectors requiring signals from several fire signatures before a fire alarm is initiated.

The technique known as 'coincidence' or 'double knock' will automatically withhold or limit a fire signal given by a detector until the presence of the fire is confirmed by the response of a second detector in an independent circuit at a different location. 'Gating' is another technique for automatically withholding a fire signal given by a single detector until the presence of fire is confirmed by its second response within a pre-determined time period. Techniques such as those mentioned above might reduce the frequency of false alarms but may, at the same time, cause undue delays in the operation of a detector in a genuine fire.

Advanced computer-controlled addressable detection systems such as analogue and multi-state types can provide improved capability for discriminating between a genuine fire and a false or unwanted signal from a non-fire source or a small fire with a local influence such as a puff of cigar or pipe smoke. An analogue system gives a 'pre-alarm warning' if the signal from a detector exceeds a certain threshold level and a 'fire warning' if the signal exceeds a higher threshold level. At a very low threshold level, a fault signal is given. The signal level represents the amount of heat, smoke or flame that is being sensed. In a multi-state system, each detector is capable of transmitting several states such as fault, normal, pre-warning and fire.

It is necessary to design a cost-effective AFD system which is capable of achieving the right balance between detection of genuine fires and suppression of false alarms. An acceptable level needs to be determined for the ratio between number of false alarms and number of fires attended by fire brigades.

8.2 Alarm

8.2.1 Conventional alarms

The alarm, when raised, is a signal for occupants to evacuate or to be alert in preparation for evacuation. The alarm should result in the fire service being summoned, so that they can start fire fighting and if necessary assist in evacuation. The most common form of fire alarm sounder used is the electric bell. Its sound is able to carry through a building. It can ring as an intermittent pulse or a continuous sound signifying respectively alert and evacuation. Where a large site is involved, it may be necessary to use a siren. Where there will be difficulty in hearing an alarm due to disabled occupants or noisy equipment, then a visual signal, for example, flashing lights, will be necessary. The alarm may be required to wake sleeping occupants. There have been a number of fires where lives have been lost through the lack of an alarm system.

In buildings which have resident staff, visitors, patients or inmates, the fire alarm is required to alert the staff and to commence a pre-arranged fire evacuation plan. Low-note buzzers or electronic sounders can be particularly useful in situations where it is necessary to warn staff, rather than the occupants, of fire. In shops and shopping complexes, the alarm can take the form of a coded message to alert staff to the existence of a fire and its location. In hospital premises and prisons, it is normal to evacuate initially the zone of fire origin as part of a staged evacuation plan. Visitors are usually asked to leave the building while other occupants make their own way or are assisted to a safe refuge. Conventionally, in these instances, fire alarm bells ringing continuously for evacuation and intermittently for alert are used. Coded messages are also sometimes used.

The fire alarm system can also be used to activate other systems such as door hold-open devices, motorised ventilation dampers/controls, powered smoke extraction systems, fire extinguishing systems and staff call systems. Self-contained smoke detectors with integral alarms could provide an earlier warning and so significantly reduce the number of deaths.

However, a conventional fire alarm described above or even an automatic detector can be insufficient in being able to convey an effective warning of fire (Tong and Canter, 1985). A delay in the accurate perception of fire threat would also lengthen the 'recognition time', the second component period, denoted as B, of the total evacuation time, H, discussed in Section 3.7.3.2. During this period, from perception to commencement of evacuation, people try to gain enough information to confirm the existence of a fire. Even direct information of fire (such as smoke) may be insufficient to motivate evacuation due to inaccurate perceptions of the rate of fire growth (Canter *et al.*, 1988).

8.2.2 Informative fire warning systems

For the reasons mentioned above, it is necessary to communicate to the occupants of a building, timely, adequate and convincing information about the existence, location and spread of fire. Such information can be provided by computer-controlled informative fire warning systems (IFWS) which have been developed over the last few years in the UK and in other countries. Ramachandran (1991) reviewed the research carried out in the UK on IFWS and the basic features necessary in these systems for motivating rapid and safe evacuation.

The Fire Research Station (FRS), UK, developed the prototype of an IFWS known as BRESENS (Pigott, 1983). In systems such as BRESENS, simple electronic design should ensure that no switch, sounder or sensor will give rise to a fire alarm when one of these or the associated wiring develops an open or short circuit condition. The outputs of fire 'sensors' in the prototype equipment installed at FRS demonstrated the theoretical possibility of tailoring the output for an approximate sensitivity to the ratio between false alarms and genuine fires. It is possible to calibrate the sensor outputs and then compensate them for drift with time. The amount of compensation necessary can itself be used as a measure of probable serviceability, permitting demand maintenance.

The FRS specified its prototype equipment with liquid crystal 16-character text displays, with display units placed at strategic points throughout a building. These displays were programmed to show the time and date in normal use and, when relevant, to indicate that an alarm test was being carried out. During a fire alarm the system can display the word 'Fire' with two levels of detail – the general area of the fire for display everywhere, and with a room number for those in the immediate area of fire origin.

To improve the performance and quality of a system like BRESENS, under contract from FRS, Surrey University commenced, in 1981, an investigation to determine the nature and effective methods of presentation of appropriate information through automatic informative fire warning (IFW) systems. A series of studies was carried out. A summary of methods, results and conclusions relating to these studies are contained in Ramachandran (1991).

To appreciate the need for IFW systems, it was necessary to assess the deficiencies of responses to conventional fire alarm systems. For this purpose, a street survey (Tong and Canter, 1985) of randomly selected individuals and fire drills in three buildings were carried out. Results revealed that a conventional alarm provokes deficient responses such that it may be ignored or it may simply be the first alerting signal in a complex process whereby a need is created to explore the reason for the alarm sounding. An alarm will be ignored if it is interpreted as a circuit malfunction or system test. As pointed out earlier by Tong (1983), inadequate behavioural responses of occupants in many cases were due to a failure of conventional alarm systems to provide a clear and unambiguous fire warning.

Certain psychological criteria need to be met if effective fire alarm systems are to be manufactured and installed – see Tong (1983) and Canter *et al.* (1988). These include a clear meaning of a fire alarm and a valid indication of the presence of a fire and its location. Also, building occupants should be provided with information on the most appropriate response to an alarm and on the escape routes available.

Case studies (Tong, 1983), based on training and fire drills, identified the following problems in the provision of information by existing means. A visual indication about the zone of alarm origin by some form of indication equipment was highly centralised in some cases where no information was directly available to building occupants about the location of the fire. Information had to be obtained from the point at which the zone indicators were installed by telephone, personal attendance or radio paging devices. Problems were encountered in all three methods. Other problems included inadequate amount of information and locationally imprecise information.

The weaknesses of existing fire alarms and communication systems are being rectified by advanced IFW systems that improve the input of information to the central control equipment and the output of information from the central control. Information is taken directly from each individual 'addressable' detector and messages are displayed that enable people to distinguish fire alarms from other alarms. To significantly reduce delays in evacuation, IFW systems should be backed up by efficient and organised management. Large delays can occur, for example, from a lack of exits of sufficient width or lack of provision of equipment, e.g. wheelchairs for facilitating the evacuation of disabled people.

The messages to be displayed visually on an IFW system should be determined with due regard to an optimum balance between urgency and detail. To tackle this problem, it would be useful to consider the recommendations of Canter *et al.* (1988) for message length, use of abbreviations, message specificity, message formulation, type of information and message format. In a care establishment, for example, the precise location of the fire (e.g. 'linen store') should be displayed only on the affected floor while a less specific message (e.g. 'Ward G7') is displayed on a non-affected floor. Location addresses should first refer to the most general area (e.g. block or wing) followed by a more specific address (e.g. floor number) and finally by the most precise address (e.g. ward or room). Location addresses should take priority over the inclusion of any other type of information. Updated information on fire/smoke spread should be displayed along with, but separate from, the address of the fire source.

Instructions displayed on an IFW system should only be targeted at the general population of building occupants. Instructions to individuals should be communicated by other means. If information is displayed on a VDU-like node, the use of grouping as a means of emphasising specific information is preferable to organising text in a long string. More detailed and specific messages would only be appropriate for buildings such as hospitals where there is a stable population and not for sport areas, shopping centres and hotels where the population is more transient and will only be able to react to very simple instructions.

Under a contract from the Fire Research Station, Technica carried out an investigation to explore the possibility of using microprocessor technology more sophisticated than an IFW system such as BRESENS (16-character LCD display) – see Bellamy (1989). Technica performed two laboratory experiments by presenting to subjects (members of the public) different warning displays (lists of choices) for information acquisition, message identification and action decision. The subjects' responses were recorded automatically on a computer. Five building types were included – residential block, hotel, hospital, department store and office building. The complete details including statistical design and analysis of these experiments are described in the final report (Technica Ltd, 1990). Ramachandran (1991) presented a summary of this report.

The first experiment was based on a $(3 \times 2 \times 2 \times 2)$ factorial design with three modes each with high and low quality, two levels of building familiarity and two levels of threat. The three modes were:

1   Graphic – computer-generated colour mimics based on building floor plans
    and elevations; isometric (i.e. not perspective)
        High quality – 3-dimensional
        Low quality – 2-dimensional
2   Visual display of text messages
        High quality – computer VDU
        Low quality – LCD (BRESENS)
3   Auditory (speech)
        High quality – computer*-generated
        Low quality – fire bell
    *Commodore Amiga PC model A1000

Subjects familiar with building plans were considered 'trained'; otherwise they were 'untrained'. 'High threat' referred to a fire located on the ground floor directly below the specified location of the subject. A 'low threat' fire was a fire located on the third floor above the location of the subject. The 24 combinations provided by the factors and their levels were 'replicated' by using 48 males and 48 females (four subjects to each combination) with repeated messages for each subject across building types. Each subject was tested on one mode, one level of familiarity and one level of threat.

The results of the first experiment are given in Table 8.1. As shown in this table, the three component response times were analysed separately. The ranks of each of the six modes for the mean of each of the response times are also shown in Table 8.1, together with the ranks of the probability (%) for the three cases indicated in columns 2 to 4. The results are shown in brackets. For each of the modes, the mean of the ranks over the six attributes (columns 2 to 7) is shown in the last column.

The results in Table 8.1 reveal that 3D graphic and Amiga speech are the best modes. With these two modes, the second experiment (7 × 2 factorial design) was performed with seven multi-mode combinations, specified in Table 8.2, constituting the main factor and familiarity as the second factor. 'IFWS speech' referred to modified computer (Amiga) generated speech while 'speech' was the unmodified mode. 'Alarm' referred to computer (Amiga) generated alarm sound. The subjects were 56 males and 56 females with four males and four females for each of the 14 factorial combinations.

The overall mean for the total response time for all factorial combinations was 43 seconds in the first experiment and 38 seconds in the second. The second experiment (not the first) revealed that mode had a significant effect on the total time. The longest total response time of 50 seconds was for the combination modified 3D/IFWS speech/alarm and the shortest time of 31 seconds was for 3D/ speech.

The main implication of the first experiment was that IFW systems could produce as much as a six-fold increase in the probability (proportion) of occupants evacuating immediately when compared with a conventional fire warning. The evacuation initiation probability could be further enhanced by adopting the mode combination 3D/IFWS speech. Familiarity with a building increased the

Table 8.1 Experiment 1: Summary of results and overall rankings (best to worst) for modes (ranks are shown in parenthesis)

| Mode | Dependent variable | | | | | | | | | | | | |
	Genuine fire (location correct) warning interpretation %		Genuine fire warning interpretation %		Immediate evacuation %		Warning acquisition time (secs)		Warning interpretation time (secs)		Action decision time (secs)		Mean of ranks
Graphic 3D	73	(1.5)	81	(1)	64	(1)	18	(5)	14	(1)	12	(3)	2.1
Graphic 2D	73	(1.5)	75	(2)	45	(2.5)	19	(6)	16	(3)	14	(5)	3.3
BRESENS	56	(4)	61	(4)	44	(4)	12	(3.5)	18	(4.5)	12	(3)	3.8
Amiga Text	48	(5)	52	(5)	28	(5)	10	(2)	24	(6)	16	(6)	4.8
Fire bell	5	(6)	13	(6)	11	(6)	6	(1)	18	(4.5)	12	(3)	4.4
Amiga speech	70	(3)	72	(3)	45	(2.5)	12	(3.5)	15	(2)	11	(1)	2.5

Table 8.2 Experiment 2: Summary of results and overall rankings (best to worst) for modes (ranks are shown in parenthesis)

Mode	Dependent variable						
	Genuine fire (location correct) warning interpretation %	Interpretation as a genuine fire warning %	Immediate evacuation %	Warning acquisition time (secs)	Warning interpretation time (secs)	Action decision time (secs)	Mean of ranks
Fire bell	3 (7)	16 (7)	8 (7)	6 (1.5)	16 (6)	10 (1.5)	4.3
3D/IFWS speech	91 (1)	92 (1.5)	77 (1)	13 (4)	10 (1.5)	11 (3)	2.0
BRESENS/ IFWS speech	88 (3)	89 (3)	63 (2)	18 (5)	12 (4)	12 (5)	3.7
Modified 3D/ IFWS speech/ alarm	70 (5)	77 (5)	42 (5)	20 (6.5)	14 (5)	15 (7)	5.6
3D/IFWS speech/alarm	84 (4)	84 (4)	50 (4)	20 (6.5)	11 (3)	12 (5)	4.4
3D/speech	89 (2)	92 (1.5)	58 (3)	11 (3)	10 (1.5)	10 (1.5)	2.1
Amaiga alarm	6 (6)	33 (6)	23 (6)	6 (1.5)	18 (7)	12 (5)	5.2

probability of interpreting a warning as a genuine fire emergency but such a correct interpretation did not guarantee immediate evacuation, which depended largely on the mode of warning.

Conventional fire detectors would reduce the delay in discovering a fire but may not reduce the 'recognition time' from discovery to commencement of evacuation. IFW systems have the potential to reduce both the two time components and hence to reduce the total time before the commencement of evacuation. Consequentially, IFW systems have the potential to reduce life risk significantly – see Ramachandran (1993).

8.3 Suppression

8.3.1 First-aid fire fighting

If a fire is discovered soon after ignition and its size is small, the occupants of a building attempt to fight and extinguish the fire using first-aid methods. Such an initial attack on a fire usually involves 'sundry means' such as buckets of water or sand, smothering, garden hose, immersion and beating. According to a sample of fires in dwellings analysed by Ramachandran *et al.* (1972), 'sundry means' were used by occupants in about 44 per cent of the fires. About 43 per cent of these were successfully extinguished; fire brigades had to extinguish only the remaining 57 per cent. Occupants used portable fire extinguishers of the types such as dry powder, water and carbon dioxide in only about 3 per cent of the dwelling fires. Only 27 per cent of these fires were extinguished, with the fire brigades putting out the remaining 73 per cent.

If a fire is attacked by the occupants with first-aid methods, even if the fire is not extinguished, the fire brigade can be expected to bring such a fire under control quicker than a fire which is not attacked by first-aid methods. According to the analysis of Ramachandran *et al.* (1972), the mean (average) 'control time' of fire brigades was 6.7 minutes for all fires with initial attack by first-aid methods and 9 minutes for all fires with no initial attack. The mean control time was 6.5 minutes for all 'sundry means' and 8.9 minutes for all portable fire extinguishers.

The analysis discussed above did cast some doubts about the effectiveness of portable fire extinguishers. It is possible that the extinguishers in dwellings were located at considerable distance from places of fire origin, e.g. in cars or garages. An analysis by Sime *et al.* (1981) indicated that people have inadequate knowledge of the location of extinguishers. According to Chandler (1978), the success rate of extinguishers used in hospital fires was less than the success rate of other methods. This conclusion was confirmed by Canter (1985) who suggested that people (especially staff in hospitals, hotels, etc) should be made aware of the location of extinguishers and trained in the use and capabilities of different types and sizes of extinguishers.

It should be pointed out, however, that a number of small fires extinguished by portable fire extinguishers were not reported to the fire brigade. The percentage of such fires was 70 per cent according to the UK Fire Extinguishing Trade Association

(FETA) – see *Fire Prevention*, March 1990. The percentage of fires extinguished by portable fire extinguishers was 80 per cent in 2002, according to a later (March 2003) report of FETA. This report on a survey into portable fire extinguishers and their use contains a detailed analysis of the effectiveness of extinguishers in the UK and in five other EU member countries – Austria, Belgium, Germany, France and Netherlands. Across these six countries, it was found that in 81.5 per cent of incidents, fires were successfully put out by extinguishers. This report (March 2003) of FETA provides some guidance and recommendations on the siting (location), maintenance and the use of different types of extinguishers. Fire brigades should be called quickly even if an attempt is made by occupants to fight the fire by first-aid methods.

8.3.2 Sprinklers

8.3.2.1 Performance

In essence, an automatic sprinkler system is a fire-fighting system designed to be operated by the fire itself, so as to dispense water in the area where it is needed to ensure rapid suppression of the fire with minimum damage to property. The salient feature of the system is an adequate water supply which can be pumped through a network of pipes, usually at ceiling level, to a series of sensitive 'sprinkler heads' which are designed to respond to the thermal conditions created by the fire. Thus, only those heads which have been affected by the fire will operate and allow water to flow from them to be distributed in the form of a spray onto the fire below. Sprinklers are generally required to operate at an average temperature of 68°C but there are special requirements for certain occupancies.

All sprinkler systems can be categorised as one of four basic types; they differ in terms of how the water is put into the area of the fire. Wet pipe systems and dry pipe systems use automatic sprinklers, while deluge systems, instead of automatic sprinklers, use open sprinklers. The fourth type is similar to a deluge system, except that automatic sprinklers are used.

Several factors cause uncertainties in the activation and operating times of sprinklers in actual fires, although deterministic models using experimental data have been developed to estimate the response times of different types of sprinklers. Based on factors such as rate of temperature rise, height of upper fire surface above the floor and height of the premises, Bengtson and Laufke (1979/80) estimated sprinkler operation times varying from 2.5 minutes for 'extra high hazard' occupancies to 16.8 minutes for 'light hazard'.

According to statistics on actual fires attended by UK fire brigades, in a sprinklered building, there is a 55 per cent chance that a fire may not produce sufficient heat to activate the system such that it is either self-extinguished or extinguished by first-aid methods. In the remaining 45 per cent of fires requiring sprinkler intervention, the system operates in 87 per cent of cases and does not operate in 13 per cent of cases. According to an investigation carried out by the Fire Research Station, UK, quoted by Rogers (1977), one-third of fires in sprinklered buildings are extinguished by the system and are not reported to the brigade. Hence, fire brigades attend only two-thirds of fires in sprinklered buildings such that, on the whole, sprinkler intervention is required in 63 per cent (= (2/3 × 0.45) + 0.33) of fires. Also, sprinklers operate in 59 per cent (= (2/3 × 0.45 × 0.87) + 0.33) of fires. Sprinklers, therefore, operate in 94 per cent (= 0.59/0.63) of fires in which their action is required.

Rutstein and Cooke (1979) estimated, for various types of occupancies in the UK, the percentages of fires in which sprinklers operate satisfactorily, which range from 92 per cent to 97 per cent. Based on data for a hundred years, Marryatt (1988) estimated for Australia and New Zealand a success rate of over 99 per cent for sprinklers. The success rate for sprinklers in the USA was about 96 per cent for the period 1897–1964 according to the National Fire Protection Association (NFPA), 85 per cent for the period 1970–1972 according to the Factory Mutual Research Corporation (FMRC) and 95 per cent for the period 1966–1970 according to the US Navy. The figures mentioned above were quoted in a study by Miller (1974) who also estimated success rates of 86 per cent for wet systems, 83 per cent for dry systems and 63 per cent for deluge systems.

The reasons for sprinkler failure (non-operation) in the UK were investigated by Nash and Young (1991). The main causes were shut valves (55 per cent), system fault (7 per cent) due to problems in design or manufacture and other and unknown causes (38 per cent). According to the NFPA investigation mentioned above, out of the 4 per cent of failures, 36 per cent were due to system shut down; of these, 85 per cent could probably be attributable to human error. Sprinkler stop valves were shut in one out of every 74 fires, water supply inadequate in one out of every 276 fires and pipework blocked in one out of every 550 fires.

8.3.2.2 Effectiveness – property protection

The effectiveness of a sprinkler system in controlling fire spread or extinguishing a fire can be assessed in terms of number of heads operating. According to an analysis by Baldwin and North (1971), 75 per cent of fires are controlled or extinguished by four heads or less, 80 per cent by five heads or less and 98 per cent by 35 heads or less. The corresponding figures for Australia and New Zealand (Marryatt, 1974) are 90, 92 and 99 per cent respectively. According to Rees (1991), 69 per cent of fires are controlled by 5 heads or less, 83 per cent by 10 heads or less and 94 per cent by 25 heads or less; these results are based on data published by the FMRC for the years 1978–1987. The figures mentioned above demonstrate that only sufficient sprinkler heads required to control a fire will activate, thus reducing the amount of water damage and fire loss.

Consequences of water damage and accidental leakage have been used as arguments against the installation of sprinklers in certain areas, e.g. computer centres, art galleries and libraries. Case studies of some individual fires reveal that losses due to water damage were not appreciable and the chance of a leakage occurring is very small. Additional loss due to water damage is likely to be smaller than that which would result from further fire spread in the absence of sprinklers.

Effectiveness of sprinklers in reducing property damage has been well documented, discussed and established in fire protection literature. Some of these studies were mentioned in Chapter 3 (Tables 3.12, 3.13 and 3.14) in regard to financial loss and in Section 3.3.1 and Chapter 5 (Table 5.7 and Figures 5.1 and 5.2) in regard to extent of fire spread and area damage. Fire loss data compiled by FMRC for the period 1980–1989, for a wide range of occupancies, indicate that the average fire loss for an unsprinklered building is approximately 4½ times the average loss for an adequately sprinklered building (Rees, 1991). An analysis of NFPA data for the period 1980–1990 has shown that the reduction in average loss per fire ranges from 43 per cent for stores and offices to 74 per cent for educational establishments (Hall, 1992).

Based on Home Office statistics for the years 1981 to 1987, Beever (1991) found that the probability of fire damage exceeding a given area is very much smaller with sprinkler protection than without. According to research carried out by Morgan and Hansell (1984/85), one in ten fires in offices will exceed 16m^2 in a sprinklered building but 47m2 in a building without sprinklers.

As discussed in Section 3.3.2, with reference to the exponential model of fire growth, sprinklers would reduce the rate of fire growth in a fire developing beyond the stage of 'established burning'; this action would increase the 'doubling time' of the fire. Sprinklers would also reduce the probability of flashover occurring in a fire in a compartment – see Section 5.3 and Table 5.6. Melinek (1993a) assessed the effectiveness of sprinklers in reducing fire severity, expressed in terms of area damage. He estimated that sprinklers would reduce the probability of fire size in industrial and commercial buildings reaching 100m^2 by a factor of five. He also found that damage to the structure of a building would be reduced by a factor of 2.5.

8.3.2.3 *Effectiveness – life safety*

Although sprinklers have been designed primarily for reducing property damage in industrial and commercial buildings, they have the potential to reduce life risk in non-industrial buildings, particularly those with large numbers of people at risk. Such non-industrial buildings include retail shops and department stores, office buildings, cinemas, theatres, clubs, pubs, restaurants, hotels, hospitals, flats and apartment buildings. These buildings may, perhaps, require the installation of fast response sprinkler systems which operate and detect fires quicker than those designed for industrial properties. Such systems may provide extra time, particularly for occupants not in the room of fire origin, to escape to safe places within or outside the building of fire origin.

Even if a fire is discovered soon after the start of ignition by automatic detection or sprinkler systems, it may be too late for people in the room of fire origin to attempt any fire fighting or escape. According to UK fire statistics, most of the fatal and non-fatal casualties in dwellings were found in the room of fire origin. Statistics reveal that the fatality rate per fire, for fires discovered at ignition, is higher than the fatality rate for fires not discovered at ignition, but discovered within five minutes after the start of ignition – see Ramachandran (1993). Fast-response domestic sprinklers can, perhaps, save some of the casualties in the room of fire origin by extinguishing the fire quickly. According to the specifications of the NFPA, USA, residential sprinklers should act five times quicker than commercial and industrial sprinklers.

Sufficient statistical data are not available for estimating the number of lives that could be saved by installing sprinklers in large non-industrial buildings. However, tentative results are available from some statistical studies. Comprehensive data for Australia and New Zealand for a 100-year period up to 1986 show that, during that period, there were 11 deaths in 9,022 fires in sprinklered buildings (Marryatt, 1988). This represents a fatality rate of 0.0012 per fire. According to Hall (1992), if sprinklers are installed in one- and two-family houses in the USA, 63 to 69 per cent reduction in death rates per thousand fires can be achieved. Under some assumptions Melinek (1993b) estimated that, if all the buildings in the UK were sprinklered, the number of fatal casualties in fires would be reduced by half and the number of non-fatal casualties by about 20 per cent. He has also shown that sprinklers can significantly reduce the number of multi-casualty fires.

Ramachandran (1993) attempted to estimate the probable reduction in life risk if sprinklers are installed in all single and multiple occupancy dwellings in the UK. For this analysis, he expanded the regression model in Equation (3.46), by considering also the parameter K given by

$$K = \lambda (B + E - F) \tag{8.1}$$

where, as defined in Section 3.7.2.2, B is the 'recognition time' and E the 'design evacuation time'. The time period F is the time taken by a combustion product, e.g. smoke, to travel from the place of fire origin and produce an untenable (lethal) condition on an escape route.

By retarding the rate of fire growth, sprinklers would increase the value of F in addition to decreasing the value of D in Equation (3.46) by acting as an automatic detection system. Ramachandran assumed that sprinklers would reduce the average value of the discovery time D to 3 minutes from 15.5 minutes with a saving of 12.5 minutes and increase the average value of F by 4 minutes. He also assumed that sprinklers will not affect the values of B and E. Under the above assumptions, with a total saving of 16.5 minutes and $\lambda = 0.0007$, applying Equations (3.46) and (8.1), Ramachandran estimated that sprinklers would reduce the fatality rate per fire in dwellings to 0.0009 from the level of 0.0124 for unsprinklered dwellings revealed by fire statistics. The fatality rate of 0.0009 for sprinklered dwellings is not significantly different from the rate of 0.0012 estimated by Marryatt (1988) for all sprinklered buildings.

In a report (unpublished) to the Fire Research Station, UK, Ramachandran (1999) investigated the effectiveness of sprinklers in reducing life risk in non-industrial buildings. According to his results, reproduced in Table 8.3, sprinklers would reduce the probability of flashover in fires in the five occupancies considered

Table 8.3 Effectiveness of sprinklers – non-industrial buildings

Occupancy type	Probability of confinement to item first ignited		Probability of flashover		Probability of spreading beyond room		Average area damage (m²)	
	Sprinklered	Unsprinklered	Sprinklered	Unsprinklered	Sprinklered	Unsprinklered	Sprinklered	Unsprinklered
Office buildings	0.58	0.37	0.25	0.53	0.03	0.15	5.69	13.22
Hotels	0.85	0.30	0.07	0.50	Less than 1%	0.14	2.75	21.29
Pubs, clubs restaurants	0.59	0.26	0.26	0.62	0.07	0.17	8.95	23.98
Hospitals	0.87	0.54	0.12	0.30	0.06	0.04	0.57	7.59
Flats	0.90	0.53	0.04	0.37	0.01	0.07	1.85	2.91

by a factor of two or more. Consequently, sprinklers considerably reduce the probability of a fire in these buildings (except hospitals) spreading beyond the room of origin. Sprinklers also significantly reduce the average area expected to be damaged in fires in non-industrial buildings.

Some communities in the USA have adopted ordinances which promote the use of residential sprinklers. The leading ones are Scottsdale, Arizona; Prince George's and Montgomery Counties, Maryland; Greenburgh, New York; and Cobb County, Georgia. According to a recent investigation (Butry *et al.* 2007) in benefit-cost analysis of residential sprinkler systems, in terms of fire-risk mitigation, multipurpose network systems achieve greater cost-effectiveness over alternative systems.

8.3.2.4 *Reliability*

As discussed in Section 8.3.2.1 there is a small chance that, for various reasons, a sprinkler system may not operate when it is required to act in an actual fire. This probability for failing to operate 'on demand' can vary from 15 per cent to 1 per cent depending on the type of the system and the manufacturer. Combining all the data discussed in Section 8.3.2.1, the following estimates appear to be reasonable for conditional probabilities (percentages) if failure occurs:

- 55 per cent due to system shut off for maintenance;
- 17.5 per cent due to shut sprinkler stop valves;
- 7.5 per cent due to defective system;
- 5 per cent due to inadequate water supply;
- 2.5 percent due to blocked pipe work; and
- 12.5 per cent due to other causes.

Consider now, as an example, a reliability of 0.92 for sprinkler operation when a fire occurs. The unreliability or failure probability is 0.08. If, on average, 200 fires occur every year in a group of buildings with 2000 sprinkler systems, the overall annual failure rate per system is likely to be 0.008 (= (200/2000) × 0.08). The mean time between failures (MTBF) for each system is hence, approximately, 125 years (= 1/0.008). Based on the conditional probabilities estimated earlier, the annual failure rates per system for different causes are likely to be:

- 0.0044 due to system shut off for maintenance;
- 0.0014 due to shut sprinkler stop valves;
- 0.0006 due to defective system;
- 0.0004 due to inadequate water supply;
- 0.0002 due to blocked pipe; and
- 0.001 due to other causes.

The reciprocals of these annual failure rates are the corresponding MTBFs in years.

The probabilities discussed above provide some 'global' estimates for the reliability of a sprinkler system installed in a group of buildings. This reliability can be improved by controlling (reducing) or eliminating the unreliabilities associated with some or all the causes of failure. The system reliability of sprinklers installed in a particular building within the group would, however, depend on the reliabilities for the different component parts of the system. The main components are sprinkler heads (considered as a group), pipe work and water supply from public mains or elevated private reservoirs, gravity tanks etc. The reliabilities (success probabilities) for these three components may be denoted by R_H, R_p and R_w respectively. One minus the reliability is the unreliability or probability of failure.

The failure of any of the three main components would result in the system failing to operate when a fire occurs. For such a 'series' arrangement, according to reliability theory, the system reliability, R_s, is given by the product

$$R_s = R_H \cdot R_p \cdot R_w \qquad (8.2)$$

A group of sprinkler heads located in a specific area may be considered to be in 'parallel' arrangement, since the failure of one head to operate will not lead to system failure; another head located close to the failed head may operate. Hence, the reliability, R_H, of a group of sprinkler heads may be estimated by

$$R_H = 1 - (1 - R_{H1}) (1 - R_{H2}) (1 - R_{H3}) \dots \qquad (8.3)$$

where R_{H1}, R_{H2}, R_{H3}, ... are the reliabilities of individual heads. Equation (8.3) follows from the reliability theorem that, for a parallel arrangement, the unreliability of a system is the product of the unreliabilities of the components. It should be noted, however, that failure of one or more heads would reduce the effectiveness of the system in controlling fire spread.

The model discussed above is a simple framework for evaluating the reliability of a sprinkler system for a particular building, based on the reliabilities of its components. This framework needs to be expanded to consider other components and sub-components such as water pressure. This is a complex problem on which practically no research has been carried out so far.

Satisfactory operation to release water is the primary function of a sprinkler system. Raising an alarm is the secondary function. The alarm may not sound in the event of a fire but this will not affect the primary function. Due to some causes, a sprinkler head may operate in a non-fire situation, raise an alarm and discharge water until the time the sprinkler stop valve is shut. Such a 'false' operation of sprinklers is a rare occurrence.

Blockages of a sprinkler system can occur due to debris entering the pipe work. This problem applies mainly to systems being installed in new buildings where sections of the system are erected during the early stages of construction when building materials and debris are widespread on the site. Poor maintenance can cause the failure of a sprinkler system to operate when a fire occurs.

References

Appleby, D and Ellwood, SH (1989), A fire detection system using distributed processing, *Proceedings of the 9th International Conference on Automatic Fire Detection*, NIST, Gaithersburg, MD

Baldwin, R and North, M A (1971), *The Number of Sprinkler Heads Opening in Fires*, Fire Research Note 886, Fire Research Station, Borehamwood.

Beever, P (1991), How fire safety engineering can improve safety, Institution of Fire Engineers Meeting, Torquay, UK, October.

Bellamy, L (1989), Informative fire warning systems: a study of their effectiveness, *Fire Surveyor*, August, 17–23.

Bengtson, S and Laufke, H (1979/80), Methods of estimation of fire frequences, personal safety and fire damage, *Fire Safety Journal*, 2, 167–180.

Bukowski, R W *et al.* (1987), *Hazard 1 Vol 1. Fire Hazard Assessment Method*. Report NBSIR 87-3602, National Bureau of Standards, Gaithersburg, MD.

Bukowski, R W and Istvan, S M (1980), *A Survey of Field Experience with Smoke Detectors in Health Facilities*, NBSIR 80-2130, National Bureau of Standards, Gaithesburg, MD.

Butry D T, Brown M H, and Fuller S K (2007), *Benefit-Cost Analysis of Residential Fire Sprinkler Systems*, Report NSTIR 7451, National Institute of Standards and Technology, Office of Applied Economics, Building and Fire Research Laboratory, Gaithersburg, MD.

Cafolla, D (1997), The impact of unwanted fire alarms on the provision of fire safety in healthcare facilities, Dissertation for Master of Science in Fire Safety Engineering, University of Ulster.

Canter, D (1985), *Studies of Human Behaviour in Fires: Empirical Results and Their Implications for Education and Design*. Fire Research Station, Borehamwood.

Canter, D, Powell, J and Booker, K (1988), *Psychological Aspects of Informative Fire Warning Systems*, Fire Research Station, Borehamwood.

Chandler, S E (1978), *Some Trends in Hospital Fire Statistics*, Current Paper CP 67/78, Building Research Establishment, Watford.

Cholin, R R (1975), Reappraising early warning detection, *Fire Journal*, 69, 2, 54–58.

Custer, R L P and Bright, R G (1974), *Fire Detection: The State of the Art*, Technical Note 839, National Bureau of Standards, Gaithersburg, MD.

Davies, D (1984), Means of cutting down false alarms in automatic systems, *Fire*, 77, 9–14.

Finucane, M and Pinkney, D (1988), *Reliability of Fire Protection and Detection Systems*, Report SRD R431. Safety and Reliability Directorate, Atomic Energy Authority, Warrington.

Fry, J F and Eveleigh, C (1975), *The Behaviour of Automatic Fire Detection Systems*, Current Paper CP 32/75, Building Research Establishment, Watford.

Gupta, Y P (1984/85), Automatic fire detection systems: aspects of reliability, capability and selection criteria, *Fire Safety Journal*, 8, 105–117.

Hall, J R (1992), *US Experience with Sprinklers: Who Has Them, How Well Do They Work*, National Fire Protection Association, Quincy, MA.

Helzer, S G, Buchbinder, B and Offensend, F L (1979), *Decision Analysis of Strategies for Reducing Upholstered Furniture Fire Losses*, NBS Technical Note 1101. National Bureau of Standards, Gaithersburg, MD.

Japanese Association of Fire Science and Engineering (1982). *Bulletin*, 32, 6.

Marryatt, H W (1974). The Australian experience with automatic sprinklers. *Australian Fire Protection Association Journal*, September, 2–7.

Marryatt, H W (1988), *Fire: A Century of Automatic Sprinkler Protection in Australia and New Zealand, 1886–1986*, Australian Fire Protection Association, Boxhill, Victoria.

Melinek, S J (1993a), Effectiveness of sprinklers in reducing fire severity, *Fire Safety Journal*, 21, 299–311.

Melinek, S J (1993b), Potential value of sprinklers in reducing fire casualties, *Fire Safety Journal*, 21, 275–287.

Military Standardization MIL-HDBK.217B (1974), *Reliability Prediction of Electronic Equipment*, Naval Publications and Forms Centre, Philadelphia, PA.

Miller, M J (1974), Reliability of fire protection systems, *Chemical Engineering Progress*, 70, 4, 62–67.

Morgan, H P and Hansell, G O (1984/85), Fire sizes and sprinkler effectiveness in offices: implications for smoke control design, *Fire Safety Journal*, 8, 187–198.

Nash, P and Young, R A (1991), *Automatic Sprinkler Systems for Fire Protection*, Paramount Publishing, London.

National Health Service (NHS) Estates (1996), *Fire Practice Note 9 – NHS Healthcare Fire Statistics*, Her Majesty's Stationery Office, London.

Peacock, S T and Sutcliffe, N (1982). *Report on False Alarms for Fire Alarm Systems in the Lincolnshire Area Health Authority*, University of Bradford, Bradford.

Pearce, N (1986), Fire alarm systems in healthcare premises, *Fire Surveyor*, April.

Pigott, B B (1983), Computer-based analogue fire detection systems, *Fire Engineers Journal*, 43, 128, 23–26.

Ramachandran, G (1985), The human aspects of fires in buildings – a review of research in the United Kingdom, *Fire Safety: Science and Engineering*. ASTM STP 882, American Society for Testing and Materials, Philadelphia, PA.

Ramachandran, G (1990), Human behaviour in fires, *Fire Technology*, 26, 2, 149–155.

Ramachandran, G (1991), Informative fire warning systems, *Fire Technology*, 27, 1, 66–81.

Ramachandran, G (1992), *Statistically-Determined Fire Growth Rates for a Range of Scenarios: Part 1: An Analysis of Summary Data. Part 2: Effectiveness of Fire Protection Measures – Probabilistic Evaluation*, unpublished report to the Fire Research Station, Borehamwood.

Ramachandran, G (1993), Early detection of fire and life risk, *Fire Engineers Journal*, December, 33–37.

Ramachandran, G (1999), *Reliability and Effectiveness of Sprinkler Systems for Life Safety*, unpublished report to the Fire Research Station, Borehamwood.

Ramachandran, G, Nash, P and Benson, S P (1972), *The Use of Fire Extinguishers in Dwellings*, Fire Research Note 915, Fire Research Station, Borehamwood.

Rees, G (1991), Automatic sprinklers; their value and latest developments, *Fire Surveyor*, October, 9–13.

Rogers, F E (1977), *Fire Losses and the Effect of Sprinkler Protection of Buildings in a Variety of Industries and Trades*, Current Paper CP 9/77, Building Research Establishment, Watford.

Rutein, R and Cooke, R A (1979), *The Value of Fire Protection in Buildings*, Report 16/78. Scientific Advisory Branch, Home Office, London.

Sime, J, Canter D and Breaux, J (1981), University team studies use and success of extinguishers, *Fire*, 73, 909, 509–512.

Tong, D A (1983), Human response to fire alarms, *Fire Surveyor*, 12, 4, 4–8.

Tong, D and Canter, D (1985), Informative warnings: in situ evaluations of fire alarms, *Fire Safety Journal*, 9, 267–279.

US Consumer Product Safety Commission (1993), *Smoke Detector Operability Survey: Report on Findings*, US Government Printing Office, Washington, DC.

9 Performance and reliability of human response and evacuation

9.1 Recognition

As defined in Section 3.7.2.2, the total time taken by an occupant or group of occupants, say, on a particular floor of a building to reach a safe place within or outside the building is the sum of three time periods. In sequential order, the first period is the time taken to discover a fire by occupants or detect the fire by automatic detection systems after the start of the fire. The second period is the 'recognition time' or 'gathering phase' which has been discussed in several research studies concerned with human behaviour in fires. The third is the evacuation time (Section 9.3) relating to the period from the commencement of evacuation to reaching a safe place. The actions of the occupants during the three periods mentioned above would depend, to some extent, on whether they are in the room of fire origin or in some other room – see Figures 9.1 and 9.2 reproduced from Ramachandran (1993a).

Consider the second period which is the subject matter of this section. This period is the elapsed time from detection or discovering fire (Figure 9.1) or receiving fire information (Figure 9.2) to commencing evacuation (Figures 9.1 and 9.2). It is not possible for occupants to put into motion any actions to cope with the fire until someone has identified that the fire is present and has acted upon that identification by informing others. As pointed out in several studies on human behaviour in fires, the early stage of fire recognition is typically characterised by ambiguity – see Canter (1980). It is clear that early acceptance of the fact that the unusual circumstances present constitute a fire of some severity is frequently delayed to a dangerous extent. Once a fire has been recognised as such, there is then the possibility for a range of actions, including first-aid fire fighting and commencement of evacuation (Section 9.3).

Statistics on fires attended by fire brigades provide some information on the first period of total evacuation time, the fire discovery time by occupants. Laboratory experiments with different types of automatic detection systems can provide estimates for this period, if any such systems have been installed in a building. But neither statistics on fires nor experiments can provide any estimates for the second period, recognition time, for any type of occupancy. Some data for this period for a few occupancy types can be obtained from case studies discussed in the papers of

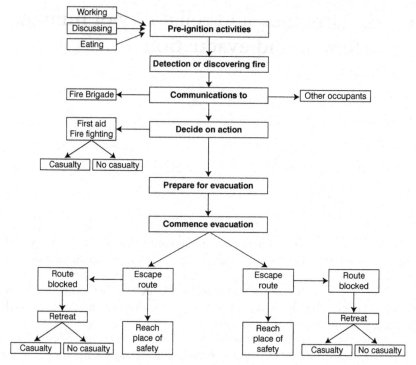

Figure 9.1 Occupants' response – room of fire origin

some authors on human behaviour in fires – see Canter (1980). For several types of occupancies, on average, the recognition time is likely to be about 2 minutes – see Canter (1980) and Ramachandran (1993a, 1993b).

To collect satisfactory data on recognition time it would be necessary to carry out evacuation exercises or computer simulated laboratory experiments with 'subjects' (members of the public) as discussed with reference to the results produced in Tables 8.1 and 8.2.

Sime (1991) defined recognition time as 'time to start to move' from the onset of an alarm or discovery of a fire by someone in a building or device such as a smoke alarm. A delay in this time exceeding two minutes is a fundamental problem which characterises large-scale fire disasters involving injury, including those at the Bradford City Football Club ground (1985), Kings Cross Underground Station (1987), Isle of Man Summerland Recreation Complex (1973), Kentucky Beverley Hills Supper Club (1977) and Woolworths (1979).

9.2 Response

Reviewing research in the UK on the human aspect of fires in buildings, Ramachandran (1985) discussed occupants' response to learning about a fire. As pointed out by him, initial response may be triggered by four senses: sight, hearing,

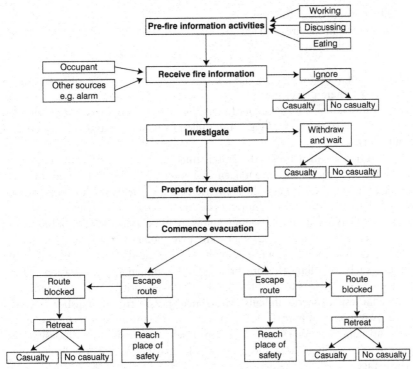

Figure 9.2 Occupants' response – room other than room of fire origin

smell and touch. Studies of actual fires suggest that sight is the most common, followed by sound, with smell and touch much less often. Visual detection of smoke, and less frequently of flames, predominates in domestic fires. Several group-residential buildings in which fires were detected by sight had audible alarm systems installed and in others flames rather than smoke were seen. All group-residential buildings in which fires were heard had audible alarm systems, though in one case the sound of a bang preceded the sounding of the alarm bell. Sometimes, the alarm bell is ignored due to the 'normality' of false alarm sounding. In a domestic fire detected by smell, a sleeping occupant may be awakened by a choking sensation from smoke emanating from downstairs.

Responses of those discovering a fire independently of the first person, simultaneously or after, are less well documented. The discovery is usually by sight or sound as it is if knowledge of the fire is received via the initial respondent. A noise stemming from a fire may be ignored if it is thought to be attributable to some other source, e.g. slamming the door.

In general terms, occupants of residential buildings respond to real fire situations by one or more of the following actions:

- seeking to confirm the existence of the fire;
- fighting the fire;

- telling others about the fire;
- escaping from the fire; and
- assisting others to escape.

Their choice of these reactions or the priority they place on each varies according to their perception of danger, responsibility and location. In cases where an individual is incapacitated mentally, physically, by sleep or by alcohol, the ability to respond normally is often dulled or frustrated as a consequence of impaired senses.

If a fire occurs in a hotel, the initial human response is usually to seek it out. Hotel staff (particularly management) seem more likely to carry out this investigation than their guests. In domestic fires it is the man who confronts the fire and delegates the job of phoning the fire brigade to his wife or others. A nursing auxiliary's response to seeing a glow in an occupied ward of a hospital is usually to enter the ward and attempt to remove the blazing object, e.g. a chair.

People do not often fight a fire immediately, perhaps because of more pressing priorities such as calling the fire brigade and commencing evacuation. Those whose initial response is to fight the fire are in the minority. Hotel staff and those with similar responsibilities in halls of residence operate fire extinguishers only as a secondary response. Where domestic fires evoked the fire-fighting response, it was usually an immediate one, using a garden hose or water supply with easy access rather than an extinguisher which, if available, may be located at a distance, say, in the garage.

Escaping from the fire seems to be the most common response. Generally, those threatened by a fire elsewhere seem more willing to stay in their room and undergo the risk of leaping from windows. In domestic fires, some occupants insist on returning to their bedrooms to retrieve their cherished belongings, even though such an action would put them at risk of inhaling smoke and dying eventually. Mentally handicapped people and people who have been drinking heavily do not respond positively and quickly to escaping from a fire and hence are likely to sustain fatal or non-fatal injuries. In some hotel fires, guests following a signalled escape route faltered at a bolted emergency exit, unable to work out how to operate the lock.

Telling others about a fire can be for the purpose of warning potential victims or informing the fire brigade directly or by delegation. Warning potential victims can be from a neighbour or passer-by who noticed smoke coming from a window.

Those with responsibility for people in normal day-to-day life seem to apply that role equally in fire situations to assisting them to escape. People with such a responsibility include duty staff in a hospital, the husband of a bed-bound woman and the staff of a hotel.

9.3 Evacuation

Once an occupant or group of occupants of a building recognise the existence of a fire, they will decide to commence evacuation and escape to a safe place within or outside the building. This course of action has to be achieved before the available escape routes become untenable due to the build-up of excessive or lethal levels of heat, smoke or toxic gases. The occupants should move quickly to the safe place from the places in the building where they were located at the time when the existence of the fire was recognised. This period of occupants' movements is the evacuation time defined in Section 9.1 and Section 3.7.2.2 as the third component period of the total evacuation time. For a successful evacuation, the total time, which is the sum of discovery or detection time, recognition time and evacuation time, should be less than the time taken by a combustion product, e.g. smoke, to travel from the place of fire origin and produce an untenable condition on the escape route.

The mobility of occupants during evacuation depends on a wide range of ambulancy from mobile to totally immobile. Considering a hospital, for example, some patients may only need minimum assistance but the non-ambulant may require maximum assistance in preparing for evacuation and during evacuation. Factors contributing to non-ambulancy include physiological deficiencies, whether through mental capability limitations or temporary reduction of ability because of sleep, drugs or alcohol. A factor enhancing the escape potential is training, through fire drills, in the use of first-aid fire fighting devices such as portable fire extinguishers and in following safe escape routes. Training and publicity through sign boards are particularly important if routes to be used normally are blocked temporarily or permanently for some reason.

Evacuation of disabled people from fire has been the subject of intensive code activities and research during the past three decades.

National concern for life safety of people with disabilities was first marked by a seminar, 'Fire Safety for the Handicapped' held in Edinburgh in 1975 – see Marchant (1975). The 1985 edition of the Life Safety Code of the National Fire Protection Association, USA, focused on life safety for people with disabilities in situations where many such people were found and where there was a record of serious life loss from fire. Special requirements for disabled people were introduced in the 1985 edition of the National Building Code of Canada. Based on some substantial early effort of the Home Office, the British Standards Institution, in 1988, issued BS 5588, Part 8, Code of Practice for Means of Escape for Disabled Persons, which included extensive guidance on the use of elevators (lifts) for egress during fires.

Reviewing the above-mentioned code activities concerned with the evacuation of disabled people, Pauls and Juillet (1993) raised significant questions about the adequacy of building codes and fire codes which existed at that time to deal with the life safety of people with disabilities. The authors concluded that the codes were deficient in relation not only to people with disabilities but also to all building users. Other important research studies on the topic of evacuation of disabled people include those of Dunlop (1993), Rubadiri *et al.* (1993), Shields *et al.* (1996, 1998), Walsh (1998) and Yoshimura (1998).

While fire is a major threat to occupants in its immediate vicinity, it is generally smoke and toxic gases which pose the greatest threat to occupants who are remote from the fire. Hence, as suggested in Figures 9.1 and 9.2, the following two categories should be distinguished for evaluating the evacuation success of occupants of a building:

1   occupants in the room of fire origin;
2   occupants in other rooms.

The second category can be further sub-divided into

1   occupants in other rooms in the floor of fire origin;
2   occupants on other floors.

Occupants on floors above the floor of fire origin are at greater risk than those on floors below the fire floor.

Occupants in the room of fire origin may become aware of the fire immediately and directly after it started, whereas those in other rooms may receive information about the fire either directly from those in the room of fire origin or indirectly through fire alarm or other communication systems. If the room of fire origin is unoccupied or unprotected by automatic fire detectors or sprinklers, the fire discovery time will be long and highly variable. If the room is occupied, someone in the room will discover the fire, perhaps, earlier than a detector system, if it has been installed. Fire detectors are only designed to reduce the delay in discovering a fire, particularly in an unoccupied area. On the other hand, computer based informative fire warning systems (Section 8.1.2) are capable of reducing the delays in both discovery time and recognition time.

As discussed above, for evaluating the evacuation or egress success of occupants of a building involved in a fire, it is important to consider the location of the occupants in relation to the location of the fire, mobility of the occupants, training through fire drills, delay in response or recognition and whether effective detection and other communication systems and suppression systems such as sprinklers have been installed or not. The behaviour of occupants will be significantly influenced by whether they are alone or with a group. Generally, group reactions will exhibit a greater inertia than individuals. Other factors include the size, density and distribution of occupants. People with mobility handicaps should have their work spaces on the ground floor, if possible.

The last, but not the least, important factor affecting escape potential is the escape system designed for a building including the number, location and capacity of the escape route network and continuing ease of use of the routes throughout the duration of the fire. The clarity and simplicity of the egress system and dead-end corridors could also affect the number of persons ultimately escaping. Uncertainty in the successful evacuation of a building can also be caused by a variety of escape routes available to the occupants and the need to choose the safest route.

As fire, smoke and toxic products continue to spread, the number of routes available for escaping will diminish and vary with each group of the escaping population. Continuing degradation of the environment and reduction in the choice of escape routes could change the mode of escape for groups of the population not yet evacuated. Hence, in parallel to the movement of people involved in evacuation, it is necessary to analyse the uncertainties in the development of fire and combustion products.

9.4 Design evacuation time

It is necessary to evaluate the evacuation time for determining an appropriate design for a given building with a specified number of occupants and their mobility characteristics. The design parameters include number of floors, number of rooms on each floor and number, location and capacity (widths etc) of the escape routes. The values of these parameter depend on the design value estimated for the evacuation time.

The time it takes an occupant to move from their starting point to a location of safety is simply a function of travel speed and distance.

Travel time (s) = distance (m) / travel speed (m/s)

Distance is a function of the escape route selected by the occupant. This selection process is affected by the occupant's familiarity with the building layout, the availability of the escape route, segments (corridors, staircases etc) on the escape route and the degree of difficulty in moving through each segment.

Travel speed is affected by the occupant's mobility and by the mobility of any accompanying occupants; by crowding, light levels, presence of smoke, quality of the floor and wall surfaces (e.g. roughness, unevenness etc), width and riser height of stairs, width of doorways and corridors etc. The travel speed is also affected by training through fire drills.

Some of the above-mentioned factors are occupant characteristics; others are building characteristics. The designer or engineer must deal with each factor explicitly or be able to justify why a factor is not relevant to the analysis.

A range of values need to be calculated to predict the movement component of total evacuation time. These include clearing time for each segment of the escape route, time for a person to travel the longest (most remote) route and location of individuals so that exposure levels can be calculated. The calculation method chosen will depend to a great extent on the values needed for the evaluation of a design.

To estimate the evacuation time, information is needed on the following people movement characteristics:

- Speed: rate of travel along a corridor, ramp, stair;
- Floor: number of persons passing through a particular segment of the escape route per unit of time (e.g. persons/sec passing through a doorway or corridor);
- Specific flow: flow per unit of width of the egress segment (e.g. persons/sec/m of corridor width).

Most of the information on people movement has been collected in fire drills.

The parameters have been investigated in many research studies for the movement of able-bodied and disabled people on stairs, in corridors and through doorways. Calculation methods have been suggested for estimating the evacuation times for specific types of buildings – see, for example, Pauls (1987) for tall buildings. Watts (1987) discussed computer models for evacuation analysis.

Speed has been shown to be a function of the density of the occupant flow type of egress component and mobility capabilities of the individual (Nelson and Mowrer, 2002).

For a density greater than 0.55 pers/m^2:

$$v = k - a k D \tag{9.1}$$

where
v = speed (m/s)
a = constant, 0.266m^2/pers
k = velocity factor (m/s)
D = density of occupant flow (pers/m^2).

For a density less than 0.55 pers/m^2, too few other people are present to impede the walking speed of an individual. In this case, maximum walking velocities for level walkways and stairways are

$$v = 0.85k \tag{9.2}$$

The velocity factor, k, varies from 1.00 m/s to 1.40 m/s.

At lower densities, people have a greater freedom to move at their own pace. As the crowd density increases, they become more controlled by others in the moving stream. At a density of about 1.88 persons per m^2, the combination of closeness of individuals and speed of movement are indicated to be the maximum. Such a high density is not comfortable. Higher densities not only slow the flow but also can reduce movement to a shuffling gait and, in the extreme, a crushing condition. People movement is faster for level segments of an escape route than stairs. It is faster for stairs with lower risers and deeper treads than stairs with higher risers and shallower treads.

The speed correlations presented in Equations (9.1) and (9.2) principally relate to adult, mobile individuals. Proulx (1995) indicates that the mean speed on stairs for children under 6 and the elderly was approximately 0.45 m/s in unannounced drills in multi-storey apartment buildings. The speed for an 'encumbered' adult is 0.22–0.79 m/s, also appreciably less than the maximum speed expressed in Equation (9.2). (An encumbered adult is an individual carrying packages, luggage or a child.)

Mean velocities for impaired individuals were estimated by Shields *et al.* (1996). These velocities depend on whether an individual uses an electric wheelchair, manual wheelchair, crutches, walking stick, walking frame/walker or rollator. For level walkways, the mean speed ranges from 0.51 m/s for walking frame/walker to 0.94 m/s for crutches. For stairs-down, the mean speed is 0.22 m/s for crutches and

0.32 m/s for walking stick. For stairs-up, the mean speed is 0.22 m/s for crutches and 0.34 m/s for walking stick. For people with no disability, the means speeds are 1.24 m/s for level walkways and 0.70 m/s for stairs, up or down.

Typical densities of people range from 1.0 to 2.0 persons per m^2 – see Fruin (1987), Pauls (1995), Predtechenskii and Milinskii (1978) and Frantzich (1996).

The specific flow is the product of the density and the speed:

$$F_s = Dv \qquad (9.3)$$

where F_s is the specific flow, persons/s/m. Expressions for F_s as a function of density can be obtained by substituting in Equation (9.3), the relationship for speed, v, given in Equation (9.1) or (9.2).

The width referenced in the units for the specific flow, Equation (9.3), relates to the 'effective width' as defined by Pauls (1995). The concept of effective width is based on the observation that people do not generally occupy the entire width of an egress segment, staying a small distance away from the walls or edge of the segment. Nelson and Mowrer (2002) refer to this small distance as a 'boundary layer'.

Maximum specific flow is achieved at a density of

$$D_{max} = \tfrac{1}{2}\,a \qquad (9.4)$$

where a is the constant defined with reference to Equation (9.1). Because a is independent of the type of egress segment, according to this correlation, the specific flow is maximised at the same density for all types of segments. Predtechenskii and Milinskii (1978) provide results from their data indicating differences in the density at which the specific flow is maximised for different types of egress segments or components.

The 'specific flow' provides a measurement of the flow capability of an egress component on a per unit basis (e.g. per metre). In an evaluation of an egress component or multiple egress components the total flow can be calculated and related to the affected population:

$$\text{Flow capacity} = F_c = F_s \cdot W_e \qquad (9.5)$$

where F_s is given by Equation (9.3) and W_e is the effective width. The evacuation time for a populated area through one exit element is the population, P, divided by the flow capacity of the exit element, plus the travel time through the exit element.

The method (Equations (9.1) to (9.5)) described above was used by Nelson and Mowrer (2002) to obtain a first order approximation of the egress time in buildings. The method involves determining the maximum flow rate for each of the egress components in the egress system. Proulx (1995) described in detail building evacuation models, including simple first-approximation methods and movement assumptions used in these calculations. She highlighted the effective-width model for evacuation flow, especially in relation to prediction formulae for the total evacuation time of large buildings.

Melinek and Booth (1975) carried out an analysis of evacuation times and of the movement of crowds in buildings. Using published data on crowd flows along corridors, the authors estimated that the normal capacity of corridors was about 1.5 people per metre per second; this value can be exceeded for a large rush of people. At low crowd densities (under one person per square metre), the speed of individuals is equal to their free walking velocity. The flow rate is a maximum when the crowd density is between one and five persons per square metre. For densities greater than five persons per square metre, congestion exists and the flow rate decreases as the crowd density increases.

Also, for movement up or down staircases or both, as the crowd density increases, the flow reaches a maximum and then decreases. There is no reduction of flow for bends in the staircase. The capacity of a staircase decreases if a large distance has to be traversed and reduces markedly by two-way flow, which should be avoided where possible. Velocities of 1.3m per second along corridors and 0.5m per second on stairs (along the line of stairs) can be assumed for unimpeded flow. Free flow at the head of the staircase can be assisted by a suitable system of barriers.

For a multi-storey building with n floors, Melinek and Booth (1975) derived the following formula for the minimum total evacuation time, T_r, for the population of floor r and above:

$$T_r = \left(\sum_{i=r}^{n} Q_i \right) / \left(N' b_{r-1} \right) + r t_s \tag{9.6}$$

where

Q_i = population of floor i.
b_r = staircase width between floor $r - 1$ and r.
N' = rate of flow of people per unit width down the stairs
t_s = time for a member of an unimpeded crowd to descend one storey.

The minimum total evacuation time for the whole building is equal to the highest of the n values of T_r ($r = 1$ to n). If there are several staircases, b_r will be the total width of the staircases considered. Using the formula in Equation (9.6), evacuation times were calculated assuming that the populations and staircase widths were the same for all floors and $N' = 1.1$ persons per metre per second. In most of the eleven office buildings considered in the study, the observed evacuation times were greater than the calculated values, the average difference being about 2 minutes. The calculated times ranged from 1.8 min for a six-floor building to 9.9 min for a 25-floor building.

In a multi-storey building designed for 'total evacuation', the escape stairs should have the capacity to allow the whole building to be evacuated simultaneously. For large buildings, over 30m high, complete evacuation would be lengthy and difficult and, hence, the fire safety codes envisage evacuation of only part of the building, usually the fire floor and the floor above. This is referred to as 'phased evacuation' which is applicable to a large building in which every floor is a floor with fire resistant compartments and which is fitted with an appropriate fire warning system.

According to means of escape provisions in fire safety codes, building occupants should be able to reach the entrance of an escape route, such as a protected staircase, within a reasonable period of time in the event of a fire. This period known as the 'design evacuation time', is used to determine the number and widths of escape routes required to facilitate an unimpeded flow of people. For this purpose, the following formula has been suggested in British Standard BS 5588:

$$M = 200b + 50 \ (b - 0.3) \ (n - 1) \tag{9.7}$$

where M is the maximum number of people who can enter a staircase of width b (in metres) serving n floors.

Equation (9.7) was derived by substituting the value of u given by

$$u = 50 \ (b - 0.3) \tag{9.8}$$

in the following equation which assumes a uniform population density:

$$M = N \, b \, t + (n - 1) \, u \tag{9.9}$$

where the flow rate N is approximately 80 persons per metre width per minute. The parameter, u, is the number of people who can be accommodated on the stairway between one storey and the next. The parameter t is the design evacuation time, usually taken as 2.5 minutes, which is considered to be an acceptable period within which all the occupants of a floor should be able to enter a protected staircase after leaving their places of occupation. For buildings with two or more staircases, it is assumed that any one staircase may be unusable in a fire.

For buildings without compartmental floors between storeys, the maximum number of people who can enter a staircase is recommended to be $222b$. For a building designed for phased evacuation, British fire safety codes recommend that the minimum width of a stair should be based on the formula $[(P \times 10) - 100]$ mm, where P is equal to the number of people on the most heavily occupied storey.

For the parameter u, the following general formula can be used:

$$u = (5.2b + 4b^2)\rho \tag{9.10}$$

where ρ is the density of people per m^2 in the staircase, a calculation which may be determined with due regard to the characteristics of the evacuating population and the presence of active fire protection systems such as detectors and sprinklers. Melinek and Booth (1975) assumed that $\rho = 3.5$ but Fruin (1971) suggested a lower density of 1.5 to avoid psychological discomfort and to minimise the risk of panic over prolonged delays. The absence of a b^2 term in Equation (9.7) implies that for wide staircase widths it makes less allowance for the standing capacity of the landing.

In sequential order, the design evacuation time, t defined in Equation (9.9), is the third component period of the total evacuation time, H, defined in Section 3.7.2.2. It is inappropriate to assign for E an arbitrary value such as 2.5 minutes without taking into consideration the first two component periods, discovery or detection time D and recognition time B. In determining the value of E, it is also necessary to consider the duration of two other phases of evacuation movement of occupants – movement for down staircases and movement through exits. The evacuation time E is the sum of the three phases of occupants, movement – time (t) to enter a staircase, movement down the staircase and movement through an exit. The value of E may be adjusted to take account of the presence or absence of active fire protection measures such as detectors, communication systems, sprinklers, emergency lighting and smoke ventilation systems.

It is also necessary that the total evacuation time H ($= D + B + E$) is less than the time, F, taken by a combustion product to travel from the place of origin and produce an untenable condition on any segment or component of an egress route. The deterministic design criterion is H is less than F. To satisfy this condition, the approach based on partial safety factors, discussed in Section 3.7.2, may be adopted – see Equation (3.91).

The partial safety factors approach is only a semi-probabilistic approach based on the expert judgement of a fire safety engineer and the quality of information available to them for estimating the values of the partial safety factors α_f and α_h.

A more rigorous probabilistic method is the 'Beta method', discussed in Section 3.7.4, which takes into account the uncertainties in the evaluation of F and H. In this method, the deterministic design criterion is modified to the probabilistic criterion specified in Equation (3.77) or (3.78). To apply this method, it is necessary to determine the probability distributions of H and F, together with their parameters such as mean and standard deviation. This might be possible by performing simulations for several scenarios, based on mathematical (deterministic) or computer models for evacuation and for spread of heat, smoke and other toxic products. Probability distributions can also provide more realistic values for the partial safety factors α_f and α_h in Equation (3.91). This method based on the 'design point' was described in detail by Ramachandran (1998) with reference to the determination of fire resistance for a compartment.

As discussed in Section 3.7.3, a normal probability distribution can be assumed for the total evacuation time H. This assumption may not be realistic for buildings with a mixture of able-bodied and disabled occupants. For such occupancies the distribution may be long-tailed such as log normal. An appropriate distribution for H can be determined, if possible, by collecting and analysing necessary data provided by case studies, fire drills, evacuation exercises and computer simulations of evacuation models. Such a method can also be adopted to determine the probability distribution of F for any combustion product, although a normal distribution has been assumed in Section 3.7.3.

Egress failure would occur with a probability Q if $H > F$. If this undesirable event occurs, one or more people may die due to visual obscuration caused by smoke, incapacitation and other reasons. As discussed by Ramachandran (1993a), the probability of death in a fire or fatality rate per fire, P_d, is the product of the probability of egress failure, Q, and the conditional probability, P_p, of death if egress failure occurs:

$$P_d = Q \times P_p \qquad (9.11)$$

As illustrated by Ramachandran, it would be possible to estimate the values of P_d and P_p for any type of occupancy by analysing data on real fires attended by the fire brigades.

By specifying an acceptable level for the product P_d, the value of Q, probability of egress failure, can be adjusted to take account of the presence or absence of active fire protection measures such as detectors, communication systems, sprinklers and smoke ventilation systems. It should be mentioned again in this context that, in addition to acting as a detection system, sprinklers, even if they fail to extinguish a fire, would retard the rate of growth of heat and smoke, thus increasing the value of F and providing more time for the occupants to escape to a safe place.

By following the above probabilistic method, the design value of the evacuation time E can be determined for any acceptable levels for the fatality rate, P_d, and probability of egress failure, Q. This procedure is explained in Table 9.1, reproduced

Table 9.1 Design evacuation time and fatality rate for single and multiple occupancy dwellings

E	H	θ	β	Probability of success	Probability of failure (Q)	Fatality rate per fire (P_d)
(min)	(min)					
2	14	1.07	0.32	0.6255	0.3745	0.0097
3	15	1.00	0.00	0.5000	0.5000	0.0130
4	16	0.94	−0.29	0.3859	0.6141	0.0160
5	17	0.88	−0.60	0.2743	0.7257	0.0189
6	18	0.83	−0.87	0.1922	0.8078	0.0210
7	19	0.79	−1.10	0.1357	0.8643	0.0225
8	20	0.75	−1.33	0.0918	0.9082	0.0236
9	21	0.71	−1.58	0.0571	0.9429	0.0245
10	22	0.68	−1.76	0.0392	0.9608	0.0250

$P_d = Q \times P_p$; $P_p = 0.026$

Note: $H = D + B + E$
$D = 10$ min
$B = 2$ min
$F = 15$ min
$\theta = F/H$
$\beta = (\theta - 1) / c\,(\theta^2 + 1)^{1/2}$; $c = 0.15$

$$\theta = \frac{1 + \beta c(2 - \beta^2 c^2)^{1/2}}{1 - \beta^2 c^2}$$

H and F have normal probability distributions.

from Ramachandran (1993a). In this table, the parameter E is the average or expected time for occupants to travel from the places where they are located and reach the entrance to a protected staircase. It does not include the duration concerned with the occupants' movement down the staircases and movement through exits.

Hasofer and Odigie (2001) developed a stochastic model for evaluating the interaction between the spread of untenable conditions and occupant egress. The authors measure safety by the expected number of deaths. The building is represented by a network for modelling fire spread and by another network for modelling occupant egress. A major innovation is the introduction of the concept of discrete hazard function. It allows the interaction between the various factors involved in the spread of untenable conditions and occupant egress to be taken into account.

References

Canter, D (1980), *Fires and Human Behaviour*, Wiley, Chichester.

Dunlop, K (1993), Real fire emergency evacuation of disabled people, *Proceedings of the CIB W14 International Symposium on Fire Safety Engineering*. University of Ulster, Jordanstown.

Frantzich, H (1996), *Study of Movement on Stairs During Evacuation Using Video Analysing Techniques*, Lund Institute of Technology, Lund, Sweden.

Fruin, J J (1971), *Pedestrian Planning and Design*, Metropolitan Association of Urban Designers and Environmental Planners, New York.

Fruin, J J (1987), *Pedestrian Planning and Design*, revised edition, Elevator World Educational Services Division, Mobile, AL.

Hasofer, A M and Odigie, D O (2001), Stochastic modeling for occupant safety in a building fire, *Fire Safety Journal*, 36, 269–289.

Marchant, E W (1975), Fire safety for the handicapped, *Proceedings of Seminar on Fire Safety for the Handicapped*, University of Edinburgh, Edinburgh.

Melinek, S J and Booth, S (1975), *An Analysis of Evacuation Times and the Movement of Crowds in Buildings*, Current Paper CP96/75, Building Research Establishment, Fire Research Station, Borehamwood.

Nelson, H E and Mowrer, F W (2002). Emergency movement, Section 3, Chapter 14, *SFPE Handbook of Fire Protection Engineering*, 3rd edn, National Fire Protection Association, Quincy, MA.

Pauls, J (1987), Calculating evacuation times for tall buildings, *Fire Safety Journal*, 12, 213–236.

Pauls, J (1995), Movement of people, Section 3, Chapter 13, *SFPE Handbook of Fire Protection Engineering*, 2nd edn, National Fire Protection Association, Quincy, MA.

Pauls, J and Juillet, E (1993), Life safety of people with disabilities, *Proceedings of the CIB W14 International Symposium on Fire Safety Engineering*, University of Ulster, Jordanstown.

Predtechenskii, V and Milinskii, A (1978). *Planning for Foot Traffic Flow in Buildings*, Amerina Publishing, New Delhi.

Proulx, G (1995). Evacuation times and movement times in apartment buildings. *Fire Safety Journal*, 24, 229–246.

Ramachandran, G (1985), The human aspects of fires in buildings – a review of research in the United Kingdom, *Fire Safety: Science and Engineering*, ASTM STP 882, American Society of Testing and Materials, Philadelphia, PA.

Ramachandran, G (1993a), Probabilistic evaluation of design evacuation time, *Proceedings of the CIB W14 International Symposium on Fire Safety Engineering*, University of Ulster, Jordanstown

Ramachandran, G (1993b), Early detection of fire and life risk, *Fire Engineers Journal*, December, 33–37.

Ramachandran, G (1998), *Probabilistic Evaluation of Structural Fire Protection – A Simplified Guide*, Fire Note 8, Building Research Establishment, Fire Research Station, Borehamwood.

Rubadiri, L, Roberts J P and Ndumu, D T (1993), Towards a coherent approach to engineering fire safety for disabled people, *Proceedings of the CIB W14 International Symposium on Fire Safety Engineering*, University of Ulster, Jordanstown.

Shields, T J, Dunlop, K and Silcock, G W H (1996). *Escape of Disabled People From Fire. A Measurement and Classification of Capability for Assessing Escape Risk*. BRE Report 301. Building Research Establishment, Fire Research Station, Borehamwood.

Shields, T J, Smyth, B, Boyce, K E and Silcock, G W H (1998), Evacuation behaviours of occupants with learning difficulties in residential homes, *Human Behaviour in Fire: Proceedings of the First International Symposium*, University of Ulster, Jordanstown.

Sime, J D (1991), *Human Behaviour in Fires*, Publication 2/91, Fire Research and Development Group, Home Office, London.

Walsh, C J (1998), A rational fire safety engineering approach to the protection of people with disabilities in or near buildings during a fire or fire related incident, *Human Behaviour in Fire. Proceedings of the First International Symposium*, University of Ulster, Jordanstown.

Watts, J M (1987), Computer models for evacuation analysis, *Fire Safety Journal*, 12, 237–245.

Yoshimura, H (1998), Sounding out the disabled in the lower extremities on their escape behaviour in building fire for safer fire escape design, *Human Behaviour in Fire: Proceedings of the First International Symposium*, University of Ulster, Jordanstown.

10 Performance and effectiveness of fire service intervention

10.1 Introduction

The demand for fire protection required by a community takes two forms: a potential demand and a realised demand. The first form is concerned with 'hazards' which have the potential to initiate the occurrence of fires and cause damage to life and property. When fires actually occur, the potential demand is converted into realised demand. Fire prevention activities and protection measures, passive and active, are aimed at reducing the adverse effects arising from potential demand. The fire services are also involved in these activities to some extent but their main responsibility is to meet the realised demand effectively by putting out fires, rescuing people and carrying out salvage operations.

Three factors primarily determine the level of resources needed by a fire department to cope with the realised demand. Firstly, the types of hazards involved (e.g. apartment houses, industrial properties) determine the types and numbers of fire fighting equipments needed, location of fire stations, with due regard to their proximity to areas with high risk in terms of frequency of fires and damage and number of personnel that should be assigned to engines, ladders and special units. Secondly, the geography of the area protected affects the travel time required by fire fighting equipments to reach various sites of fires. Extra protection may be needed for isolated locations, e.g. ridges, or where access may become difficult or restricted, e.g. tunnels, mountain roads. Thirdly, peak-period alarm rates determine the extra personnel and equipment needed to handle the simultaneous demand for service.

Subject to the factors mentioned above and economic and other constraints, a fire department has to consider alternative deployment policies and identify a policy or set of policies which will ensure 'optimum' performance in terms of reduction in damage to life and property. The central parameter in obtaining an optimum solution to this problem is the travel time to fire incidents in a region, which depends on the region's area, alarm rate, number of fire stations and other characteristics. Travel time is the major component of attendance time or response time. Several 'descriptive models' are available for estimating the travel time directly or indirectly by converting results based on models for travel distance – see Walker *et al.* (1979).

Models discussed in this chapter are concerned with the evaluation of the relationship between attendance time and damage to property and life. Other aspects discussed relate to percentage of fires requiring fire brigade intervention, methods of extinction, control time and its interaction with attendance time and the effectiveness of fire brigade intervention.

10.2 Attendance time and travel time

10.2.1 Definition

Attendance time is the elapsed time from the moment a fire brigade is called until fire fighting personnel and equipment arrive at the fire scene. After arrival, a certain time is required for preparing and positioning equipment at the scene before commencing fire fighting. This period, 'set up time', is not included in the attendance time defined in the United Kingdom fire statistics; it is part of the 'control time' elapsing from arrival until the fire is brought under control by the fire brigade. In United States fire statistics, the 'set up time' is included in the 'response time' which is the elapsed time from the moment the fire department is notified until a fire company is on the scene and is ready to operate.

An important component of the attendance time, or response time, is the 'travel time', the time between the start of a fire unit from its quarters and its arrival at the fire scene. An analysis of fire station location should focus on changes in travel time but attendance time, or response time, is often the most important measure since changes in fire cover policy might affect other components, 'dispatching time' and 'turnout time'. The first of these two components is the time between the receipt of an alarm and the dispatch of a unit. The second is the time required for the unit to leave its quarters and start moving once it has been dispatched. Although our discussion in this section concentrates on travel time, in most cases, attendance time or response time can be substituted for travel time and the statements will remain correct.

10.2.2 Travel time vector

Different types of units (engines and ladders, for example) perform different functions at a fire. Hence, it is necessary to distinguish travel times for each type of equipment. In addition, since two units of the same type working together may be able to take some action that neither could take alone, the travel time of each arriving piece of equipment is important. Therefore, in evaluating alternative deployment policies, it would be useful to consider the list or vector of travel times for engines, ladders and other fire fighting units that would respond to each incident under each policy.

The travel time vector for engines, for instance, gives the time of arrival (relative to the time of departure from the fire station) of the first engine, second engine etc. It is, however, difficult in practice to use individual travel or response time to compare the attractiveness of alternative deployment policies. For example, the travel time for the first engine under one policy may be longer than the travel time for the first engine under another policy. On the other hand, the travel time for the second engine under the first policy may be shorter than the travel time for the second engine under the second policy. One way of comparing policies is to calculate the average of the travel times of the first and second engines. From a fire cover point of view, the first engine response time is more important than the second engine response time.

10.2.3 Travel time measures

A wide variety of travel time measures may be used in different situations. One measure is based on the frequency distribution of first-engine (or ladder) travel time to incident locations throughout a city or in different regions of a city. Consider, for example, a long travel time of 4 minutes. The frequency or probability of exceeding this time may be 10 per cent under one deployment policy while it may be 20 per cent under another policy. The first policy would certainly be preferred. Another approach would be to compare alternative policies on the basis of how well they satisfy requirements for maximum acceptable travel times for different incident locations.

Another set of travel time measures relates to average travel times to groups of incidents or incident locations. Incident locations can be grouped, for example, by region of the city or company response area. The average travel time to incidents can be found by weighting the travel time to each location by the structural fire rate at that location (and not by alarm rate which includes false alarms).

10.3 Travel distance

10.3.1 Methods of estimation

Travel time of an engine (or ladder) generally increases with distance travelled. In many situations it may be inconvenient, impossible or even unnecessary to measure or estimate travel times directly. In an analysis of deployment policies, travel distances can be used directly or can be converted to travel times. An advantage of direct use of distances is that they are more stable than times. They do not change with weather, time of day etc. Moreover, there is likely to be broad agreement on distances and how they are measured.

Consider the distance d_{ij} between two specific points i and j that have grid coordinates (x_i, y_i) and (x_j, y_j) respectively. If the streets are laid out as a rectangular grid, then

$$d_{ij} = |x_i - x_j| + |y_i - y_j|$$

If the travel is 'as the crow flies' in a straight line connecting the two points, then

$$d_{ij} = [(x_i - x_j)^2 + (y_i - y_j)^2]^{1/2}$$

In practice, neither of the two conditions mentioned above may hold exactly and hence an approximation is needed. In many cities throughout the USA the following formula has provided good estimates of the travel distance (Walker *et al.*, 1979):

$$d_{ij} = 1.15 [(x_i - x_j)^2 + (y_i - y_j)^2]^{1/2}$$

Another approach to estimating travel distances was used in The New York City – Rand Institute's Fire Operations Simulation Model. In the simulation, the distance between two points was computed on the following basis: If the distance travelled is short, the fire company is presumed to travel on a right-angle grid of streets oriented at a specified angle with respect to the coordinate axes. If the distance travelled is long, it is assumed that a straight line is followed. If the distances are of intermediate length, a combination of right-angle and straight-line distances is used. Carter (1974) explained the details of these computations.

The third possible approach is based on a network model which will provide estimates of travel distances with the greatest accuracy if the model has a 'node' for every intersection and an 'arc' for every street segment. One-way streets could be modelled explicitly in the network by using directed arcs. Parks, lakes, railroads, highways and other barriers to direct travel are automatically taken into account, since permissible paths in the network would avoid these obstacles. The US Bureau of the Census has created a file of data for computerised map generation that can be used to construct such a network. It exists for several Standard Metropolitan Statistical Areas (SMSA) in the USA and is called the Dual Independent Map Encoding (DIME) file. In the UK, some of the Geographic Information System (GIS) computer packages can enable the creation of networks to evaluate travel distances. Efficient algorithms are available to compute the length of the shortest path between every pair of points in a network – see, for example, Dreyfus (1969), Dreyfus and Law (1977).

Although a network can be used to estimate travel distances which can be converted to times, the result will not necessarily be the same as when using a network to estimate travel time directly. Timed measurements take into account hills and traffic lights and rarely the convenient set of rectilinear streets. Hence, the shortest-distance path between two points may not be the fastest path. For example, part of the trip could be made on a limited-access divided highway which may make the trip longer but faster. When a network is used to estimate travel times, the computer program assumes that the fire companies follow the fastest route, even if some other route would have a shorter distance. For the reasons mentioned above, very specific and accurate timed measurements over the station grounds are made in the UK instead of a generalised fire model.

The network method is particularly useful in cities that are irregularly shaped, divided by having a limited number of crossover points, or that contain large areas (parks or airports, for example) through which fire companies cannot travel. However, many cities in the USA are sufficiently regular that much simpler methods are appropriate for estimating travel times by conversion from travel distances. Such a conversion would be easy if fire companies travelled at a constant speed but this is not supported by data collected in several cities. These data show that average speed is higher on longer trips and an appropriate time–distance relation should reflect this.

10.3.2 Time–distance relationship

A statistical analysis (Walker *et al.*, 1979) of empirical data, collected in four cities (Denver, Trenton, Wilmington and Yonkers) revealed that travel time and travel distance were related according to a curve. The individual travel time functions for the four cities were remarkably similar to each other. This was the reason for combining the data obtained in the four cities and estimating the parameters of a single function. These data were collected during experiments in which selected units measured their travel times with stop watches and their travel distances with odometers. An analysis of the data revealed that a square-root relationship is applicable for distances up to 0.38 miles and a linear relationship for greater distances.

The results of an analysis in a range of US cities (Walker *et al.*, 1979) did not indicate any practical differences between travel speeds under conditions of daylight and darkness. While speeds were somewhat slower during rush hours, they were closer to speeds at other hours than anticipated. The reduction in average speed (about 20 per cent) was greatest during the 8am to 9am period. The results of an additional analysis indicated clearly that response speed did not change markedly by time of day.

10.3.3 Square-root law

Suppose we wish to compute the average travel distance of engine companies to fires in a particular region of a city. We will be computing this average over all the engines and alarm boxes in the region. Let D denote the distance travelled by an available engine to a fire scene and $E(D)$ the average (or expected) travel distance. According to square-root law (Walker *et al.*, 1979)

$$\bar{D}_1 = E(D) = k(A / N)^{\frac{1}{2}} \tag{10.1}$$

where k is a constant of proportionality, A is the area of the region and N is the number of firehouses in the region that have engines available to respond. This inverse relationship between $E(D)$ and N implies, as one would expect, that the average travel distance would decrease if the number of firehouses in the region is increased.

In deriving Equation (10.1) it has been assumed that each firehouse has a company available and ready to respond when an alarm is received. But the number of companies available will vary from time to time and hence, it is not realistic to use the number of firehouses as the value for N if Equation (10.1) is used to estimate the average travel distance over a period of time. However, as discussed by Walker *et al.* (1979), if the number of available engines in a region is not too small (say, more than 2.0), a reasonable approximation is to replace N in Equation (10.1) by the expected (long-run average) number of firehouses having at least one engine company available.

The square-root law is also based on the following assumptions:

- Alarms are distributed randomly but with uniform probability density throughout the region of interest.
- Firehouses are spread either in a regular pattern or randomly throughout the region.
- Boundary effects are insignificant.
- Fire companies travel either on a straight line between two points or on a dense rectangular grid of streets.

In the real world, none of these assumptions is strictly true. Hence, the validity of the square-root law for any region needs to be examined by performing computer simulations. In the simulation experiment, conditions such as alarm rates and number of active ladder companies should be varied over a broad range, reflecting real situations, although the number of engine companies may be fixed at an appropriate level. Alarm locations and patterns should be so chosen that they imitate reality. The locations of engine and ladder companies may be fixed at their actual locations adding, if required, new locations at appropriate spots.

Walker *et al.* (1979) described a simulation experiment performed for the Bronx region of New York City. (The reader should refer to this book for details.) The results of this exercise supported the validity of the square-root law and showed a general consistency between the parameters for engines and ladders. Hence, the analysis was repeated, with the data grouped from various simulations and for engines and ladders.

Walker *et al.* (1979) also examined the results from two independent studies: one to determine fire station locations in Bristol, England (Hogg, 1968) and the other to determine locations for ambulances in a suburban county near Washington DC (Berlin and Liebman, 1974). In Hogg's study neither the site locations nor the fires were evenly distributed spatially, both being more dense in the centre of the rectangular region under study. In her analysis, variations in the alarm pattern by time of day were considered but possible unavailability of fire companies was ignored. She calculated travel times from knowledge of the distances involved and from estimates of travel speed. Berlin and Liebman used a 'set-covering model' and computed the relationship between the maximum response time and the number of locations occupied. Their computations were based on a linear relationship between time and distance. Both the studies mentioned above confirm the validity of the square-root model in terms of travel or response time but the results can be converted to travel distances.

10.3.4 *Average travel times in a region*

The average travel time in a region is the average of many point-to-point travel times. The frequency of each potential trip is determined by the frequency of alarms at each alarm box, the dispatching rules used to assign fire companies to respond and the availability of the fire companies assigned to respond. Using data on actual travel times for a series of alarms, the average travel time for a region can be estimated by dividing the sum of travel times for all trips by the number of trips. The computation should include repetitions of the same trip, company m in travelling to alarm box b, as many times as that run was made during the period of interest.

An alternative approach is to use other historical data and convert travel distance to time for each possible trip by applying one of the methods discussed earlier. This estimate can be combined with information on the fraction of all trips made by all first arriving companies that involve company m in responding to box b. Then, by considering all the m to b combinations, the average regional travel time can be estimated.

Both the two methods mentioned above involve tedious computations. A simple approximation is obtained by combining the square-root law for average distances (e.g. Equation (10.1)) with a function for converting travel distances into travel times based on an analysis of data such as the one mentioned in Section 10.3.2. With the resulting function, average travel times can be estimated for regions where comparatively little is known about the details of travel patterns, alarm distributions etc.

10.4 Attendance time and fire damage

10.4.1 *Introduction*

Fire brigades are required to provide fire cover to different categories of fire risk by meeting certain standards, specified in terms of attendance or response time. In the United Kingdom, for example, there are four risk categories: A, B, C, D. These correspond approximately to commercial and industrial city complexes, centres of large towns, built-up areas of towns and rural areas. Exceptionally high risks and remote rural areas are treated as special cases. For each risk category there is a recommended first attendance. For A category it is two pumps (first attendance fire appliance) within a maximum period of five minutes and one further pump within eight minutes. It is one pump within five minutes plus a second within 8 minutes for B category, one in 8 to 10 minutes for C and one in 20 minutes for D. For known small fires, such as those on waste ground or in derelict buildings, brigades use their discretion.

The safety levels provided by standards such as those mentioned above are practically unknown quantities. It would be useful to evaluate these levels in terms of life loss and property damage in order to assess changes in these levels due to a re-location of fire stations or re-allocation of fire fighting equipments, personnel and other resources to existing stations. For this purpose, it is necessary to evaluate the relationship between the attendance or response time and damage due to fires, life and property. This is a difficult task, particularly due to the random nature of fire spread and interactions between various factors. However, some researchers have attempted to model such a relationship as described briefly in the next two subsections.

10.4.2 Attendance time and fire loss

In the Untied Kingdom, for example, Hogg (1973) carried out a study of fire losses in relation to attendance time. She modelled a fire as growing in stages: confined to object first ignited, confined to room, beyond room and beyond building. Fire spread from stage to stage was governed by transition probabilities as functions of (dependent on) time. The starting probabilities at discovery, together with the transition probabilities of fire spread, generated the probability that a fire would be in a given stage according to the discovery to arrival time. For each stage, i.e. extent of spread category, the observed mean area burned was obtained according to whether or not the fire was fought before the arrival of the fire brigade. Hogg's model took account of the first pump at each station and the first at each fire and provided output in terms of arrival times of the first pump.

Hogg estimated the relationship between attendance time and fire loss for four major types of occupancies, each subdivided into two categories: fires fought before the arrival of the fire brigade and fires not so fought. Using loss per unit area, Hogg converted area damage to financial loss. As one might expect, the losses in the first category were much smaller than those in the second. The overall results obtained by Hogg are given in Table 10.1. In the USA a similar study was carried out by Halpern, Isherwood and Wand (1979).

Maclean (1979) attempted to evaluate the loss-attendance time relationship for nine occupancy groups by applying 'longitudinal' analysis and comparing the size of a fire at one known time with the size of the same fire at a different time. He used area damage to describe fire size. Converting area damage to financial loss, Maclean estimated that an increase in attendance time of one minute beyond the level at that time would result in an overall increase of £1570 per fire (at 1978 prices) to the total loss to the economy.

10.4.3 Attendance time and life loss

In Section 3.3.3, the relationship between fatality rate and delays in discovering fires was estimated with the aid of fire statistics published by the Home Office, UK. Based on this relationship, a parameter λ was estimated to have values of 0.0008 and 0.0006 for single-occupancy and multiple-occupancy dwellings. This parameter, measuring the average increase in fatality rate per minute can, perhaps, be regarded as applicable to the period following the discovery of a fire until the arrival of the fire brigade at the fire scene. Under this assumption, it can be approximately estimated that one additional life per thousand fires would be lost for every minute of delay in attendance or response time.

Table 10.1 Direct loss in relation to attendance time – non-dwellings, UK
(£, 1967 values)

Attendance time (mins)	Single storey 21,806 fires	Multi-storey 40,419 fires
1	3,420	4,878
2	3,865	5,187
3	4,312	5,489
4	4,764	5,784
5	5,226	6,073
6	5,700	6,355
7	6,189	6,632
8	6,696	6,903
9	7,224	7,170
10	7,776	7,431
11	8,354	7,687
12	8,960	7,939
13	9,599	8,187
14	10,271	8,432
15	10,981	8,672
16	11,730	8,909
17	12,521	9,143
18	13,358	9,374
19	14,244	9,601
20	15,181	9,827

The value of λ can be expected to increase with time since the start of the fire, although it has been assumed as a constant independent of time in the analysis mentioned above. Hogg (1973) attempted to evaluate this relationship for the period from discovery of fire to the arrival of the brigade at the fire scene. She estimated, for this period, that the increase in the probability of one or more deaths for every minute delay in arriving at the scene would be 0.015 on average over a duration of 21 minutes. According to this result, the average value of λ would increase 15 times, from 0.001 to 0.015, from the first period up to discovery of fire to the second period after discovery. It is difficult to explain the reason for such a significant increase in λ from one period to a subsequent period but it should be pointed out that Hogg's results were based on a much smaller (7,818) number of fires.

Using a 'square-root law', Hogg (1973) estimated that if the number of fire stations covering the UK in 1967 had been double the actual number of stations, the fatalities predicted would have been fewer by only 51, a fall from 584 to 533, while a fourfold increase in the number of stations would probably have saved a further 37 lives. She concluded that any change in fire cover would only have a marginal effect on the number of lives lost. It ought to be mentioned in this connection that the increasing installation of smoke alarms in dwellings in the UK may affect the attendance time/life loss relationship.

Corman *et al.* (1976) investigated the effects on fire casualties of reducing response distances of fire companies. According to their analyses, based on a 'pairing scheme' of fatal and non-fatal fires, none of the means of the differences of response distance for fatal pairs was significantly different from zero. For injury data, however, the mean of engine distance difference was significantly non-zero at the 87 per cent probability level and for ladders it was significantly non-zero at the 92 per cent level. The important conclusion of this study was that the effect of fire company response distance (for average distances typical of New York City) on fire casualties was very small compared with the effects of other factors, such as time of day, season, type of building construction and floor of fire origin.

10.5 Fire fighting

10.5.1 Fire brigade intervention

Some fires in buildings burn out themselves for a variety of reasons or are extinguished by first-aid methods such as portable fire extinguishers, buckets of water or sand and smothering. Fire incidence data contain information on some of these 'small' fires which were reported to the fire brigade but were out on (or before) the arrival of the brigade. In sprinklered buildings, some 'small' fires do not produce sufficient heat to activate these fire protection systems, such that they are either self-extinguished or extinguished by first-aid means.

According to UK fire statistics, about 55 per cent of the fires in sprinklered buildings attended by fire brigades or reported to them are 'small' fires as defined in the previous paragraph. The remaining 45 per cent are 'big' fires requiring intervention by sprinklers, with sprinklers operating in 87 per cent of these cases (39 per cent of all reported fires) and not operating in 13 per cent of the cases (6 per cent of all fires). Some of the fires in which sprinklers operate are extinguished by the system before the arrival of the brigade and some are extinguished by the brigade. The Home Office Research and Planning Unit has conducted surveys in which a question on first-aid fire fighting was asked. They produced figures of the order of 1 in 10 fires being reported to the fire service.

Fire brigades deal with 'big' fires which are neither self-extinguished nor extinguished by first-aid means or sprinklers. Based on UK fire statistics for the period 1984 to 1986, the proportion of fires extinguished by fire brigades is shown in Table 10.2 for seven types of occupancies and the two cases, sprinklered and non-sprinklered buildings. Also shown in this table are proportions of fires extinguished by sprinklers and proportions of small fires self-extinguished or extinguished by first-aid methods. In the sprinkler case, the figures include estimated numbers

Table 10.2 Percentage of small fires and fires extinguished by sprinklers and by fire brigade

	Sprinklered building			Non-sprinklered building	
		Extinguished by			Extinguished by fire brigade
Occupancy	Small fires	Sprinklers	Fire brigade	Small fires	
Textile industry	37	39	24	33	67
Chemical and allied industries	37	44	19	52	48
Paper, printing and publishing industries	37	42	21	47	53
Timber, furniture, etc. industries	37	38	25	13	87
Retail distributive trade	37	42	21	21	79
Wholesale distributive trade	37	39	24	17	83
Office buildings	37	49	14	28	72

of fires which are neither reported to the fire brigade nor any insurance claims are made. Such 'unreported fires' extinguished by sprinklers, according to an unpublished research study quoted by Rogers (1977), constitute about one-third of fires in sprinklered buildings. Fire brigades only attend two-thirds of fires in these buildings. Also, in this case, 4 per cent ($= 6 \times \frac{2}{3}$) of reported fires in which sprinklers did not operate have been included in fires extinguished by the brigade.

10.5.2 Methods of extinction

The growth of a fought fire depends not only on the burning characteristics of the fire and fire fighting before the arrival of the brigade, but also on the method used by the fire brigade to extinguish the fire. These methods include hose reels, main jets and other means. According to UK fire statistics for 1991, fire fighting by brigades is not required in 29 per cent of the fires. In 71 per cent of fires requiring fire brigade action, no hose reels or main jets are used in 17 per cent of the fires, only hose reels in 68 per cent of fires, 1 or 2 main jets in 12 per cent of fires, 3 or 4 main jets in 2 per cent of fires and 5 or more jets in 1 per cent of fires. This distribution of number of fires according to hose reels and main jets indicates, to some extent, the sizes of fires which have to be tackled by the fire brigades. Until 1965, in the absence of data on financial losses, fires extinguished by 5 or more jets by the fire brigades in the UK were regarded as large fires.

However, there is a counter-acting factor affecting the sizes of fires at the time of extinction. The sizes reflected in the extent of fire spread or damage also depend on the number of jets used and the times at which the jets are applied. A simulation method, based on data for a large number of fires, can be applied to estimate the general relationship between average spread or damage and the arrival times of pumps. Rutstein (1975) attempted this method of analysis and found that, on average, non-dwelling fires were growing at the rate of about 10 per cent per minute before the arrival of the first pump. Once fire fighting commenced, the average rate of spread was about 1–2 per cent per minute before the arrival of the second pump and about 0.25 to 0.5 per cent per minute before the arrival of the third pump. These results indicate that fire damage depends mainly on the arrival time of the first pump.

10.5.3 *Control time*

As discussed in the previous section, the time taken by a fire brigade to bring a fire under control and extinguish it depends, to some extent, on the number of hose reels or main jets used and the build-up of jets over the fire fighting period. Control time also depends on the size of the fire at the time of arrival of the brigade at the fire scene. This size is related to the elapsed period T_A , from the start of the fire until fire brigade arrival. A fire discovered or detected soon after ignition, and reported to the fire brigade, will be in its early stage of growth when fire fighting commences and hence can be controlled quickly.

A reduction in T_A will, therefore, shorten the control time, T_B, as well, thus reducing the total fire duration (*T*) and damage to an appreciable extent, according to the exponential model in Equation (3.42). The relationship between T_A and T_B may be assumed to be approximately linear and can be estimated with the aid of raw data on individual fires, such as those available in the UK. Ramachandran (1992) carried out such an analysis and obtained the results shown in Table 10.3.

The regression coefficient given in Table 10.3 indicates the additional control time (mins) required for every additional delay of one minute in arriving at the fire scene due to delays in discovery of fire or calling the fire brigade or attendance (response). This coefficient also indicates the extent to which the control time (mins) will be reduced for every minute of early arrival of the Fire and Rescue Service at the fire scene due to early detection of fire or early attendance. The coefficient was statistically significant in most of the cases. Figures such as those in Table 10.3 can be used in conjunction with an exponential (or any other) model of fire growth to estimate the additional damage likely to occur due to delays in discovering fires or responding to them. Long delays due to these factors would increase the chance of a fire spreading beyond the room of origin, particularly in a building without sprinklers. Figures 10.1 and 10.2 are examples showing the extent to which damage and probability of spread would increase if the fire brigade commences its attack when a fire is in its second or third stage of growth. Such figures were produced by Ramachandran (1992) for some types of industrial occupancies and scenarios by performing simulations based on an exponential model of fire growth (Equation (3.42)) in real fires attended by fire brigades.

Table 10.3 Control time (minutes) and arrival time (minutes): regression coefficients

Industry	Area of fire origin		
	Production	Storage	Other areas
Food, drink, tobacco	0.458	0.359	0.405
Chemicals and allied industries	0.417	0.560	0.370
Metal manufacture	0.334	0.325	0.118
Mechanical, instrument and electrical engineering	0.336	0.367	0.218
Textiles	0.439	0.198	0.117
Clothing, footwear, leather and fur	0.318	0.172	0.351
Timber, furniture, etc. industries	0.249	0.311	0.299
Paper, printing and publishing industries	0.465	0.440	0.187

10.5.4 Number of jets

According to an analysis carried out by the Fire Research Station, UK on behalf of the Swedish Fire Protection Association (Bengtson and Hagglund, 1986) the number of hose reel jets (J_h) required to fight a fire of size A_f square metres is given by the following equation:

$$J_h = 2.71 + 4 \log_e A_f \tag{10.2}$$

In this analysis, five hose reel jets were considered to be equivalent to one main jet in regard to discharge rate of water. Hence, in terms of main jets (J), Equation (10.2) may be rewritten as

$$J = 0.54 + 0.8 \log_e A_f \tag{10.3}$$

If, for example, A_f is equal to 80m2, $J_h = 20$ or $J = 4$. In Equations (10.2) and (10.3) A_f denotes damage in terms of fire area and not total area including smoke and water damage.

Data for all fires were utilised in deriving Equations (10.2) and (10.3) and hence this result was weighted heavily in the direction of small fires. It is unwise to apply this result to large fires for which Thomas (1959) obtained the following equation based on 48 fires ranging in ground plan area from 28m² to 56,000 m²:

$$J = 0.1\sqrt{A}$$

where A is the area in square feet, or

$$J = 0.33\sqrt{A} \tag{10.4}$$

Non-sprinkled building — Intervention by fire brigade
Occupancy type — Retail distributive trade
Area of fire origin — Assembly areas

Intervention stage	Spread stage	Extinction by SM/FB	Calculated time		Damage m²	Probability of		
			Elapsed (mts)	Cumulative (mts)		Intervention	Extinction (conditional)	Extinction
1	1	SM	0	0*	1.15	0.00	0.00	0.00
	1	FB	9.4	9.4	1.63	1.00	0.33	0.33
	2	FB	24.0	33.4	3.99	1.00	0.27	0.27
	3	FB	52.4	85.8	27.96	1.00	0.27	0.27
	4	FB	42.0	127.8	133.29	1.00	0.13	0.13
	Average damage				26.49			
2	1	SM	0	0	1.15	0.00	0.00	0.00
	2	SM	13.5	13.5*	1.67	0.00	0.00	0.00
	2	FB	29.4	42.9	4.99	1.00	0.40	0.40
	3	FB	57.8	100.7	42.85	1.00	0.40	0.40
	4	FB	47.4	148.1	249.88	1.00	0.20	0.20
	Average damage				69.11			
3	1	SM	0	0	1.15	0.00	0.00	0.00
	2	SM	16.2	16.2	1.67	0.00	0.00	0.00
	3	SM	81.8	95.3*	15.98	0.00	0.00	0.00
	3	FB	90.5	185.8	463.10	1.00	0.67	0.67
	4	FB	80.1	265.9	9114.87	1.00	0.33	0.33
	Average damage				3316.18			

* Time of commencement of fire brigade attack
SM Small fire
FB Fire brigade

Stages
1: Confinement to item first ignited
2: Spreading beyond item first ignited but confinement to contents of room of fire origin (structure not involved)
3: Spreading beyond item first ignited and other contents but confinement to room and involvement of structure
4: Spreading beyond room of origin

Figure 10.1 Effectiveness of early intervention by fire brigade – example based on UK fire statistics, non-sprinklered buildings

Sprinkled building   Intervention by fire brigade
Occupancy type   Retail distributive trade
Area of fire origin   Assembly areas

Intervention stage	Spread stage	Extinction by SP/FB	Calculated time		Damage m²	Probability of		
			Elapsed (mts)	Cumulative (mts)		Intervention	Extinction (conditional)	Extinction
1	1	SP	0	0*	1.80	0.70	1.00	0.70
	1	FB	21.5	21.5	4.60	0.30	0.31	0.09
	2	FB	15.0	36.5	9.36	0.30	0.44	0.13
	3	FB	23.2	59.7	27.96†	0.30	0.19	0.06
	4	FB	26.4	86.1	133.29†	0.30	0.06	0.02
	Average damage				7.14			
2	1	SP	0	0	1.67	0.79	0.89	0.70
	2	SP	4.6	4.6*	2.67	0.79	0.11	0.09
	2	FB	16.8	21.4	5.90	0.21	0.64	0.13
	3	FB	25.0	46.4	19.20	0.21	0.27	0.06
	4	FB	28.3	74.6	94.32	0.21	0.09	0.02
	Average damage				5.21			
3	1	SP	0	0	1.67	0.92	0.87	0.80
	2	SP	4.6	4.6	2.67	0.92	0.11	0.10
	3	SP	17.1	21.7*	14.97	0.92	0.02	0.02
	3	FB	31.9	53.6	67.47	0.08	0.75	0.06
	4	FB	35.1	86.7	420.06	0.08	0.25	0.02
	Average damage				14.35			

* Time of commencement of fire brigade attack
† Figure for non-sprinled building used as estimate
SP Sprinklers
FB Fire brigade

Stages
1: Confinement to item first ignited
2: Spreading beyond item first ignited but confinement to contents of
    room of fire origin (structure not involved)
3: Spreading beyond item first ignited and other contents but confinement
    to room and involvement of structure
4: Spreading beyond room of origin

Figure 10.2 Effectiveness of early intervention by fire brigade – example based on UK fire statistics, sprinklered buildings

where A is the area in square metres. This sample was biased heavily towards large fires over 200 m². Equation (10.4) corresponds to the use of one jet per 10m circumference to the fire. Equations (10.3) and (10.4) give approximately the same number of jets for 200m² but the first equation underestimates the number for fires of larger sizes.

According to Thomas (1959), with A in square feet, the average control time T_4 (minutes) is given by

$$T_4 = \sqrt{A}$$

which is equivalent to

$$T_4 = 3.3\sqrt{A} \tag{10.5}$$

with A in square metres. Reviewing data on 134 fires attended by fire brigades which were published by Illinois Institute of Technology Research Institute (Labes, 1968), Baldwin (1970) obtained the following relationship between control time and fire area in square metres:

$$T_4 = 1.66\,A^{0.559} \tag{10.6}$$

Equations (10.3) or (10.4) and (10.5) or (10.6) provide an estimate of the number of jets that would be required to control a fire of a certain area within a certain time. For example, if the fire area is 100m², according to Equations (10.4) and (10.5), 3.3 or 4 jets would be required to control the fire within 33 minutes.

References

Baldwin, R (1970), *The Use of Water in the Extinction of Fires by Brigades*, Fire Research Note 803, Fire Research Station, Borehamwood.

Bengtson, S and Hagglund, B (1986), The use of a zone model in fire engineering application, *Fire Safety Science: Proceedings of the First International Symposium*. Hemisphere Publishing Corporation, New York.

Berlin, G and Liebman, J (1974), Mathematical analysis of emergency ambulance location, *Socio-Economic Planning Sciences*, 8, 323–328.

Carter, G (1974), *Simulation of Fire Department Operations: Program Description*, Report R-1188/2-HUD/NYC, Rand Corporation, Santa Monica.

Corman, H, Ignall, E J, Rider, K L and Stevenson, A (1976), Fire casualties and their relation to fire company response distance and demographic factors, *Fire Technology*, 12, 193–203.

Dreyfus, S (1969), An appraisal of some shortest path algorithms, *Operations Research*, 17, 3, 395–412.

Dreyfus, S and Law, A (1977), *The Art and Theory of Dynamic Programming*, Academic Press, New York.

Halpern, J, Isherwood, G and Wand, Y (1979), Response times and fire property losses in single and double family dwelling units, *INFOR*, 17, 373–379.

Hogg, J M (1968), The siting of fire stations, *Operational Research Quarterly*, 19, 275–287.

Hogg, J M (1973), *Losses in Relation to the Fire Brigade's Attendance Time*, Fire Research Report 5/73, Scientific Advisory Branch, Home Office, London.

Labes, W G (1968), *Fire Department Operation Analysis, Final Report.* Illinois Institute of Technology Research Institute, Office of Civil Defense, Washington, DC.

Maclean, A D (1979), *Fire Losses – Towards a Loss-Attendance Relationship*, Fire Research Report 17/79, Scientific Advisory Branch, Home Office, London.

Ramachandran, G (1992), *Statistically Determined Fire Growth Rates for a range of scenarios: Part 1: An Analysis of Summary Data. Part 2: Effectiveness of Fire Protection Measures – Probabilistic Evaluation*, upublished report to the Fire Research Station, Borehamwood.

Rogers, F E (1977), *Fire Losses and the Effect of Sprinkler Protection of Buildings in a Variety of Industries and Trades*, Current Paper CP 9/77, Building Research Establishment, Fire Research Station, Borehamwood.

Rutstein, R (1975), *Methods of Planning Fire Cover Using Cost Effectiveness Criteria*, Fire Research Report 7/75, Scientific Advisory Branch, Home Office, London.

Thomas, P H (1959), Use of water in the extinction of large fires, *Institution of Fire Engineers Quarterly*, 19, 35, 130–132.

Walker, W E, Chaiken, J M, Ignall, E J (Editors), (1979), *Fire Department Deployment Analysis: A Public Policy Analysis Case Study*, The Rand Fire Project, Rand Corporation, North Holland, NY.

11 Whole project analysis

11.1 Introduction

This chapter describes the application of full quantitative fire risk assessment. The process is called 'full' since the analysis quantifies the probabilistic and physically deterministic processes of fire events for a particular building. Chapter 3 provides detailed information on the methods for undertaking these two key processes and Chapter 4 contains information to help inform the judgement of the acceptability of any fire risk predicted.

In contrast, some methods in Chapter 3 concentrate on purely statistical methods, where the consequences of fire events are contained in the data. This has the advantage that all the consequences implicit in the analysis are from actual events. The disadvantage is that it implicitly assumes that any particular data set is from a homogenous set of buildings and that they are similar to the building being assessed. For that reason, purely probabilistic methods tend to be used to inform decisions concerning large sets of buildings such as those concerning regulation or insurance portfolios.

Full quantitative fire risk assessment allows the performance and reliability of fire precautions and the physical nature of a building, its fire hazards and occupants to be explicitly taken into account in the fire safety assessment of the building (Charters 2000). It can be used to quantify levels of fire risk to life, property, business and/or the environment and the uncertainties affecting risk in terms of probabilities (Charters 2000).

Full quantitative fire risk assessment is usually most useful in quantifying the levels of risk for an individual building either during design and/or occupation:

- in demonstrating equivalent life safety to a code-compliant building or satisfaction of fire risk criteria;
- for cost-benefit analysis of property protection/business continuity; and/or
- in assessing the environmental impact of large fires.

Therefore, circumstances where full quantitative fire risk assessment would be useful include where life safety is affected by buildings:

- containing very large numbers of people who are unfamiliar with the building
- containing large numbers of sleeping occupants
- containing large numbers of vulnerable people
- with very high fire growth rates
- with restricted means of escape; and/or
- with other unusual features which could be significantly detrimental to fire safety.

Examples where full quantitative fire risk assessment would be useful for life safety in buildings and structures may include the larger assembly buildings, hotels, hospitals, industrial and commercial premises, tunnels, offshore installations and ships.

- Property protection/business continuity is affected by buildings with:
  - a high value;
  - high value operations; and/or
  - high value images.
  Premises where full quantitative fire risk assessment would be useful for property protection/business continuity may include certain landmark, headquarters, leisure, assembly and heritage buildings as well as some industrial, commercial and transport premises (Charters and Wu 2002).
- Environmental protection is affected by buildings with a high potential impact on the environment. Premises where full quantitative fire risk assessment would be useful may include industrial and transport premises particularly where hazardous goods may be present in significant quantities (Charters and McGrail 2002).

Fire risk assessment is essential where deterministic fire safety engineering cannot adequately address all the fire scenarios of concern. This tends to occur when the consequences in terms of life safety, loss prevention and/or environmental protection of a fire may be intolerable.

Factors that might indicate where quantitative fire risk assessment would be essential include:

- When the reliability and performance of protection systems is critical. Fire risk assessment should be used to assess the defence in depth of a design that relies heavily on a single fire safety system.
- When the variability of and uncertainties governing input parameters have significant impact on the results. Fire risk assessment should be used where there are significant variations in variables like the number of people, their characteristics, fire growth rates etc and deterministic analysis shows that credible combinations of the variables are not acceptably safe (Charters, McGrail, Fajemirokum and Wang 2002).
- When a wide range of fire scenarios is deemed to be necessary. Fire risk assessment should be used in complex, high value, mission critical, buildings where high levels of defence in depth (multiple fire safety systems) are required.

11.2 Risk parameters

Since fire risk assessment is one of the tools of risk management, it is clear that there is a need to measure risk on the basis that if it isn't measured, it can't be managed. A measure of risk is often called a risk parameter. There are several ways of measuring any particular risk and the optimum risk parameter is therefore related closely to the type of risk being analysed and the decision being taken. For example:

- If the objective is life safety, and there is concern about injuries (or fatalities) to people in a building, the risk parameter could be measured in terms of the number of injuries (or fatalities) per building year.
- If the objective centres on property protection and/or business continuity, the risk parameter could be measured or monitored in terms of the 'financial value of losses per year'.

In general terms, the risk parameter is measured in 'outcomes per unit of activity', where the 'outcome' is the unwanted event and the 'unit of activity' is a measure of time, number or frequency of occurrences, etc. The unit of activity can have a significant impact on the perceived level of fire risk. For example, risk parameters using 'per building year' may be appropriate if society's experience or perception of those risks is linked to a specific building or building type. However, 'per building year' as a measure of activity, will naturally be biased against large or high occupancy buildings. If the unit of activity is 'per area year' or 'per occupant year', this may produce a less biased comparison between buildings of different sizes.

11.3 The full quantification fire risk assessment process

Risk assessment is the appraisal of one or more risk parameters. Risk assessment can be used to implement a safety objective more cost-effectively by directing spending to address the most significant safety issue(s).

Having identified a suitable risk parameter, the full quantitative risk assessment process can begin. The full quantitative fire risk assessment process involves four main tasks (see Figure 11.1):

1   **Hazard identification** – to identify hazards that could give rise to incidents or events of concern.
2   **Frequency analysis** – to estimate how often such events are likely to occur.
3   **Consequence analysis** – to estimate the potential severity of such events.
4   **Risk evaluation** – to decide what, if anything, should be done to address the levels of risk predicted.

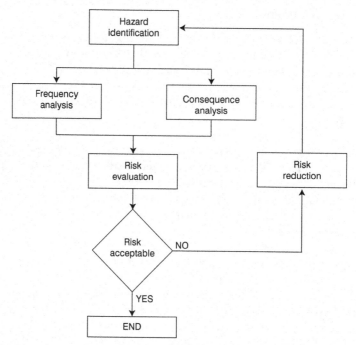

Figure 11.1 Typical quantified fire risk assessment process

11.3.1 *Hazard identification processes*

For fire safety, events of concern might be those that could lead to the outcome of concern (death/injury, asset damage and/or business interruption etc). There are many forms of hazard identification and these include:

- checklist type approaches (see Section 2.1);
- structured brainstorms (see matrix methods in Section 2.2).

In fact, a highly effective way of assessing risks is to undertake a structured brainstorm process such as HAZID (HAZard IDentification) or HAZOP (HAZard and OPerability) studies and then identify the core events of concern using full quantitative risk assessment.

11.3.2 *Frequency analysis*

In order to estimate the frequency of fire events, historical data is collected on how often fires occur in different building types. Because serious fires in buildings are relatively infrequent, there is arguably insufficient information to provide a meaningful direct estimate of the frequency of severe fires. If all the data is broken down into sub-events that could lead to the severe event of concern, this can be reconstructed using fault and event trees to provide a prediction for the frequency of rare, but severe fire events.

Frequency analysis is addressed in detail in Chapter 3.

11.3.3 Consequence analysis

There are several ways of estimating the severity of the consequences of an event of concern – by historical data, simple analytical methods, computational methods and/or small/full-scale tests and experiments. Each approach has its advantages and disadvantages.

Historical data gives details of known past outcomes, but this may not always be complete or even relevant in the context of current or future practices. Simple analytical methods can be used to estimate the severity of event outcomes cost-effectively, but the predictions are only as good as the model, its data and its application including any assumptions (implicit or explicit). Computer models can be used to estimate the severity of outcomes more closely, but such analysis can be expensive and time consuming, limiting the range of events being quantified. Full-scale testing can give the most accurate assessment of the level of fire hazard and the response of a building and its occupants, but again it is usually highly costly and time consuming as separate experiments are needed for fire hazards and occupant response.

Recently, significant advances have been made in the prediction of fire consequences using computer models. Models have been developed and improved to give more accurate and cost-effective prediction of fire consequences and again great care is needed in their application and a detailed understanding of the physical and computational basis of the model is necessary for competent use.

Consequence analysis is addressed in detail in Section 3.6.

11.3.4 Risk evaluation

The level of risk can be calculated by combining the frequency and consequence analysis for the outcomes of events. Once the level of risk has been quantified, it is necessary to decide whether or not it is acceptable, and if not, what should be done to reduce it. The two main approaches to risk evaluation are:

1 **Risk comparison** – where the risk of an activity is compared with the risks from relevant prescriptive standards for life safety. For risk comparison the Health and Safety Executive and Health and Safety Commission have also proposed risk acceptance criteria. For example, the maximum tolerable risk to an individual member of the public is a 10^{-4} probability of death per year. This equates to a one in 10,000 chance of an individual dying in one year.

2 **An economic approach** – where the cost-effectiveness of risk reduction can be assessed. This approach is usually used for property protection or business continuity risk assessments.

Risk evaluation is addressed in detail in Chapter 4. If the risks are not acceptable, then risk reduction is required.

11.3.5 Risk reduction

Overall, several approaches to risk reduction can be employed (see Section 11.1 for risk management). In terms of the full quantitative fire risk assessment process, there are three main approaches:

1   Reduce the hazard – less material, lower hazard material, control of ignition sources, etc.
2   Reduce the frequency – by reducing the hazard or introducing additional mitigation measures or improving the reliability of existing measures to reduce the frequency of severe events of concern.
3   Reduce the consequences – again by reducing the hazard or introducing additional mitigation measures to reduce the severity of event outcomes.

The full quantitative fire risk analysis can then be modified in line with the proposed changes and the predicted levels of fire risk evaluated.

11.4 Examples

The first example fire risk assessment was to satisfy the Building Regulations for a novel design. A quantitative probabilistic method was used to satisfy acceptance criteria and demonstrate equivalency with similar buildings designed to the prescriptive guidance. The second example is a financial risk assessment of the risk-cost benefit of the installation of sprinklers in schools.

BUILDING REGULATIONS

The objectives of the Building Regulations 2000, with respect to fire safety, are set out in five functional requirements under Part B:

1   To ensure satisfactory provision of means of giving an alarm of fire and a satisfactory standard of means of escape for persons in the event of a fire in a building.
2   That fire spread of the internal linings of buildings is inhibited.
3   To ensure the stability of buildings in the event of fire; to ensure that there is a sufficient degree of fire separation within buildings and between adjoining buildings; and to inhibit the unseen spread of fire and smoke in concealed spaces in buildings.
4   That external walls and roofs have adequate resistance to the spread of fire over the external envelope, and that spread of fire from one building to another is restricted.

5   To ensure satisfactory access for fire appliances to buildings and the provision of facilities in buildings to assist fire fighters in the saving of life of people in and around buildings.

Guidance on how to comply with Part B of the Building Regulations can be found in Approved Document B. This guidance is intended for some of the more common building situations. However, it recognises that there are alternative ways of complying with the requirements. The means for developing alternative design solutions is commonly known as fire safety engineering. Paragraph 0.11 goes on to state that a fire engineering approach may be the only practical way to achieving a satisfactory standard of fire safety in some large and complex buildings. Paragraph 0.13 indicates that factors that should be taken into account include:

*   the anticipated probability of a fire occurring; and
*   the anticipated fire severity.

Furthermore, alternative solutions are accepted on the basis of equivalency, that is to:

> ... demonstrate that a building, as designed, presents no greater risk to occupants than a similar type of building designed in accordance with well established codes.

Therefore, fire risk and fire risk assessment can play a major role in the design of new buildings. Further information on fire engineering can be found in BS 7974: 2001 'Code of practice on the application of fire engineering principles to the design of buildings'. Part 7 of BS 7974 contains guidance on the application of probabilistic fire risk assessment.

BS 7974 CODE OF PRACTICE ON THE APPLICATION OF FIRE ENGINEERING PRINCIPLES TO THE DESIGN OF BUILDINGS

This Code of Practice was published under the Fire Standards Policy Committee and is published as part of the PD 7974 series. Other parts published or about to be published are as follows:

*   Part 0: Introduction to fire safety engineering
*   Part 1: Initiation and development of fire within the enclosure of origin
*   Part 2: Spread of smoke and toxic gases within and beyond the enclosure of origin
*   Part 3: Structural response and fire spread beyond the enclosure of origin
*   Part 4: Detection of fire and activation of fire protection systems
*   Part 5 Fire service intervention
*   Part 6: Evacuation
*   Part 7: Probabilistic risk assessment.

This Code of Practice provides guidance on the use of BS 7974:2001 and is a framework for an engineering approach to the achievement of fire safety in buildings. It gives guidance on the application of scientific and engineering principles to the protection of people and property from fire. It also gives a structured approach to assessing the effectiveness of the total fire safety system in achieving the design objectives.

It provides guidance on the design and assessment of fire safety measures in buildings. It provides some alternative approaches to existing codes and guides for fire safety and also allows the effect of departures from more prescriptive codes to be evaluated. It recognises that a range of alternative and complementary fire protection strategies can achieve the design objectives.

This Code of Practice is intended to provide a framework for a flexible but formalised approach to fire safety design that can also be readily assessed by the statutory authorities. It is intended that this Code of Practice, when used by persons suitably qualified and experienced in fire safety engineering, will provide a means of establishing acceptable levels of fire safety economically and without imposing unnecessary constraints on other aspects of building design.

BS 7974 can be used to define one or more fire safety design issues to be addressed using fire safety engineering. The appropriate Published Documents (PDs) can then be used to specific acceptance criteria and/or undertake detailed analysis.

A fire safety engineering (FSE) approach that takes into account the total fire safety package can often provide a more fundamental and economical solution than more prescriptive approaches to fire safety. It may, in some cases, be the only viable means of achieving a satisfactory standard of fire safety in some large and complex buildings.

Fire safety engineering can have many benefits. The use of BS 7974 will facilitate the practice of fire safety engineering and in particular it will:

*   provide the designer with a disciplined approach to fire safety design;
*   allow the safety levels for alternative designs to be compared;
*   provide a base for selection of appropriate fire protection systems;
*   provide opportunities for innovative design; and
*   provide information on the management of fire safety for a building.

Fire is an extremely complex phenomenon and there are still gaps in the available knowledge. When used by a suitably qualified person, experienced in fire safety engineering, this series of documents will provide a means of establishing acceptable levels of fire safety economically and without imposing unnecessary constraints on aspects of building design.

PART 7: PROBABILISTIC RISK ASSESSMENT

This PD provides guidance on the application of probabilistic risk assessment for fire safety engineering in buildings. This approach can be used to show how regulatory, insurance or other requirements can be satisfied. Probabilistic risk assessment, like fire safety engineering in buildings, is a developing field. As with all engineering and risk disciplines, models and data can never fully describe actual circumstances and so judgement is required in assessing whether a design is acceptable. This judgement should be based on the best and most appropriate facts and evidence available.

This PD may be applied to the design of new buildings and the appraisal of existing buildings. Probabilistic risk assessment may be used in conjunction with the other PDs (see Figure 11.1) and other guidance documents. It may also be used to justify approaches that differ from those in other guidance documents.

This PD provides guidance on probabilistic risk analysis in support of BS 7974, 'Application of fire safety engineering principles to the design of buildings – Code of Practice'. It sets out the general principles and techniques of analysis that can be used in fire safety engineering. This PD also outlines the circumstances where this approach is appropriate and gives examples illustrating their use.

This PD also includes data for probabilistic risk assessment and criteria for assessment. The data included is based on fire statistics, building characteristics and reliability of fire protection systems. The criteria included cover life safety and property protection, both in absolute and comparative terms.

This PD does not contain guidance on techniques for hazard identification or qualitative risk analysis.

Probabilistic risk assessment of fire in buildings (with the exception of nuclear, chemical process, offshore and transport) is not widely used and so a discussion of possible future developments is included.

11.4.1 Example 1 Application of full quantitative fire risk assessment of a shopping centre design

11.4.1.1 Introduction

The Fields shopping centre is a multi-purpose retail/leisure complex in Denmark (Charters, Paveley and Steffensen 2001). The podium is approximately 200m by 150m with linked adjacent blocks. The retail and leisure complex can be divided into the following functional areas:

- car park
- retail levels
- offices
- leisure facilities
- conference centre
- cinema complex.

A shopping and leisure complex such as Fields is not specifically addressed in the Danish Building Regulations, so a performance-based fire engineering approach has been adopted. Innovatively for a commercial development, a quantified fire risk assessment was undertaken in parallel to ensure that there was sufficient 'defence in depth' and this provided a powerful insight into the fire safety of the complex.

11.4.1.2 The fire strategy

A fire strategy was developed for the shopping and leisure complex using a performance-based fire engineering approach and incorporating international practice for shopping centres and the Danish Building Regulations.

Some of the fire safety objectives for the project were:

*   safe and efficient evacuation of high populations;
*   minimise property loss and business interruption;
*   maintain open routes and shop fronts;
*   provide flexibility for varied tenant use.

The fire safety measures for the building were designed to provide an integrated package. With fire safety systems such as sprinklers, smoke control to protect the mall, automatic fire detection and a voice alarm system, the mall forms an evacuation route through which people can escape from the fire (see Figure 11.3).

Each area of the complex is within their own fire/smoke compartment, although they will share escape routes, including stairs. Evacuation of the complex is phased, evacuating each functional area individually. The shopping centre area is sub-divided into a number of evacuation zones.

The fire strategy was developed and agreed with the local authorities and was used as the basis of the building design and formed part of the Building Regulations submission. It is also be used for the future development and management of the centre.

The provision of automatic sprinklers, smoke control to protect the malls and large shops, automatic fire detection and a voice alarm system, allows the alternative design to incorporate:

*   extended travel distances within malls;
*   travel distances in shops to be measured to the exit to the mall;
*   shops not fire separated from the mall;
*   fire resistance levels will be reduced in specific areas; and
*   open stairs at the main entry/exit point from the mall areas will be used for egress direct to the outside.

11.4.1.3 Quantified fire risk assessment

One of the more innovative aspects of the project was that a fire risk analysis was required to be undertaken in parallel with the deterministic fire engineering. This was in accordance with the principles of the Draft Common Guidelines for the Øresund region, which includes acceptance criteria.

The risk analysis examined the overall 'defence in depth' of the proposed fire safety systems and determined the level of risk that occupants were exposed to. This included sensitivity analysis and incorporated the performance, reliability and the effect of the failure of key fire safety measures (such as smoke control).

The quantified fire risk assessment was undertaken using SAFiRE (Simplified Analytical Fire Risk Evaluation). SAFiRE combines consequence and frequency analysis in the same way as traditional quantified fire risk assessment (developed in the nuclear industry). Frequency analysis is based on statistical analysis, simple fault and event trees and Monte Carlo analysis of highly variable parameters. Consequence analysis is based on statistical analysis and simplified models for time to untenability and time for evacuation. This method is used for the design process because it is significantly quicker than traditional quantitative risk assessment.

SAFiRE takes into account the regulatory paradigm. That is, it addresses the same fire safety objectives as Building Regulations and assesses adequacy using the same variables, e.g. exit width, travel distance, number of people, compartment size, etc.

The method is also consistent with the principles of traditional quantitative risk assessment. That is, it uses deterministic models based on theoretical and empirical physical models to predict the consequences of fires and probability theory to predict the likelihood of fires and their outcomes. There is also a balance between the way that deterministic and probabilistic processes are modelled.

Figure 11.2 shows a typical example of an F-n curve for life safety. The further to the upper right the higher the risk and the further to the lower left the lower

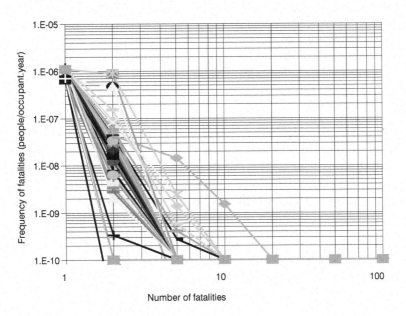

Figure 11.2 Typical output from SAFiRE

the risk. The individual curves show relatively high frequency/low consequence fire events on the upper left and relatively low frequency/high consequence events on the lower right.

The risk analysis showed that the overall risk to occupants satisfied the acceptance criteria and is the same or less than that for a department store or similar public building designed to meet the Danish Building Regulations 1995.

11.4.1 4 Conclusion

The quantified fire risk assessment showed that, for certain parts of the development, relatively simple changes to the design could significantly improve life safety. This powerful combination of deterministic fire engineering and probabilistic risk assessment provided a more rounded view of fire safety in the development and may be a taste of things to come in countries with risk-informed functional fire safety regulations.

11.4.2 Example 2: Life safety and asset protection in schools

A study of the cost/benefits of sprinklers in schools was undertaken using full quantitative fire risk assessment (Charters, Salisbury and Wu 2004). The objective of the study was to assess whether sprinkler systems in new and refurbished schools are:

- a cost-effective alternative to other fire precautions for life safety; and/or
- represent good value for money in terms of protection of property.

The study assessed the risk-cost/benefits in four schools of varying size and extent of new build/refurbishment. The types of school were:

1   refurbishment;
2   small all new build;
3   new build extension;
4   large new build.

The quantified fire risk assessments were undertaken for all four schools, occupied and unoccupied and with and without sprinkler systems.

Table 11.1 shows the results of the risk–cost benefit analysis.

The net present value (NPV) is the total (aggregate) present values of the annual benefits over the 30 years discounted at 6 per cent minus the capital cost.

In the analysis, the cost of allocating temporary accommodation was incorporated based upon £260 per 30 pupils (including toilets).

Table 11.1 Risk–cost benefit analysis

School	Capital cost of sprinkler system (£)	Average annual benefits (£)	NPV (after 30 years at 6%)	Average annual benefits with temporary accommodation included (£)	NPV with temporary accommodation included (after 30 years at 6%)
Refurbishment	53,350	2,566	+22,145	3,427	+10,300
Small new build	58,580	3,825	+10,060	5,074	-7,130
New build extension	50,550	6,194	-30,580	6,808	-39,030
Large new build	39,640	2,293	+12,210	3,173	+94

11.4.2.1 Conclusion

The results for the small new build and new build extension schools indicated relatively robust investments with pay-back periods of 22 years and 10 years respectively.

Refurbishment and large new build showed positive NPVs and as such the installation of sprinklers does not necessarily represent a cost-effective investment; this is probably due to the relatively high level of compartmentation present within the layout of the schools, coupled with the relatively low number of classrooms that are likely to be affected by any one fire.

It is interesting to note that no allowance could be made in either analysis for any excess or reduction in premium for the installation of sprinklers with respect to insurance provisions. The wider decision-making process did, however, take into account the social and educational benefits of fewer schools being lost due to fire.

References

Charters D (2000), What does quantified fire risk assessment need to do to become an integral part of design decision-making, SFPE International Conference on Performance Based Codes, San Francisco, CA.

Charters D (2004), A review of fire risk assessment methods, *Proceedings of Interflam '04*, Interscience Communications, Cambridge.

Charters D and McGrail D (2002), Assessment of the environmental sustainability of different performance based fire safety designs, *Proceedings of the 4th International Conference on Performance-Based Codes and Fire Safety Design Methods*, SFPE, Melbourne.

Charters D and Wu S (2002), The application of 'simplified' quantitative fire risk assessment to major transport infrastructure, SFPE Symposium on Risk, New Orleans.

Charters D, Paveley J and Steffensen F (2001), Quantified fire risk assessment in the design of a major multi-occupancy building, *Proceedings of Interflam '01*, Interscience Communications, Cambridge.

Charters D, McGrail D, Fajemirokum N and Wang Y (2002), Analysis of the number of occupants, fire growth, detection times and pre-movement times for probabilistic risk assessment, *Proceedings of the 7th International Symposium on Fire Safety Strategy*, International Society for Fire Safety Science, Worcester, MA.

Charters D, Salisbury M and Wu S (2004), The development of risk-informed performance-based codes, *Proceedings of the 5th International Conference on Performance-Based Codes and Fire Safety Design Methods*, SFPE, Luxembourg.

12 Interactions

12.1 Introduction

For over a century, regulatory control for fire safety has mainly been achieved through a framework of prescriptive rules for passive fire protection measures. These rules are highly empirical and depend heavily on simple standard fire tests and Codes of Practice. The rules are modified largely on the basis of experience. The rules do not take sufficient account of the effectiveness of active measures such as sprinklers, detectors and smoke ventilation systems.

A life safety level implicit in the fire regulations or any other level acceptable to the society can be achieved through combinations of passive and active fire protection measures, which are appropriate to the hazard involved, having due regard to practicability. Any combination considered for the fire safety design of a building must demonstrate an acceptable performance reflected in fatal and non-fatal casualties likely to be sustained if a fire breaks out in the building. The performance depends on the successful operation, reliability and effectiveness of the passive and active fire protection measures included in the combination.

Combinations of safety measures produce interactions between measures in their joint performance to provide a prescribed level of life and property protection. A balanced fire safety system would recognise these interactions and permit adjustments in requirements. This is commonly referred to as 'trade off' or 'equivalence'. Among combinations providing equivalent safety, a property owner may select one which is the most cost-effective. Fire safety measures also have interactions and trade-offs with fire brigade operations.

Evaluating the interactions mentioned above is a complex problem requiring the application of statistical/probabilistic methods, deterministic models and fire safety engineering techniques. Sufficient statistical and experimental data are not available at present for evaluating all the interactions.

12.2 Passive and active fire protection

12.2.1 Sprinklers and passive fire protection

12.2.1.1 Introduction

Although there is a small chance, about 5 per cent, of not operating in a fire, if maintained in a satisfactory working condition, sprinklers have a high potential to extinguish a fire before it spreads beyond the object first ignited. If a fire is not thus confined to the object first ignited, sprinklers can prevent a fire from involving all the objects in a room. For the reasons mentioned above, sprinklers would reduce the probability of 'flashover' in the room (or compartment) of fire origin which, in turn, would reduce the probability of fire spreading beyond the room to other parts of the building.

It is, therefore, reasonable to relax requirements specified in fire regulations, codes and standards for buildings which have full sprinkler protection. Sprinklered properties can be permitted to have one or more of the following concessions:

a. larger building size and height;
b. larger compartments;
c. increased travel time and distance to the entrance of a protected staircase;
d. reduced widths for exits and staircases;
e. fewer staircases;
f. reduced fire resistance.

Some of the concessions mentioned above have already been incorporated in some form or other in the fire safety codes of many countries.

12.2.1.2 Sprinklers and building or compartment size

The 'power' relationship between area damaged, $D(A)$, and total floor area, A, discussed in Section 3.3.1 (Equation (3.41)) can be applied to evaluate the interaction between sprinklers and building or compartment size. Figures 12.1 and 12.2 are examples.

According to Figure 12.1, if a damage of 2300m² is acceptable for a building of total floor area 10,000m² without sprinklers, the building size can be increased to 33,000m² if it is provided with sprinklers or to a slightly smaller size if the reliability of sprinklers is less than 100 per cent, say, 95 per cent. Likewise, as revealed by Figure 12.2, if a damage of 153m² is acceptable, the size of a sprinklered compartment can be increased from 500m² to 4,000m² or slightly less depending on the reliability of the system.

12.2.1.3 Sprinklers and travel distance

Sprinklers have the potential to reduce significantly the number of fatal and non-fatal casualties likely to be sustained if a fire breaks out in a large building occupied

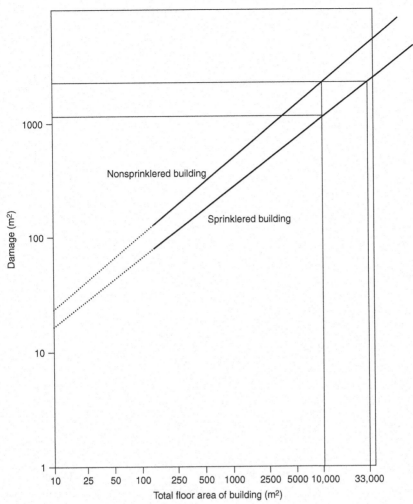

Figure 12.1 Damage and building size

by several people. Hence, for any floor of a sprinklered building, the maximum travel distance to the entrance of a staircase can be increased up to a limit such that life risk does not exceed an acceptable level.

An increase in travel distance will enable the construction of a smaller number of staircases but life risk might increase due to a reduction in the number of staircases. Sufficient data are not available, at present, for assessing the increase in life risk due to an increase in travel distance or reduction in the number of staircases for large sprinklered buildings. Hence, such a relaxation may not be justified at present but may be considered if smoke ventilation systems are also installed in addition to sprinklers.

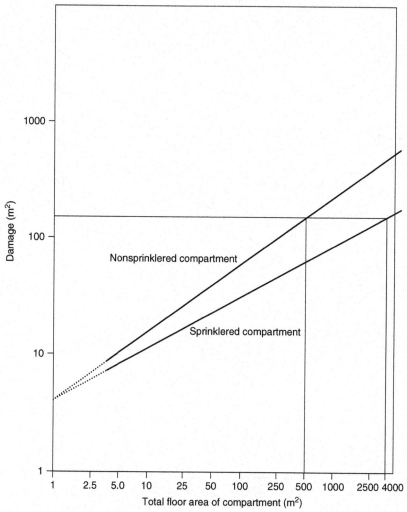

Figure 12.2 Damage and compartment size

12.2.1.4 Sprinklers and staircase width

An increase in travel distance implies an increase in the 'design evacuation time', t, for which 2.5 minutes has currently been recommended in UK fire regulations. With this value for t, the width b required for a staircase is determined according to the following equation:

$$M = Nbt + (n - 1) \, 50 \, (b - 0.3) \tag{12.1}$$

where M is the maximum number of people who can enter a staircase of width b (in metres) serving n floors. The flow rate, N, is approximately 80 persons per metre width per minute.

Consider, for example, a six-storey building with 690 people designed for total (simultaneous) evacuation. With M = 690, t = 2.5, and n = 6, calculations based on Equation (12.1) show that the total width of a stair should be at least 1.7m. Two 1.7m stairs would be necessary to meet the needs of 'discounting' of one stair assumed to be not available due to fire or smoke.

If the building is sprinklered, the design evacuation time, t, may be increased to say, 4 minutes such that, using Equation (12.1), b can be reduced to 1.34 m for M = 690 and n = 6. Two 1.34 m stairs would be necessary to allow for 'discounting' of one stair.

A reduction in the total width of staircases for a sprinklered building will produce a gain in rentable floor space but this action might increase life risk. This increase in life risk may be compensated by an increase in life safety provided by sprinklers.

12.2.1.5 Sprinklers and fire resistance

Sprinklers have the potential to reduce the severity likely to be attained in an actual fire occurring in a compartment. Hence, there is a clear case for reducing the fire resistance required for a sprinklered compartment from a level required for a compartment without sprinklers. This reduction, for any level specified for the probability of compartment failure, can be evaluated by applying the Beta method discussed in Section 3.7.4. For any specified level for β, the value of mean severity, μ_s, in Equation (3.108) for a sprinklered compartment will be less than that for an unsprinklered compartment.

For a compartment of any given size, the mean severity, μ_s, in units of time can be realistically assumed to be proportional to the logarithm of area damage expected in a fire. This assumption is based on the exponential model discussed in Section 3.3.2 – see Equation (3.42). This relationship between area damage and severity can be estimated by carrying out some further research and statistical analysis of data – see Ramachandran (1990). Sprinklers would reduce the area expected to be damaged in a fire and, hence, would reduce the mean severity, μ_s.

In the following, simpler, semi-probabilistic approach (Ramachandran, 1993), total fire safety may be defined by probability C:

$$C = A*B \qquad (12.2)$$

where
A is the probability of 'flashover'
B is the conditional probability of 'structural failure' given flashover.

Probability of flashover, A, can be estimated from statistical data on area damage – see Section 5.3.

'Compartment failure' would occur if the design fire resistance period of a structural element is exceeded by the severity actually attained in a fire. The probability of this event, B, depends empirically on the fractile value, F, of design fire load density, L, selected for a compartment:

$$B = 1 - F \tag{12.3}$$

Probability of structural failure, B, is 0.2 if F = 0.8.

L is one of the parameters in a formula for estimating fire severity in a compartment of given dimensions, ventilation factor and thermal inertia. This formula is based on the concept of equivalent time of fire exposure in which fire resistance required is equated to fire severity.

If L_0 is the design fire load density for a non-sprinklered compartment corresponding to a fractile F_0 of the load density distribution:

$$B_0 = 1 - F_0 \tag{12.4}$$

For a sprinklered compartment:

$$B_s = K * B_0; K = A_0/A_s \tag{12.5}$$

$$F_s = 1 - B_s \tag{12.6}$$

where A_s is the probability of flashover for a compartment without sprinklers and As the probability of flashover for a compartment with sprinklers.

The design fire load density L_s, for a sprinklered compartment, would correspond to the fractile F_s. L_s will be less than L_0 since F_s will be less than F_0. If K = 3.5, as on average for many occupancy types, and F_0 = 0.8, B_0 = 0.2 such that B_s = 0.7 and F_s = 0.3.

The sprinkler factor S is defined as the ratio

$$S = L_s/L_0 \tag{12.7}$$

The value of S is about 0.6 for office buildings, retail premises and hotels. Hence, the fire resistance of a sprinklered compartment of these occupancies can be about 60 per cent of the resistance specified for a compartment without sprinklers – see Ramachandran (1993).

12.2.1.6 Sprinklers and smoke ventilation systems

The interaction between sprinklers and smoke ventilation systems has not, so far, been clearly evaluated by fire scientists and engineers. This interaction arises from the fact that water sprays from activated sprinklers remove buoyancy from combustion products and generate air currents which counter the outflow from the vents, and additionally, transport combustion products to the floor. Use of vents may reduce the total water demand by the sprinkler system but this benefit due to vents has not yet been clearly established. Sufficient statistical information needs to be collected to assess the effects of the interactions on damage to life and property and the performance of the fire service.

It is arguable whether a vent should operate before the operation of a sprinkler or not. There are indications from current research that the effect of venting on the opening of the first sprinklers and their capacity to control the fire is likely to be small. There are also indications that the earlier vents are opened, the more likely they would be effective in preventing smoke-logging of a sprinklered building. The controversy between sprinklers and vents can, perhaps, be resolved by deciding, for any type of building considered, whether property protection or life safety is the main objective. In the initial stage of fire growth, a vent should operate before a sprinkler if life safety is the dominant objective, e.g. hotels, shopping centres, office buildings. In industrial buildings, the first sprinkler may operate before the opening of any vent (Ramachandran 1998).

12.2.1.7 Sprinklers and automatic fire detectors

For discovering the existence of a fire, sprinklers installed in industrial and commercial buildings are generally less sensitive in the sense that their operating times are longer than those of automatic detectors. Normally, it would take about 6 minutes before the heat generated by a fire is sufficient to activate a sprinkler head, whereas a detector is designed to operate in 2 or 3 minutes after the start of the ignition. Consequently, if both of these systems are installed in a building, early discovery of a fire by the detector is likely to be followed by first-aid fire fighting such that the sprinkler system is not brought into action.

12.2.1.8 Automatic fire detectors and means of escape facilities

By discovering the existence of a fire very soon after the start of the ignition, automatic detection systems provide extra time for the escape of occupants of a building. Consequently, detectors have a great potential for reducing life risk. Hence, for a building fitted with detectors, particularly advanced types such as computer-controlled addressable systems, design evacuation time and travel distance to the entrance of a staircase can be increased or the total width of stairs reduced.

This concession appears to exist in some form in the fire regulations of some countries. In the UK, for example, a building can be designed for 'phased evacuation' if it has been fitted with an appropriate fire warning system. This would enable narrower stairs to be built than would be the case if the building had to be designed for total (simultaneous) evacuation.

12.3 Smoke movement and evacuation of occupants

As defined in Sections 3.7.3, 3.7.4 and 9.3, the evacuation time of occupants of a building is the third main component period of the total evacuation time. The first and second component periods are concerned with the detection or discovery time of a fire and recognition time (Section 9.1). For a successful evacuation, the total evacuation time should be less than the time taken by a combustion product to travel from the place of fire origin and produce an untenable condition on any segment or component of an egress route. Otherwise, egress failure would occur, with consequences leading to fatal or non-fatal injuries to one or more occupants.

The probability of occurrence of egress failure should be reduced by reducing the total evacuation time and the actual evacuation time, which is the time taken by an occupant or group of occupants to leave their places of occupation in the building and reach safe places inside or outside the building. To achieve this objective, for any target level specified for the probability of egress failure, a semi-probabilistic approach (safety factors) was recommended in Section 3.7.3 and a probabilistic approach (Beta method) in Section 3.7.4. Following one of these methods, appropriate design values can be determined for the total evacuation time and the actual evacuation time. To achieve these design values, the building may be equipped with necessary detection and communication systems and means of escape facilities which include staircases, emergency lighting and smoke ventilation systems.

As discussed in Section 9.3, smoke is the major combustion product threatening occupants, particularly those who are not in the immediate vicinity of a fire. The time taken by smoke to produce an untenable condition on an escape route depends on the rate (as a function of time) at which large quantities of smoke are produced and the location of the escape route with reference to the location of the fire. Egress failure or success also depends on the location of an occupant or group of occupants with reference to the location of the escape route. It also depends on the mobility conditions (disabled etc) of the occupants, as discussed in Section 9.3.

Sufficient statistical data are not available, at present, for evaluating quantitatively the interaction between smoke movement and evacuation of occupants. Simulations based on deterministic models may provide necessary data on smoke movement. Likewise, data on the movement of occupants can be obtained by performing simulations, based on computer models, for evacuation. It may not be necessary to carry out evacuation exercises or drills which are costly and time consuming.

12.4 Fire protection measures and fire brigade

12.4.1 Building design and fire brigade

One of the objectives of providing fire resistant compartments to a building is to ensure that a fire is contained within the compartment of origin, for a length of time which is sufficient for the fire brigade to arrive at the fire scene and commence fire fighting, before the fire spreads beyond the compartment. The fire resistance of a compartment should, therefore, be determined with due regard to the attendance or response time of the brigade, which is a function of the travel distance or time from a fire station to the building.

Fire spread beyond a compartment depends on the size of the compartment, apart from its fire resistance and other factors such as fire load and area of ventilation. Fire spread in a building also depends on the design and size of the building, particularly the building height. Compartments should not, ideally, extend to more than two or three floors in low-rise buildings and not more than one floor in high-rise buildings. Each floor (including basements) in a high-rise building should be a fire-resistant compartment. Fire brigades usually find it difficult to fight fires in high-rise buildings and basements.

The factors mentioned above should be taken into account for specifying or relaxing requirements in fire regulations for size and fire resistance of compartments and for size and height of buildings. These factors have been considered to some extent in fire regulations but their effects have not, so far, been assessed quantitatively with due regard to interactions.

12.4.2 Sprinklers and fire brigade

Sprinklers extinguish several fires which are not even reported to the fire brigades. Some fires to which the brigades are called are also put out by the system before the brigade arrives at the premises involved. If they fail to extinguish a fire, sprinklers would slow down the rate of fire growth and restrict fire spread until the arrival of the brigade.

The performance characteristics of sprinklers mentioned above, together with their capability for giving early fire warning, would enable a brigade to attend a smaller number of fires in sprinklered premises and to control and extinguish fires quickly in these premises. Statistical data are available in the UK for estimating the reduction in brigade control time due to sprinklers. For the reasons mentioned above, sprinklers would reduce considerably the time (and cost) spent by fire brigades in responding to fire incidents.

The interaction between sprinkler performance and fire brigade operation might affect the number of fire stations required to cover a particular city or area. In the USA, for example, a fire station in Fresno had to be closed down after almost all buildings in two districts in this city were fitted with complete automatic sprinkler protection in 1970. For providing protection to life, sprinklers are required to be installed in apartment buildings in many cities in the USA and Canada. Scotsdale in Arizona is an example where sprinklers have been installed in residential buildings.

12.4.3 Automatic detection systems and fire brigade

Early discovery or detection of a fire would enable the commencement of fire fighting by first-aid methods and/or fire brigade when the fire is small in size. This would reduce the time required by the brigade to bring the fire under control. Statistical studies indicate that, on average, the control time will be reduced by half a minute for every minute of early arrival of the brigade at the fire scene (see Table 10.2). The 'attendance time' is the period from the time when information about the fire is received by a fire brigade, to the arrival of the brigade at the scene of fire.

Since automatic fire detection systems have the potential to reduce, indirectly, the average control time for fires in an area, the average attendance time for this area can be increased up to a limit, such that risk to life and property does not exceed an acceptable level. An increase in attendance time would lead to a reduction in the number of fire stations required for an area.

The number of fire stations required for an area protected (or not) by detectors depends on the relationship between 'service time' and fire damage (to property and life). Attempts have been made to establish this relationship, which can provide inputs for cost-benefit comparisons and trade-offs between the effectiveness of detector systems and fire brigades.

Automatic fire detection systems are known to produce false alarms by operating when there are no real fires. False alarms cause wastage of time and money for a fire brigade.

12.4.4 Smoke ventilation systems and fire brigade

One of the objectives of installing smoke ventilation systems is to facilitate fire brigades by enabling fire fighters to enter a building involved in a fire and locate easily the seat of the fire. The extent to which this object is achieved can be assessed by comparing the average control time of fire brigades for fires in buildings with ventilation systems with that for fires in buildings without these systems. Data available at present are not sufficient for carrying out this investigation.

However, tentative conclusions can be drawn by analysing data on fires collected by leading manufacturers of smoke ventilation systems. These data can be combined with those contained in reports on fires furnished by fire brigades.

Such combined data were analysed some years ago using *Fire Case Histories* published by Colt International in October 1975 – see Ramachandran (1998). These case histories related to 73 fires which occurred during 1957 to 1974. Fire brigade reports were identified for 55 of these fires which occurred in the UK. According to the information contained in these reports, the average control time was 126 minutes for 19 fires in buildings without ventilation systems or sprinklers. The average control time was 57 minutes in 26 fires in buildings with only ventilation systems and 64 minutes in 7 fires in buildings with both vents and sprinklers.

The figures mentioned above, although based on small samples of data, indicate that fire brigades will be able to bring under control fires in buildings with ventilation systems quicker than fires in buildings not equipped with these systems.

References

Ramachandran, G (1990), Probability-based fire safety code, *Journal of Fire Protection Engineering*, 2, 3, 75–91.

Ramachandran, G (1993), Fire resistance periods for structural elements – the sprinkler factor, *Proceedings of the CIB W14 International Symposium on Fire Safety Engineering*, University of Ulster, Jordanstown.

Ramachandran, G (1998). *The Economics of Fire Protection*. E & F N Spon, London.

13 Combining data from various sources – Bayesian techniques

13.1 Introduction

The results of a risk assessment are sensitive to the probabilities attached to parameters such as the probability of a fire starting and probable loss if a fire occurs. Sufficient information to estimate these quantities for a particular building is unlikely to be available. In such cases, a 'prior' or initial assessment of the probabilities can be made by consulting an expert such as a fire safety engineer or by analysing statistical data for a group of buildings with similar fire risk.

A prior assessment of the probability estimated for any parameter can be revised in the light of expert opinion and/or data which may be available for the building considered. These data may be provided by a quantitative assessment of fire risk in that building and a sample of fires which occurred in the building. Bayesian techniques can be applied to revise a 'prior' estimate of the probability and obtain a 'posterior' estimate (Ramachandran 1998).

Bayesian statistical techniques provide a mechanism for combining subjective and objective information from two or more sources: national data, sample data, expert opinion, experimental results and simulations based on deterministic (zone or field) models.

13.2 Bayes' theorem

According to Bayes' theorem (La Valle, 1970), the 'posterior' probability, $P(H_i/A)$, of occurrence of event H_i given observation A is proportional to the 'prior' probability, $P(H_i)$, times the likelihood of A, given H_i, $P(A/H_i)$:

$$P(H_i / A) \propto P(A / H_i).P(H_i) \tag{13.1}$$

If $H_1, H_2, \ldots, H_p$ are a set of (exhaustive, mutually exclusive and independent) events,

$$\sum_{j=1}^{p} P(H_j) = 1 \tag{13.2}$$

To replace the proportionate sign, $\propto$, in Equation (13.1) by an equal to sign, the equation is 'normalised' with respect to the set of events in Equation (13.2). This process is achieved by dividing the right-hand side of Equation (13.1) by

$$\sum_{j=1}^{p} P(A / H_j).P(H_j) \qquad (13.3)$$

We can therefore write

$$P(H_i / A) = \frac{P(A / H_i).P(H_i)}{\sum_{j=1}^{p} P(A / H_j).P(H_j)} \qquad (13.4)$$

13.3 Probability of fire recurrence

13.3.1 Example 1

For a particular industrial building of a specified type and size, a power function (Section 3.3.1) based on national fire statistics provided a certain estimate for the annual probability of fire occurrence. This 'global' figure was adjusted for the building considered by a statistician who carried out a risk assessment and took into account the human (e.g. careless disposal of cigarettes) and non-human (e.g. faulty electrical appliances) sources of ignition actually present in the different parts of building. The adjustment followed the method described in Section 5.1.2 and Equation (5.2) and produced an estimate of 0.45 for the annual probability of fire occurrence for the property considered.

During a recent five-year period, four fires occurred in the building considered, indicating an annual probability of 0.8 for fire occurrence. This figure is considerably higher than 0.45. Statistically, however, with a higher variance, the estimated figure of 0.8 provided by a small sample is likely to be less reliable than the adjusted estimate of 0.45 based on a much larger sample provided by the national fire statistics. It would be appropriate, however, to combine both the above estimates using Bayes' theorem and obtain a composite estimate for the probability. The following procedure may be adopted.

The figure of 0.45 may be considered as the prior probability $P(H_1)$ for the occurrence of a fire during a year with $P(H_2) = 0.55$ for the probability of non-occurrence, with only two events $p = 2$ in Equations (13.2) to (13.4). If the information provided by the recent fires in the building is denoted by the letter A, $P(A/H_1) = 0.8$ and $P(A/H_2) = 0.2$. Then applying Equation (13.4):

$$P(H_1 / A) = \frac{0.8 \times 0.45}{(0.8 \times 0.45) + (0.2 \times 0.55)} = 0.77 \qquad (13.5)$$

The 'prior' probability of 0.45 has been revised upwards to the 'posterior' probability of 0.77 to give weight to the data on fire occurrence in the property considered. The 'posterior' probability for non-occurrence of a fire, $P(H_2/A)$, is hence 0.23 which is lower than the prior. However, according to the fire safety engineer who carried out a risk assessment using his own method, the annual probability of fire occurrence was only 0.4. In this case, with $P(A/H_1) = 0.4$ and $P(A/H_2) = 0.6$, a calculation as in Equation (13.5) would show that the prior probably 0.45 for fire occurrence needs to be revised downwards to the posterior probability of 0.35; the revised posterior probability for non-occurrence is 0.65.

It will be apparent that a value for $P(A/H_1)$ exceeding 0.5 will produce a posterior probability $P(H_1/A)$ greater than the prior while a value for $P(A/H_1)$ less than 0.5 will produce a posterior less than the prior. A calculation would also show that $P(H_1/A)$ and $P(H_1)$ would be equal to 0.45 for the example considered, if $P(A/H_1) = 0.5$. In this case, there is no need to revise the initial (prior) assessment of the probabilities $P(H_1)$ and $P(H_2)$ which are 0.45 and 0.55 for the example.

The prior annual probability of 0.45 and the revised annual probability of 0.77 (Equation (13.5)) for fire occurrence may be considered as 'objective' estimates based on statistics from real fires, while the value of 0.4 for this probability estimated by the fire safety engineer, although based on expert opinion, is a subjective assessment, not based sufficiently on facts and figures. Instead of the 'objective' prior probability of 0.45, the revised 'objective' probability of 0.77 may be considered as the prior since it has taken into account the recent data on fire occurrence in the property considered. In this case, $P(H_1) = 0.77$, with $P(H_2) = 0.23$. This prior may be combined with the subjective estimate 0.4 to give weight to the opinion of the fire safety engineer. Combining 0.77 and 0.4, with $P(A/H_1) = 0.4$ and $P(A/H_2) = 0.6$, a calculation as in Equation (13.5) would show that a more realistic estimate of annual probability of fire occurrence, $P(H_1/A)$, is 0.69. The corresponding annual probability of non-occurrence of fire, $P(H_2/A)$, is 0.31. These estimates are based on information from three sources: global statistics, data for the property considered and expert opinion.

13.3.2 Example 2

Bayesian techniques can be applied to the probability of financial loss or area damage exceeding a specific level. According to national fire statistics, for example, the probability of loss in a fire in an industrial building exceeding £500,000 may be 0.05. The probability of area damage exceeding 100m^2 may be 0.08. These national estimates can be used as prior probabilities for any particular industrial building and revised on the basis of expert opinion, a sample of fires in that building and estimates provided by deterministic models.

Consider, as another example, the probability of a fire spreading beyond the room of origin in an industrial building. This prior probability, according to national statistics, may be 0.07 if the room has no sprinklers and 0.02 if it has sprinklers. According to a deterministic model, however, the probability for this undesirable event may be 0.6 and 0.3 in the unsprinklered and sprinklered cases (respectively). An application of Bayes' theorem would provide 'posterior' probabilities of 0.101 if unsprinklered and 0.009 if sprinklered.

13.3.3 Example 3

One may be interested to estimate the probability that a certain number of fires would occur in a particular type of building during a given period of time, say a year. It is reasonable to assume that the fires would occur independently of one another and with a constant tendency to occur. Under this assumption, the frequency of fire occurrence is likely to follow the Poisson distribution (Ramachandran 1998).

Several factors can cause uncertainties in the value of the parameter θ of the Poisson distribution providing estimates of the probabilities of different frequencies of fire occurrence during a given period. The probabilities and hence the value of the parameter are likely to vary from one period to another. The value of the parameter needs to be revised or updated continuously as more and more information becomes available on the frequencies of fire occurrence and the factors affecting them. This task can be accomplished easily by applying, as discussed in the Sections 13.3.1 and 13.3.2, the discrete form of Bayes' theorem presented in Equation (13.4). In this equation, the letter H_i will be replaced by θ_i denoting a particular value for the Poisson parameter θ. $P(\theta_i)$ is the probability that one assigned, prior to obtaining the new information A, to θ_i being the correct value. $P(A/\theta_i)$ is the probability that the new information A would have been observed given that θ_i is true. $P(\theta_i/A)$ is the posterior probability that θ_i is the correct value after obtaining the new information A. The parameter θ may have p different values θ_j with $j = 1, 2, ..., p$.

A more complex problem is concerned with the estimation of the posterior probability distribution of fire occurrence given that the prior distribution is Poisson. This problem was discussed by Johansson (2003) who considered the Gamma distribution which is a 'conjugate' distribution of Poisson. If one considers the Gamma distribution for the probabilities of frequencies of fire occurrence, the resulting posterior distribution will also be a Gamma distribution but with different parameter values.

Another example referring to the probability distribution of occurrence of an event is concerned with uncertain situations involving only two possible, mutually exclusive events such as a sprinkler system extinguishing or not extinguishing a fire or an automatic fire detector operating or not operating when a fire occurs. For such cases, the binomial distribution can be used to denote the probability, p, of a sprinkler extinguishing a fire or of a detector operating in a fire. Such a situation was studied by Sui and Apostolakis (1988) who used Bayes' theorem to combine indirect evidence with direct evidence for the demand availability of a sprinkler system.

The Beta distribution is the 'conjugate' distribution that is applicable when the probability of obtaining the evidence, $P(A/p)$, can be calculated using a binomial distribution (Ang and Tang 1984). Using a prior distribution in the form of Beta distribution, one can simplify the use of Bayes' theorem considerably. This problem was discussed in detail by Johansson (2003). He showed how Bayes' theorem can be used to update the parameters of the prior Beta distribution to provide estimates of posterior values of the parameters.

Johansson (2003) also discussed the application of a multi-nomial distribution for situations involving more than two possible, mutually exclusive events. He used the Dirichlet distribution as a 'conjugate' distribution for the multi-nomial distribution. A Dirichlet distribution with only two parameters is the same as a Beta distribution. Thus following the method developed by him for the Beta distribution, Johansson discussed the calculation of posterior values of the parameters of the Dirichlet distribution. He has also proposed a Bayesian network model for the continual updating of fire risk measurement.

13.4 Probable loss in a fire

The probable or expected loss in a fire, whenever it occurs, can be estimated by identifying and using the probability distribution of loss and its parameters – such as mean and standard deviation. Research studies have shown that this distribution can be Pareto or log normal, as discussed in Section 3.3.4. One of these two distributions can be identified as a 'prior' distribution for any occupancy type by collecting and analysing past data on fire losses occurring on buildings of this type. Estimation of the 'posterior distribution' when the 'prior distribution' is Pareto or log normal is a complex statistical problem.

Statistical techniques are available for ascertaining whether the probability distribution of fire loss for any occupancy type is Pareto or log normal or any other distribution such as Weibull. However, none of these distributions can be expected to fully account for the uncertainties involved in the development and spread of a fire in a building and the physical or financial damage sustained.

To achieve a substantial reduction in the uncertainties it would be worthwhile developing a composite model which incorporates all suggested plausible probability distributions for fire loss. Such a model was attempted by Shpilberg (1974) using weights for the distributions before observing the data by a prior assumption regarding the likelihood of any one of the postulated distributions being the 'true' one. After observing the data, the composite model weighs the different distributions by a posterior probability that is a function of the prior assumption and the data, thereby modifying prior knowledge by observed information.

13.4.1 A heavy goods vehicle (HGV) fire in a tunnel

Carvel *et al.* (1999) considered HGV fires in tunnels from a probabilistic viewpoint. The authors described the effect of forced longitudinal ventilation on fire size by defining the coefficient k, based on the heat release rate (HRR):

$$k = \frac{\text{HRR of fire with forced ventilation}}{\text{HRR of fire with natural ventilation}}$$

Bayes' theorem was expressed as follows:

$$P(k = k_i / I) = \frac{P(I / k = k_i).P(k = k_i)}{\sum_{i=1}^{n} P(I / k = k_i).P(k = k_i)} \qquad (13.6)$$

where

P($k = k_i$) is the prior estimate of the probability that $k = k_i$

P($I/k = k_i$) is the likelihood of evidence I given $k = k_i$ and

P($k = k_i/I$) is the posterior probability, that is the probability of hypothesis k_i updated in the light of evidence, I.

In the Bayesian method, it is necessary to consider a range of hypotheses, k_1, k_2, ..., k_n only one if which can be true for a specific situation.

A 'Delphi' like panel of eight independent fire engineering and firefighting experts were presented with a description of the tunnel and the HGV. The experts were asked to provide their estimates of the HRR for two scenarios, a time t after ignition (growth phase) and after a further period of time when the fire becomes fully developed. The estimates collected from the experts were converted into values of k and collated to give a 'prior' probability distribution for k in order to keep to a minimum the errors in individual judgements. Originally, it was hoped that each expert would be able to provide estimates of the likelihood values as well as prior estimates. However, finally the principal author of this study estimated the likelihood values as he was considered to be independent of all the prior estimates and also very familiar with all of the evidence data.

For updating the prior probabilities as estimated above, results of four experiments providing values of k were considered. These experiments were:

- Hammerfest HGV test
- Hammerfest wooden crib test
- Buxton simulation of HGV test and
- Buxton wooden crib test.

The estimates of the likelihoods for the first test in the growth phase were applied to the prior probabilities using Bayes' formula (Equation (13.6)). The series of posterior probabilities produced were used as the new prior probabilities and the likelihoods for the second test were applied. This process was repeated for the fourth and third tests. It was observed that the third test was not appropriate for the calculation of k values in the growth phase of a HGV fire. These data were discounted and the posterior probabilities recalculated. The same process was repeated for k values of fully involved HGV fires. The expectations of k for each airflow velocity was calculated from the posterior probabilities using the Bayesian estimator:

$$E(k) = \sum k_i P(k = k_i / I) \tag{13.7}$$

The Bayesian analysis revealed that if an HGV fire in a tunnel is subjected to a $2\,\mathrm{ms}^{-1}$ ventilation airflow, then the fire may increase in severity seven times faster than under natural conditions and the maximum HRR may be up to four times greater. If a $10\,\mathrm{ms}^{-1}$ airflow is used, it may grow up to 22 times faster and ultimately be eight or nine times more severe, and so on. It would be interesting to compare these predictions with estimates made from the data directly.

References

Ang A H S, and Tang W H, (1984), *Probability Concepts in Engineering, Planning and Design, Volume 2 – Decision, Risk and Reliability*, Wiley, New York.

Carvel, R O, Beard, A N and Jowitt P W (1999), The effect of forced longitudinal ventilation on a HGV fire in a tunnel, *Proceedings of the International Conference on Tunnel Fires*, Independent Technical Conferences, Lyon.

Johansson, H (2003), Decision Analysis in Fire Safety Engineering – Analysing Investments in Fire Safety, PhD Thesis, Lund University, Institute of Technology, Department of Fire Safety Engineering.

La Valle, I H, (1970), *An Introduction to Probability, Decision and Inference*, Holt, Rinehart and Winston, New York.

Ramachandran, G (1998), *The Economics of Fire Protection*, E & F N Spon, London.

Shpilberg, D C (1974), *Risk Insurance and Fire Protection: A Systems Approach. Part 1: Modelling the Probability Distribution of Fire Loss Amount*. Technical Report, FMRC, Serial No. 22-431, RC74-TP-23, Factory Mutual Research Corporation, Norwood, MA.

Sui, N and Apostolakis, G (1988), Uncertainty, data and expert opinions in the assessment of the unavailability of suppression systems, *Fire Technology*, 24, 2, 138–162.

Index

CPSIA information can be obtained
at www.ICGtesting.com
Printed in the USA
LVOW13*2039200618

581276LV00003B/24/P